TWENTY FIRST CENTURY science

Project Directors

Angela Hall
Emma Palmer
Robin Millar
Mary Whitehouse

Editor

Emma Palmer

Authors

Helen Harden
Andrew Hunt
John Lazonby
Ted Lister
Janet Renshaw
Mike Shipton
Vicky Wong
Dorothy Warren

THE UNIVERSITY *of York*

Official Publisher Partnership

Contents

C5 Chemicals of the natural environment 134

C6 Chemical synthesis 164

C7 Chemistry for a sustainable world 194

How to use this book

Welcome to Twenty First Century Science. This book has been specially written by a partnership between OCR, The University of York Science Education Group, The Nuffield Foundation Curriculum Programme, and Oxford University Press.

On these two pages you can see the types of page you will find in this book, and the features on them. Everything in the book is designed to provide you with the support you need to help you prepare for your examinations and achieve your best.

Module Openers

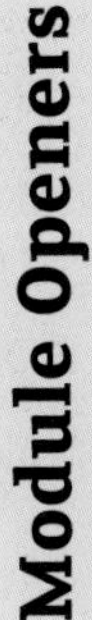

Why study?: This explains why what you are about to learn is useful to scientists.

Find out about: Every module starts with a short list of the things you'll be covering.

The Science: This box summarises the science behind the module you're about to study.

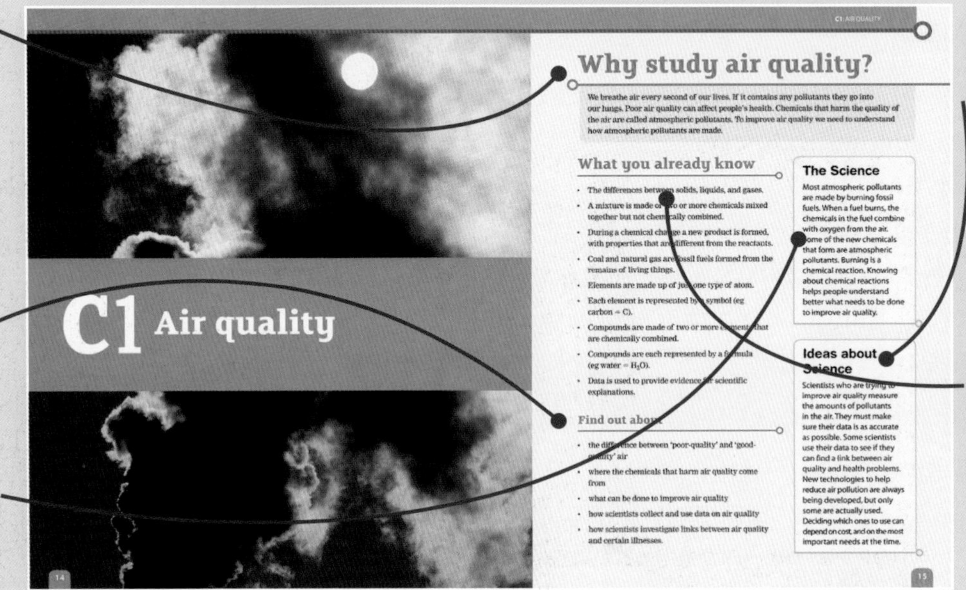

Ideas about Science: Here you can read about the key ideas about science covered in this module.

What you already know: This list is a summary of the things you've already learnt that will come up again in this module. Check through them in advance and see if there is anything that you need to recap on before you get started.

Main Pages

Find out about: For every part of the book you can see a list of the key points explored in that section.

Key words: The words in these boxes are the terms you need to understand for your exams. You can look for these words in the text in bold or check the glossary to see what they mean.

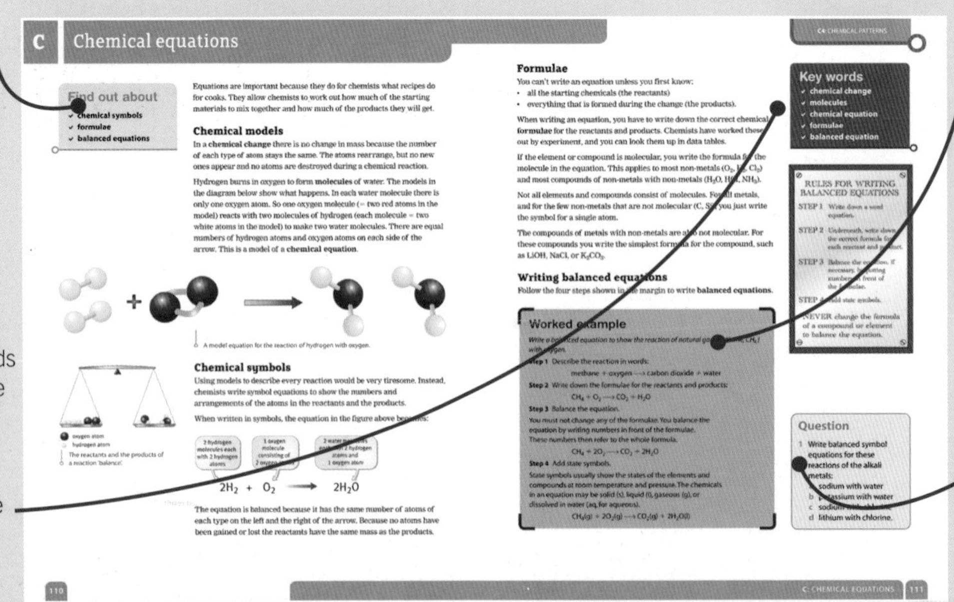

Worked examples: These help you understand how to use an equation or to work through a calculation. You can check back whenever you use the calculation in your work to make sure you understand.

Questions: Use these questions to see if you've understood the topic.

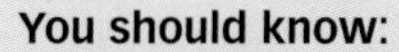

You should know: This is a summary of the main ideas in the unit. You can use it as a starting point for revision, to check that you know about the big ideas covered.

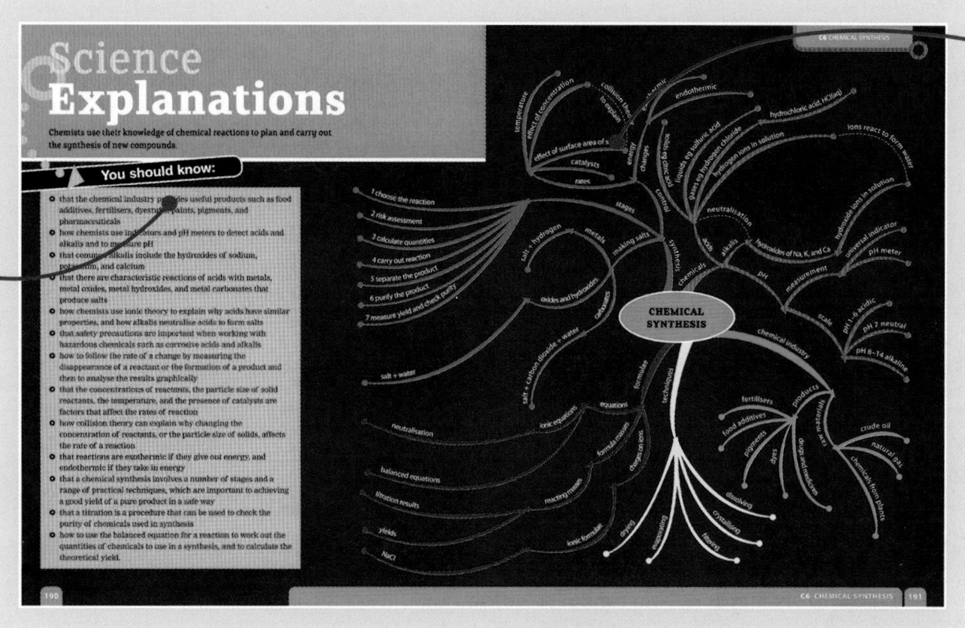
Science Explanations

Chemists use their knowledge of chemical reactions to plan and carry out the synthesis of new compounds.

You should know:

- that the chemical industry provides useful products such as food additives, fertilisers, dyestuffs, paints, pigments, and pharmaceuticals
- how chemists use indicators and pH meters to detect acids and alkalis and to measure pH
- that common alkalis include the hydroxides of sodium, potassium, and calcium
- that there are characteristic reactions of acids with metals, metal oxides, metal hydroxides, and metal carbonates that produce salts
- how chemists use ionic theory to explain why acids have similar properties, and how alkalis neutralise acids to form salts
- that safety precautions are important when working with hazardous chemicals such as corrosive acids and alkalis
- how to follow the rate of a change by measuring the disappearance of a reactant or the formation of a product and then to analyse the results graphically
- that the concentrations of reactants, the particle size of solid reactants, the temperature, and the presence of catalysts are factors that affect the rates of reaction
- how collision theory can explain why changing the concentration of reactants, or the particle size of solids, affects the rate of a reaction
- that reactions are exothermic if they give out energy, and endothermic if they take in energy
- that a chemical synthesis involves a number of stages and a range of practical techniques, which are important to achieving a good yield of a pure product in a safe way
- that a titration is a procedure that can be used to check the purity of chemicals used in synthesis
- how to use the balanced equation for a reaction to work out the quantities of chemicals to use in a synthesis, and to calculate the theoretical yield.

190 C6: CHEMICAL SYNTHESIS 191

Visual summary: Another way to start revision is to use a visual summary, linking ideas together in groups so that you can see how one topic relates to another. You can use this page as a starting point for your own summary.

Ideas about Science: For every module this page summarises the ideas about science that you need to understand.

Ideas about Science

Scientists can never be sure that a measurement tells them the true value of the quantity being measured. Data is more reliable if it can be repeated. If you take several measurements of the same quantity, for example, in a titration, the results are likely to vary. This may be because:

- the quantity you are measuring is varying, for example, the purity of separate solid samples of a product may not be uniform
- there are variations in the judgement of when the indicator colour corresponds to the end-point, or of the position of a meniscus
- there are limitations in the measuring equipment, such as burettes.

Usually the best estimate of the true value of a quantity is the mean (or average) of several repeat measurements. You should:

- be able to calculate the mean from a set of repeat measurements
- know that a measurement may be an outlier if it lies well outside the range of the other values in a set of repeat measurements
- be able to give reasons to explain why an outlier should be retained as part of the data or rejected when calculating the mean.

When comparing sets of titration results you should know that:

- a difference between their means is real if their ranges do not overlap.

To investigate the relationship between a factor and an outcome, it is important to control all the other factors that might affect the outcome, such as temperature and the concentrations of other reactants. When investigating the rates of chemical reactions you should be able to:

- identify the outcome and factors that may affect it
- suggest how an outcome might alter when a factor is changed.

In a plan for an investigation of the effect of a factor on an outcome, you should:

- be able to explain why it is necessary to control all the factors that might affect the outcome
- recognize that the fact that other factors are controlled is a positive design feature, and the fact that they are not is a design flaw.

If an outcome occurs when a specific factor is present but does not when it is absent, or if an outcome variable increases (or decreases) steadily as an input variable increases, we say that there is a correlation between the two. In the context of studying reaction rates you should:

- understand that a correlation shows a link between a factor and an outcome
- be able to identify where a correlation exists when data is presented as text, as a graph, or in a table
- understand that a correlation does not always mean that the factor causes the outcome.
- identify that where there is a machanism to explain a correlation, scientists are likely to accept that the factor causes the outcome.

192

C6: CHEMICAL SYNTHESIS

Review Questions

1. A student is developing a new sports pack. She needs to find two chemicals that will cool an injury when mixed inside the pack.
   a Should the reaction be endothermic or exothermic?
   b Which of the energy-level diagrams below represents the reaction that is likely to cool injuries more effectively?

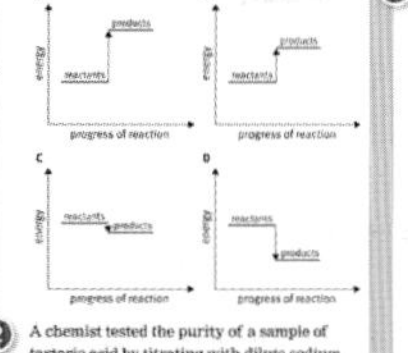

2. A chemist tested the purity of a sample of tartaric acid by titrating with dilute sodium hydroxide. The table shows his results.

	Run 1	Run 2	Run 3	Run 4
Initial burette reading (cm^3)	0.1	0.1	25.0	21.9
Final burette reading (cm^3)	25.2	25.0	50.0	46.9
Volume of acid added (cm^3)	25.1	24.9	25.0	

   a Explain why the chemist repeated the titration four times.
   b Calculate the volume of alkali added in run 4.
   c The chemist calculates the mean of his titration results. Why does he do this?
   d Suggest the range within which the true value probably lies.

3. A teacher adds 3 g of zinc granules to dilute hydrochloric acid in a flask. She uses a gas syringe to measure the volume of hydrogen gas given off.

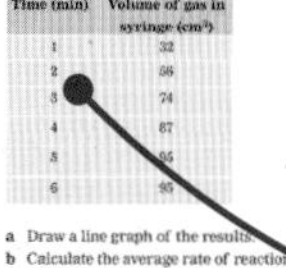

Time (min)	Volume of gas in syringe (cm^3)
1	32
2	56
3	74
4	87
5	95
6	95

   a Draw a line graph of the results.
   b Calculate the average rate of reaction between three and four minutes, in cm^3 per minute.
   c The teacher decides to repeat the experiment with 3 g of zinc powder. She keeps all other factors the same. Draw and label a line on the graph to show the expected results. Use collision theory to explain the effect of this change.
   d Explain why it was important for the teacher to keep all other factors the same.

C6: CHEMICAL SYNTHESIS 193

Review Questions: You can begin to prepare for your exams by using these questions to test how well you know the topics in this module.

Structure of assessment

Matching your course

What's in each module?

As you go through the book you should use the module opener pages to understand what you will be learning and why it is important. The table below gives an overview of which main topics each module includes.

C1	C4
• Which chemicals make up air, and which ones are pollutants? How do I make sense of data about air pollution? • What chemical reactions produce air pollutants? What happens to these pollutants in the atmosphere? • What choices can we make personally, locally, nationally, and globally to improve air quality?	• What are the patterns in the properties of elements? • How do chemists explain the patterns in the properties of elements? • How do chemists explain the properties of compounds of Group 1 and Group 7 elements?

C2	C5
• How do we measure the properties of materials and why are the results useful? • Why is crude oil important as a source of new materials such as plastics and fibres? • Why does it help to know about the molecular structure of materials such as plastics and fibres? • What is nanotechnology and why is it important?	• What types of chemicals make up the atmosphere? • What reactions happen in the hydrosphere? • What types of chemicals make up the Earth's lithosphere? • How can we extract useful metals from minerals?

C3	C6
• What were the origins of minerals in Britain that contribute to our economic wealth? • Where does salt come from and why is it so important? • Why do we need chemicals such as alkalis and chlorine and how do we make them? • What can we do to make our use of chemicals safe and sustainable?	• Chemicals and why we need them • Planning, carrying out, and controlling a chemical synthesis

C7	
• Green chemistry • Reversible reactions and equilibria • Alcohols, carboxylic acids, and esters	• Analysis • Energy changes in chemistry

How do the modules fit together?

The modules in this book have been written to match the specification for GCSE Chemistry. In the diagram to the right you can see that the modules can also be used to study parts of GCSE Science and GCSE Additional Science.

	GCSE Biology	GCSE Chemistry	GCSE Physics
GCSE Science	B1	C1	P1
GCSE Science	B2	C2	P2
GCSE Science	B3	C3	P3
GCSE Additional Science	B4	C4	P4
GCSE Additional Science	B5	C5	P5
GCSE Additional Science	B6	C6	P6
	B7	C7	P7

GCSE Chemistry assessment

The content in the modules of this book matches the modules of the specification.

The diagram below shows you which modules are included in each exam paper. It also shows you how much of your final mark you will be working towards in each paper.

	Unit	Modules Tested			Percentage	Type	Time	Marks Available
Route 1	A171	C1	C2	C3	25%	Written Exam	1 h	60
	A172	C4	C5	C6	25%	Written Exam	1 h	60
	A173	C7			25%	Written Exam	1 h	60
	A174	Controlled Assessment			25%		4.5–6 h	64

Command words

The list below explains some of the common words you will see used in exam questions.

Calculate
Work out a number. You can use your calculator to help you. You may need to use an equation. The question will say if your working must be shown. (Hint: don't confuse with 'Estimate' or 'Predict'.)

Compare
Write about the similarities and differences between two things.

Describe
Write a detailed answer that covers what happens, when it happens, and where it happens. Talk about facts and characteristics. (Hint: don't confuse with 'Explain'.)

Discuss
Write about the issues related to a topic. You may need to talk about the opposing sides of a debate, and you may need to show the difference between ideas, opinions, and facts.

Estimate
Suggest an approximate (rough) value, without performing a full calculation or an accurate measurement. Don't just guess – use your knowledge of science to suggest a realistic value. (Hint: don't confuse with 'Calculate' and 'Predict'.)

Explain
Write a detailed answer that covers how and why a thing happens. Talk about mechanisms and reasons. (Hint: don't confuse with 'Describe'.)

Evaluate
You will be given some facts, data, or other kind of information. Write about the data or facts and provide your own conclusion or opinion on them.

Justify
Give some evidence or write down an explanation to tell the examiner why you gave an answer.

Outline
Give only the key facts of the topic. You may need to set out the steps of a procedure or process – make sure you write down the steps in the correct order.

Predict
Look at some data and suggest a realistic value or outcome. You may use a calculation to help. Don't guess – look at trends in the data and use your knowledge of science. (Hint: don't confuse with 'Calculate' or 'Estimate'.)

Show
Write down the details, steps, or calculations needed to prove an answer that you have given.

Suggest
Think about what you've learnt and apply it to a new situation or context. Use what you have learnt to suggest sensible answers to the question.

Write down
Give a short answer, without a supporting argument.

Top Tips

Always read exam questions carefully, even if you recognise the word used. Look at the information in the question and the number of answer lines to see how much detail the examiner is looking for.

You can use bullet points or a diagram if it helps your answer.

If a number needs units you should include them, unless the units are already given on the answer line.

Making sense of graphs

Reading the axes

Look at these two charts, which both provide data about recycling plastics in several countries.

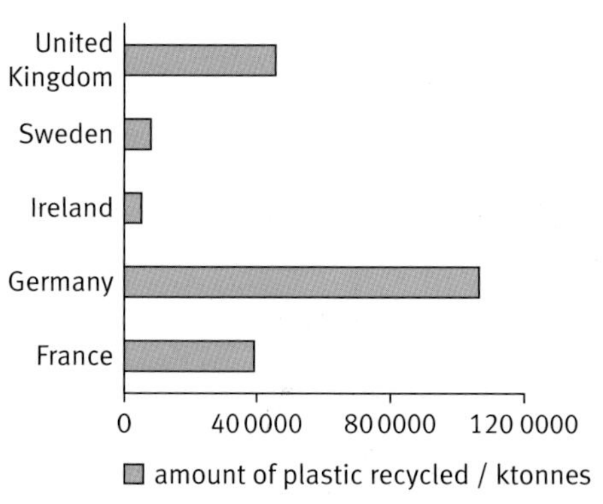

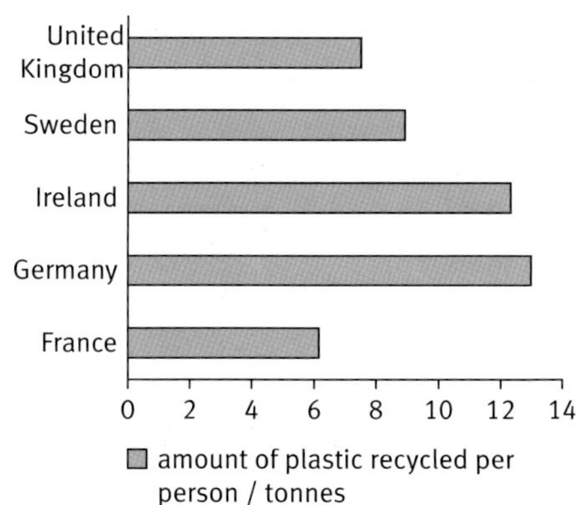

Why are the charts so different if they both represent data about recycling plastics?

First rule for reading graphs: read the axes and check the units.

Describing the relationship between variables

The pattern of points plotted on a graph shows whether two **factors** are related. Look at this scatter graph.

There *is* a pattern in the data. As the number of carbon atoms increases, the boiling point increases.

But it is not a straight line, it is quite a smooth curve, so we can say more than that. The boiling point increases quickly with each extra carbon atom when the number of carbon atoms is small. It increases, but less quickly as the number of carbon atoms gets bigger. Another way of describing this is to say that the slope of the graph – the **gradient** – gets less as the number of carbon atoms increases.

Rates of reaction

Look at the graph to the right, which shows the product of a chemical reaction being produced over time.

The shape of the graph is a curve with several different gradients. This is because the rate of the chemical reaction is changing throughout the reaction. The graph shows how the gradient, and therefore the rate of reaction, is calculated.

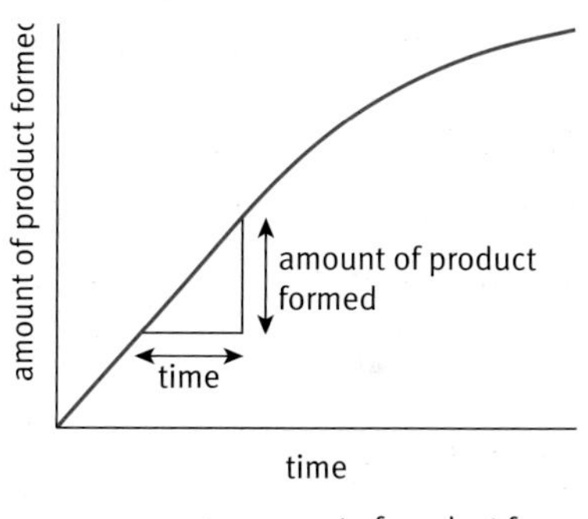

$$\text{rate of reaction} = \frac{\text{amount of product formed}}{\text{time}}$$

Graph showing product produced over time by a chemical reaction.

You should describe each part of the graph and where possible include data.

- At the start of the reaction, the amount of product is produced very quickly.
- After about 6 seconds, the rate at which it is produced starts to slow down.
- After 10 seconds, no more product is being made, showing that the reaction is over.

Second rule for reading graphs: describe each distinct phase of the graph, including ideas about the **gradient** and **data including units**.

Is there a correlation?

Sometimes we are interested in whether one thing changes when another does. If a change in one factor goes together with a change in something else, we say that the two things are **correlated**.

The two graphs on the right show how global temperatures have changed over time and how levels of carbon dioxide in the atmosphere have also changed over time. Is there a correlation between the two sets of data?

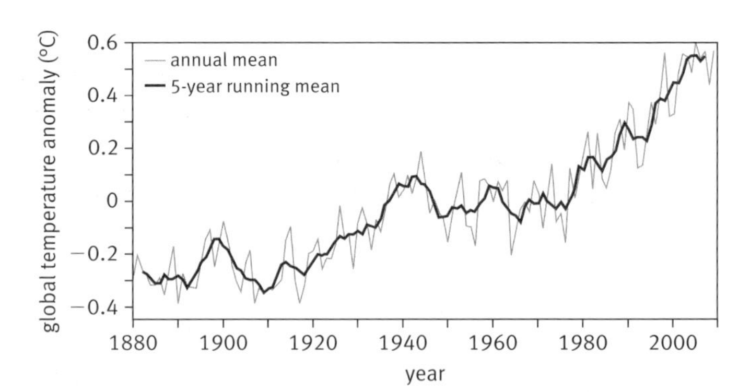

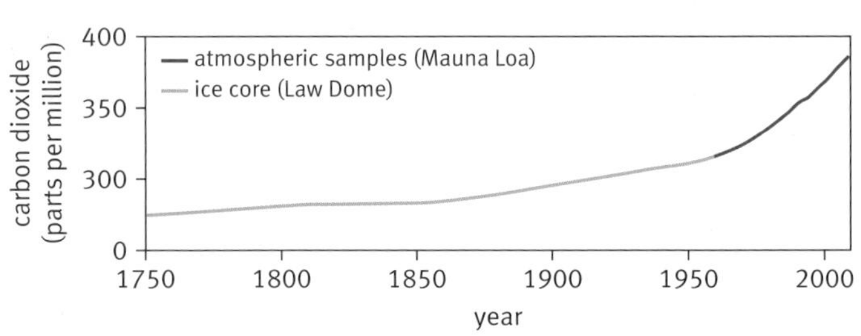

Graphs to show increasing global temperatures and carbon dioxide levels. Source: NASA.

Look at the graphs – why is it difficult to decide if there is a correlation?

The two sets of data are over different periods of time, so although both graphs show a rise with time, it is difficult to see if there is a correlation. It is easier to identify a correlation if both sets of data are plotted for the same time range and placed one above the other, or on the same axes.

When there are two sets of data on the same axes take care to look at which axis relates to which line.

Third rule for reading graphs: when looking for a correlation between two sets of data, read the axes carefully.

Explaining graphs

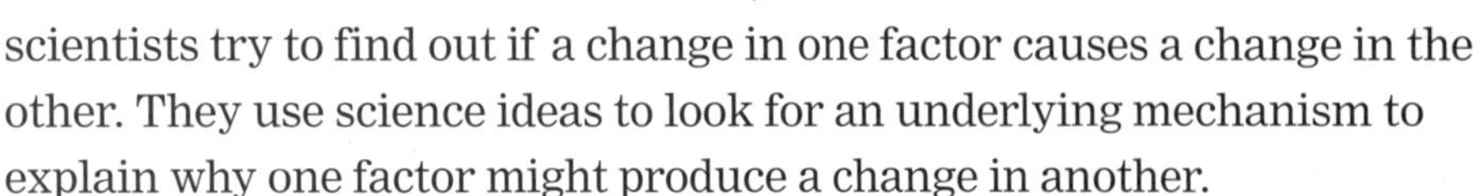

When a graph shows that there is a correlation between two sets of data, scientists try to find out if a change in one factor causes a change in the other. They use science ideas to look for an underlying mechanism to explain why one factor might produce a change in another.

Calculating reacting masses and percentage yields

Problem *How much aluminium powder do you need to react with 8.0 g of iron oxide?*

$2Al(s) + Fe_2O_3(s) \longrightarrow Al_2O_3(s) + 2Fe(s)$

Work out the relative formula masses (RFM) by using **multiplication** and **addition.**

Reactants:

RFM of Al = 27;

RFM of $Fe_2O_3 = (2 \times 56) + (3 \times 16) = 160$

Products:

RFM of $Al_2O_3 = (2 \times 27) + (3 \times 16) = 102$

RFM of Fe = 56

To find the relative masses in this reaction **multiply** the RFM by the numbers in front of each formula in the equation. Then convert the relative masses to reacting masses by including units. These can be g, kg or tonnes depending on the data in the question. The units must be the same for each of the values.

$2Al(s)$	+	$Fe_2O_3(s)$	$\longrightarrow$	$Al_2O_3(s)$	+	$2Fe(s)$
$(2 \times 27) = 54$ g		160 g		102 g		$(2 \times 56) = 112$ g

To find the quantities required, scale the reacting masses to the known quantities by using **simple ratios.** Always include the **correct units** when substituting values.

$$\frac{\text{mass of aluminium required}}{\text{reacting mass of aluminium}} = \frac{\text{mass of iron oxide used}}{\text{reacting mass of iron oxide}}$$

$$\frac{\text{mass of aluminium required}}{54\text{ g}} = \frac{8\text{ g}}{160\text{ g}}$$

Rearrange the equation, to find the mass of aluminium.

$$\text{Mass of aluminium required} = \frac{8\text{ g}}{160\text{ g}} \times 54\text{ g} = \mathbf{2.7g}$$

Problem *What is the percentage yield of iron for the same reaction, if 4.9 g of iron is actually produced?*

To work out the theoretical yield of iron, use ratios as before

$$\frac{\text{mass of iron yielded}}{\text{reacting mass of iron}} = \frac{\text{mass of iron oxide used}}{\text{reacting mass of iron oxide}}$$

$$\frac{\text{mass of iron yielded}}{112\text{ g}} = \frac{8\text{ g}}{160\text{ g}}$$

and then rearrange the equation.

$$\text{Theoretical yield of iron} = \frac{8\text{ g}}{160\text{ g}} \times 112\text{ g} = 5.6\text{ g}$$

Use this equation to calculate the percentage yield

$$\textbf{percentage yield} = \frac{\textbf{actual yield (g)}}{\textbf{theoretical yield (g)}} \times \mathbf{100\ \%}$$

Actual yield of iron given in the question = 4.9 g

Substitute the quantities.

$$\text{percentage yield} = \frac{4.9\text{ g}}{5.6\text{ g}} \times 100 = \mathbf{87.5\ \%}$$

Controlled assessment

In GCSE Chemistry the controlled assessment counts for 25% of your total grade. Marks are given for a practical investigation.

Your school or college may give you the mark schemes for this.

This will help you understand how to get the most credit for your work.

Practical investigation (25%)

Investigations are carried out by scientists to try and find the answers to scientific questions. The skills you learn from this work will help prepare you to study any science course after GCSE.

To succeed with any investigation you will need to:

- choose a question to explore
- select equipment and use it appropriately and safely
- design ways of making accurate and reliable observations
- relate your investigation to work by other people investigating the same ideas.

Your investigation report will be based on the data you collect from your own experiments. You will also use information from other people's research. This is called secondary data.

You will write a full report of your investigation. Marks will be awarded for the quality of your report. You should:

- make sure your report is laid out clearly in a sensible order
- use diagrams, tables, charts, and graphs to present information
- take care with your spelling, grammar, and punctuation, and use scientific terms where they are appropriate.

Marks will be awarded under five different headings.

Strategy

- Develop a hypothesis to investigate.
- Choose a procedure and equipment that will give you reliable data.
- Carry out a risk assessment to minimise the risks of your investigation.
- Describe your hypothesis and plan using correct scientific language.

Collecting data

- Carry out preliminary work to decide the range.
- Collect data across a wide enough range.
- Collect enough data and check its reliability.
- Control factors that might affect the results.

Analysis

- Present your data to make clear any patterns in the results.
- Use graphs or charts to indicate the spread of your data.
- Use appropriate calculations such as averages and gradients of graphs.

Evaluation

- Describe and explain how you could improve your method.
- Discuss how repeatable your evidence is, accounting for any outliers.

Review

- Comment, with reasons, on your confidence in the secondary data you have collected.
- Compare the results of your investigation to the secondary data.
- Suggest ways to increase the confidence in your conclusions.

Tip

The best advice is 'plan ahead'. Give your work the time it needs and work steadily and evenly over the time you are given. Your deadlines will come all too quickly, especially if you have coursework to do in other subjects.

Secondary data

Once you have collected the data from your investigation you should look for some secondary data relevant to your hypothesis. This will help you decide how well your data agrees with the findings of other scientists. Your teacher will give you secondary data provided by OCR, but you should look for further sources to help you evaluate the quality of all your data. Other sources of information could include:

- experimental results from other groups in your class or school
- text books
- the Internet.

When will you do this work?

Your school or college will decide when you do your practical investigation. If you do more than one investigation, they will choose the one with the best marks.

Your investigation will be done in class time over a series of lessons.

You may also do some research out of class.

C1 Air quality

Why study air quality?

We breathe air every second of our lives. If it contains any pollutants they go into our lungs. Poor air quality can affect people's health. Chemicals that harm the quality of the air are called atmospheric pollutants. To improve air quality we need to understand how atmospheric pollutants are made.

What you already know

- The differences between solids, liquids, and gases.
- A mixture is made of two or more chemicals mixed together but not chemically combined.
- During a chemical change a new product is formed, with properties that are different from the reactants.
- Coal and natural gas are fossil fuels formed from the remains of living things.
- Elements are made up of just one type of atom.
- Each element is represented by a symbol (eg carbon = C).
- Compounds are made of two or more elements that are chemically combined.
- Compounds are each represented by a formula (eg water = H_2O).
- Data is used to provide evidence for scientific explanations.

Find out about

- the difference between 'poor-quality' and 'good-quality' air
- where the chemicals that harm air quality come from
- what can be done to improve air quality
- how scientists collect and use data on air quality
- how scientists investigate links between air quality and certain illnesses.

The Science

Most atmospheric pollutants are made by burning fossil fuels. When a fuel burns, the chemicals in the fuel combine with oxygen from the air. Some of the new chemicals that form are atmospheric pollutants. Burning is a chemical reaction. Knowing about chemical reactions helps people understand better what needs to be done to improve air quality.

Ideas about Science

Scientists who are trying to improve air quality measure the amounts of pollutants in the air. They must make sure their data is as accurate as possible. Some scientists use their data to see if they can find a link between air quality and health problems. New technologies to help reduce air pollution are always being developed, but only some are actually used. Deciding which ones to use can depend on cost, and on the most important needs at the time.

Find out about

- the gases that make up air
- the atmosphere that surrounds Earth
- how other gases may be added to the atmosphere by human activity or natural processes

Key words

- molecule
- atmosphere
- mixture
- particulates

What do you know about air?

The air

Air is all around us. You cannot see the air but if you wave your hand you can feel it.

You may think that a can of fizzy drink is empty once you have drunk it, but look what happens if all the air is then removed from inside the can. The can collapses.

If you remove the air from inside a can, it collapses.

Air is made up of small **molecules** with large spaces in between. Molecules are groups of atoms joined together.

The atmosphere

An astronaut in space needs an air supply in order to breathe. A mountaineer climbing Everest feels that the air is becoming 'thinner' as he climbs. Where does our air stop?

The **atmosphere** is the layer of gases that surrounds the Earth. It is about 15 km thick. That sounds a lot but the diameter of the Earth is over 12000 km. The atmosphere is like a very thin skin around the Earth.

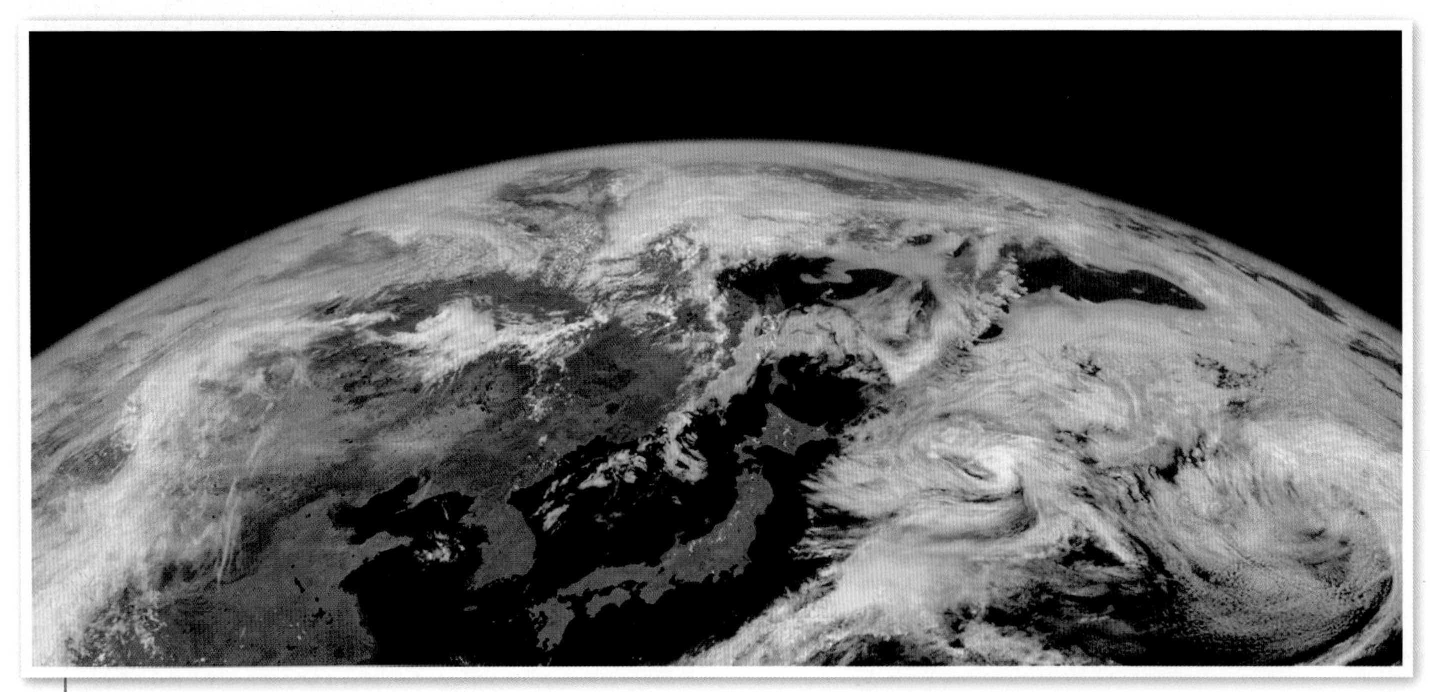

The Earth from space. White clouds of water vapour can be seen in the atmosphere.

What gases are found in the air?

Air contains the gas oxygen. Oxygen is the gas we need to breathe but air is not just made of oxygen. Air is a **mixture** of oxygen and nitrogen, with a small amount of argon plus tiny amounts of carbon dioxide and water vapour.

The tiny amount of carbon dioxide in the air is what has helped keep the Earth warm enough to support life.

However, the air contains other gases as well. Human activity has released a whole variety of different gases into the atmosphere. Many of these gases affect the quality of the air we breathe. Unfortunately, gases released in one part of the world will gradually spread through the atmosphere and can affect the air quality of people many miles away.

Some gases are naturally released into the atmosphere by volcanoes. These gases include sulfur dioxide, carbon dioxide, carbon monoxide, nitrogen dioxide, and water vapour. They also produce **particulates** in the form of smoke and ash. The particulates in volcanic ash or smoke are tiny specks of solids. They are small enough to stay suspended in the air.

oxygen molecule containing two oxygen atoms

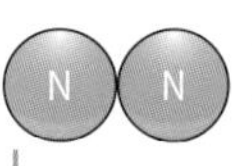

nitrogen molecule containing two nitrogen atoms

Nitrogen and oxygen make up 99% of the air.

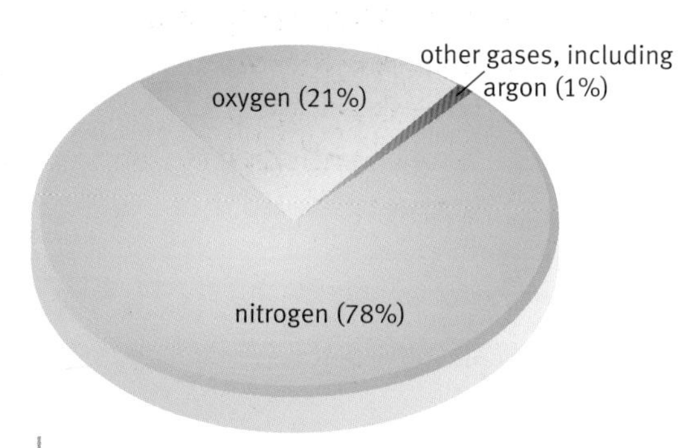

This pie chart shows the percentages of the main gases in clean air.

This geologist is working upwind of a volcano. She is wearing a gas mask to protect her from breathing in sulfur dioxide gas.

Questions

1. Explain why a can does not collapse when the drink has been finished.
2. Oxygen, carbon dioxide and water vapour are three gases found naturally in the atmosphere. Explain why each of these is important.
3. In 2010 a volcano in Iceland erupted, releasing huge quantities of volcanic ash and particulates. Explain why it was not only local people in Iceland who were affected by this eruption.

The story of our atmosphere

Find out about

- the composition of gases in the Earth's early atmosphere
- evidence for an increase in oxygen in the atmosphere
- how scientists disagree about this evidence

Early Earth

The early Earth was a violent place, constantly bombarded by meteors and covered with active volcanoes. The atmosphere consisted mainly of carbon dioxide and water vapour. These gases probably came from volcanoes, which belched huge quantities of carbon dioxide and water vapour, and also some nitrogen and methane, into the atmosphere. If it had stayed that way, Earth would have been as inhospitable as our neighbouring planet Venus.

An artist's impression of early Earth.

The temperature of the surface of Venus is almost 500 °C. Its atmosphere is largely carbon dioxide and its clouds are formed of sulfuric acid.

On Earth, temperatures began to cool. The water vapour gradually **condensed** to form the oceans. Some of the carbon dioxide began to dissolve in the oceans and later became incorporated into **sedimentary rocks**.

Exactly what happened next is less certain. Scientists are still investigating when and how the Earth's atmosphere changed from being mostly carbon dioxide to containing oxygen and just a little carbon dioxide.

How can scientists find out about the development of Earth's atmosphere?

Of course, scientists can't directly measure the composition of Earth's very early atmosphere. They have to use indirect evidence instead.

For example, the chemical composition of rocks can give scientists clues about the state of the atmosphere when the rocks were formed. So if they know the age of the rocks they find out something about the composition of the atmosphere at that time.

Scientists also look at fossil evidence of early life. The evidence suggests that oxygen levels first increased due to early plant life. These early plants used up carbon dioxide during **photosynthesis** and released oxygen. So as the oxygen level rose, the carbon dioxide level fell. Much of the carbon dioxide was removed from the atmosphere for a long time because it was trapped underground as the coal and oil that we now use as fuels.

Iron pyrite is made up of iron sulfide, which only forms if there is *no* oxygen present.

Air bubbles trapped in ice cores drilled in Antarctica allow scientists to analyse the composition of the air from hundreds of thousands of years ago. However, the Earth is about 4.5 billion years old. Ice cores don't go back far enough in time to provide evidence of the earliest atmosphere.

The details of exactly when and how fast these changes in the atmosphere happened are still a matter of scientific debate, as scientists discuss and evaluate each others' ideas and explanations.

Scientists may interpret available data in different ways. An expert in plants might interpret fossil data differently from an expert in rocks. This may lead them to come up with different scientific explanations for when and how the atmosphere changed.

Even if scientists agree on an explanation, new information may then be found that disagrees with this explanation. For example, in 2010 some American scientists reported that they had found fossils of primitive animals with shells that lived in ocean reefs 650 million years ago. These fossils are 50–100 million years older than any other known fossils of hard-bodied animals. New discoveries like this mean that scientists have to rethink their ideas and come up with new explanations to account for fresh data.

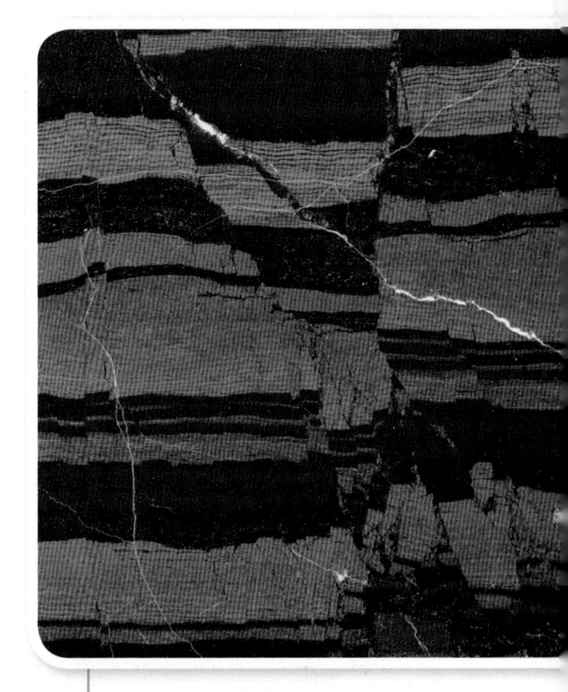

Red iron oxide rocks only form if there is oxygen present.

Questions

1. Rocks made of iron oxide have been dated to two billion years ago. What does this tell you about when oxygen first appeared in Earth's atmosphere?
2. No samples of iron pyrite have been dated as less than two billion years old. What does this tell you about when oxygen first appeared in the Earth's atmosphere?
3. What discovery could change your mind about these conclusions?

Key words

- condensed
- photosynthesis
- sedimentary rock

C What are the main air pollutants?

Find out about

- the most important air pollutants
- the problems pollutants cause
- what can influence air quality in different locations

Every time you are driven somewhere by car, or when you switch on the lights at home, new gases are made. They are released by the car, or from the power station where electricity is generated. Some of these chemicals are harmful and are called air **pollutants**. These pollutants can harm us directly by affecting our health, or harm us indirectly by affecting our environment.

The clouds coming from the cooling towers may look like pollutants, but are just harmless water vapour. There may be invisible pollutants coming out of the tall chimney.

Air pollutants spread through the atmosphere and change its composition. The biggest recent change has been in levels of carbon dioxide, which have been increasing for over a century.

The table lists the main chemicals that are released from power stations and vehicles.

Smoke is a pollutant that can be easily seen. It contains microscopic particles of carbon. Some of these are just 10 micrometres (10 millionths of a metre) in size. These are called PM10 particles. Although they are very small, they are very much bigger than atoms or molecules. Each particle contains billions of carbon atoms.

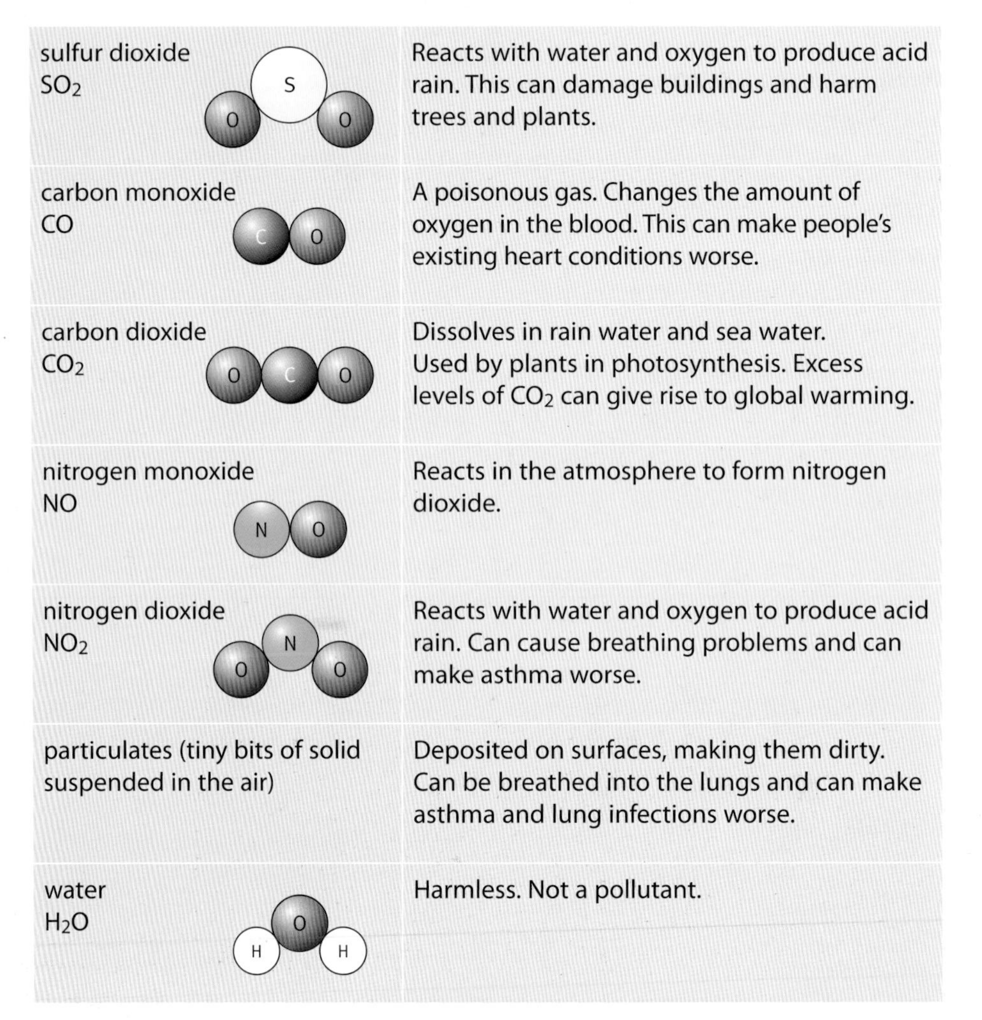

Chemical	Effect
sulfur dioxide SO_2	Reacts with water and oxygen to produce acid rain. This can damage buildings and harm trees and plants.
carbon monoxide CO	A poisonous gas. Changes the amount of oxygen in the blood. This can make people's existing heart conditions worse.
carbon dioxide CO_2	Dissolves in rain water and sea water. Used by plants in photosynthesis. Excess levels of CO_2 can give rise to global warming.
nitrogen monoxide NO	Reacts in the atmosphere to form nitrogen dioxide.
nitrogen dioxide NO_2	Reacts with water and oxygen to produce acid rain. Can cause breathing problems and can make asthma worse.
particulates (tiny bits of solid suspended in the air)	Deposited on surfaces, making them dirty. Can be breathed into the lungs and can make asthma and lung infections worse.
water H_2O	Harmless. Not a pollutant.

How can you find out about air quality?

Some people suffer from asthma or hayfever. They may be able to feel when the air quality is poor. But most people do not know whether the air quality is good or bad.

Automatic instruments collect air samples and measure the concentrations of a range of pollutants. The data is recorded automatically. Much of the data is regularly displayed in real time on websites available to the public. Newspapers and TV stations summarise the data in reports, which may give the day's overall air quality on a number scale or describe it as low, medium, or high quality.

Does it matter where you live?

Is the air quality the same all over the country? Some people live in cities. Other people live in the countryside. Will they all have air of the same quality to breathe?

The bar chart shows the concentration of NO_2 on the same day in three different places. The concentrations are clearly different. The concentration of NO_2 depends a lot on the level of human activity in the area. The amount of road traffic has a big effect.

Mace Head, in Ireland, has very pure air when the wind blows in from across the Atlantic Ocean. Scientists use it as a baseline to see what air would be like without the effects of human activities.

Most of us live in environments where the air quality is much poorer than at Mace Head.

Measuring the concentration of a pollutant

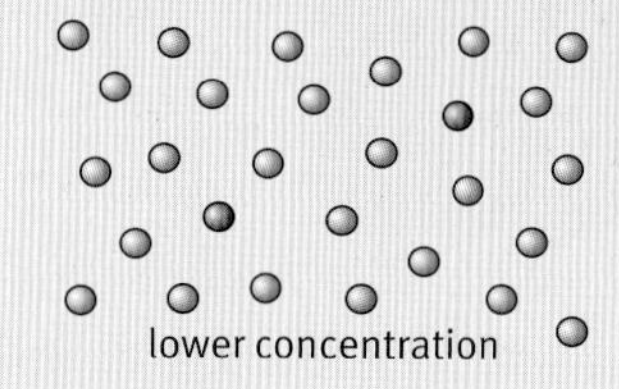

A low concentration of pollutants. There are very few pollutant molecules in a certain volume of air. This is an indication of good air quality.

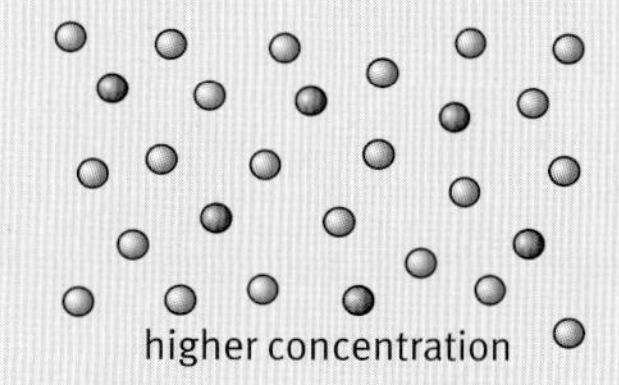

A high concentration of pollutants. There is a large number of pollutant molecules in a certain volume of air. This shows that the air quality is poor.

● molecules of pollutant
○ other molecules in air

Concentration is the amount of pollutant in a certain volume of air.

(Note: the air molecules are normally much more spread out than shown in the diagrams.)

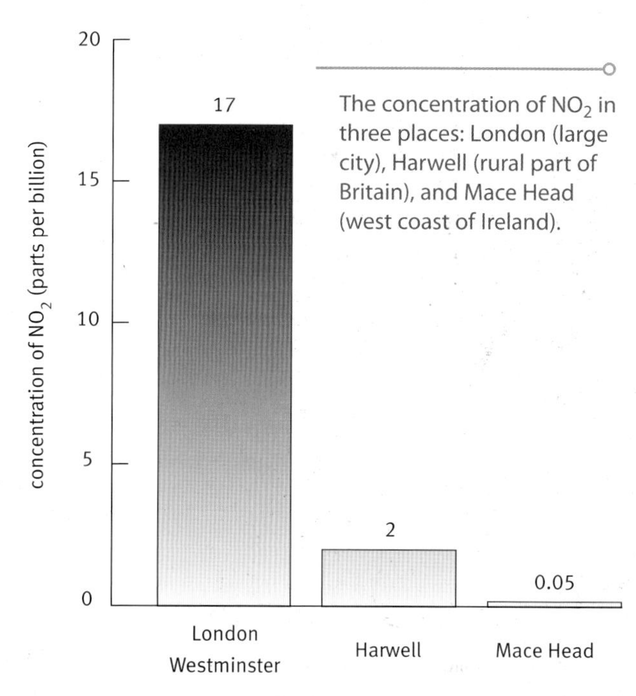

The concentration of NO_2 in three places: London (large city), Harwell (rural part of Britain), and Mace Head (west coast of Ireland).

An air-pollution monitoring station in a busy London street. It has instruments to measure particulate matter, carbon monoxide, nitrogen dioxide, and sulfur dioxide levels.

Key words

- pollutant
- emissions
- concentration

Nitrogen dioxide in London

Nitrogen dioxide levels increase when traffic is heavy. Exhaust **emissions** released from vehicles contain nitrogen dioxide. Can you see any patterns in the graph below that back this up?

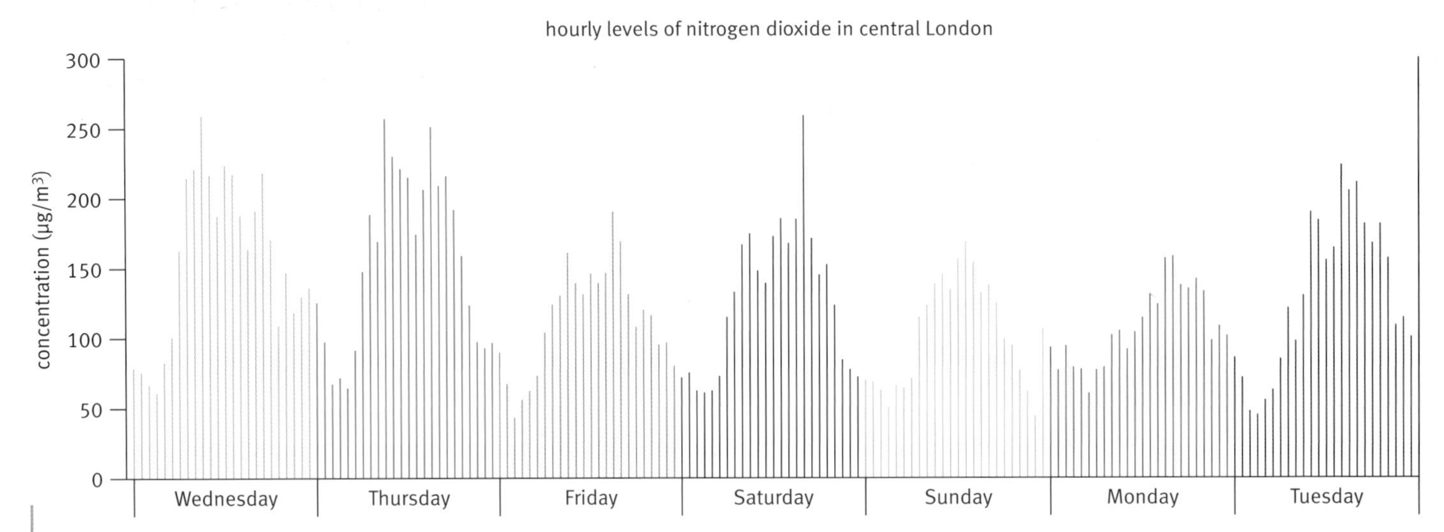

Nitrogen dioxide levels in central London over a seven-day period at the beginning of January 2009.

What influences air quality?

The quality of the air where you live depends mostly on nearby sources of air pollutants, and the weather.

- Sources: vehicles, power stations, and industry are some of the main sources of air pollutants.
- Weather: pollutants are mixed up and carried around by the winds. Wind can move pollutants many miles and even carry them from one country to another.

Questions

1 Write down one problem that can be caused by each of these air pollutants:
  a SO_2 b NO_2 c particulates

2 A newspaper article on air quality included a photograph of white clouds coming out of a power station's cooling towers. Write a note to the paper explaining why the clouds are not polluting the atmosphere.

3 Look at the chart above showing levels of NO_2 in London. Suggest reasons for the pattern of readings for Wednesday.

4 The weather moves air pollutants from one place to another. If we reduce emissions of air pollutants in our own town, we can still get pollution from other areas. Explain why it is still important to try to reduce emissions.

Measuring an air pollutant

D

Crucial data

Quantities of air pollutants are measured using a network of monitoring stations all around the country. The information is useful to individual people, as it allows them to check the air quality in their area. It is also used by the government to check whether air pollution is reaching dangerous levels anywhere.

Scientists use data to help answer questions such as 'how do pollutants travel?' or 'how do air pollutants interact with other chemicals in the atmosphere?' Data can be used to check the scientists' proposed explanation.

For example, the data from a busy street showed that the concentration of nitrogen dioxide (NO_2) was much higher on one side than the other. One explanation was that it could be caused by patterns in air movements in the street.

More data were collected to test this explanation. The computer-generated picture below shows the concentrations of nitrogen dioxide at all levels in the street. You can see that the air currents are actually circulating. This is invisible to people in the street, but obvious when you use the data.

Find out about

- measurement of air pollution
- how data is checked and used

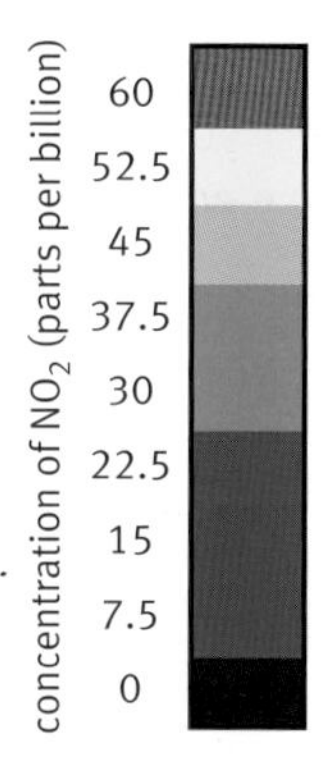

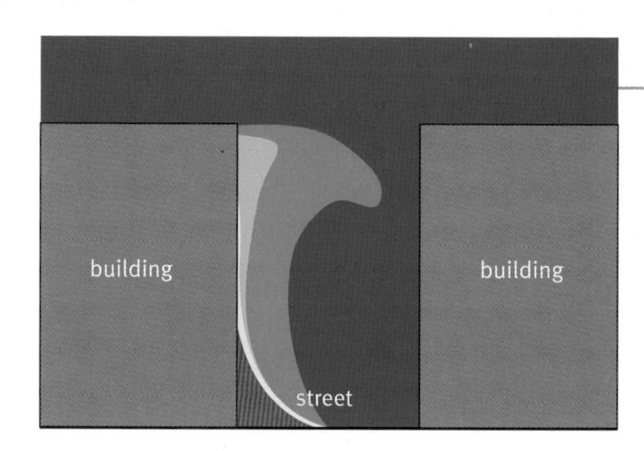

This computer-generated picture shows the concentration of the air pollutant NO_2 in a city street. The red area shows that invisible air currents have channelled the NO_2 onto just one side of the street. The tall buildings on either side make the street into a kind of 'canyon' and trap the NO_2 at street-level.

Making measurements

If you measure the concentration of nitrogen dioxide in a sample of air several times, you will probably get different results. This is because:

- you used the equipment differently
- there were differences in the equipment itself.

If you take just one reading, you cannot be sure it is very accurate. So it is better to take several measurements. Then you can use them to estimate the true value.

The true value is what the measurement should really be. The **accuracy** of a result is how close it is to the true value.

Key words

- accuracy
- outlier
- mean value
- best estimate
- range
- real difference

How can you make sure your data is accurate?

The table below shows what you should do to get a measurement of the level of nitrogen dioxide that is as accurate as possible.

The mean value is 19.1 ppb. This is the best estimate of the concentration of nitrogen dioxide in the sample of air. You cannot be absolutely sure that it is the true value. But you can be sure that:

- the true value is within the range 18.8 – 19.4 ppb
- the best estimate of the true value is 19.1 ppb.

If you had taken only one measurement, you wouldn't have been sure it was accurate. If the range had been narrower, say 19.0 – 19.3 ppb, you would have been even more confident about your best estimate of the true value.

What you do	Data	Describing what you do
Take several measurements from the same air sample. Not all the measurements will be the same.	Concentration of NO_2 in parts per billion (ppb) 18.8, 19.1, 18.9, 19.4, 19.0, 19.2, 19.1, 19.0, 18.3, 19.3	The measurements (10 in this case) are called the data set.
Plot the results on a number line. This shows that the 18.3 ppb measurement is very different from the others. Decide whether to ignore this reading.	19.6 19.5 19.4 × 19.3 × 19.2 × 19.1 × × 19.0 × × 18.9 × 18.8 × 18.7 18.6 18.5 18.4 18.3 ⊗ — this result is an outlier	A result that is very different from the others is called an **outlier**. If you can think of a reason why this result is so different (eg, you made a mistake when you took this measurement), you should ignore it.
Add the other nine results together. Divide the total by 9. The answer is 19.1 ppb of NO_2.	Total of nine readings = 171.8 $\frac{171.8}{9} = 19.1$ ppb	19.1 is called the **mean value** of the nine measurements.
You can use the mean value rather than any of the nine measurements.	The best estimate for the concentration of NO_2 is 19.1 ppb	The mean value is used as the **best estimate** of the true value.
When you write down the mean value you also record: • the lowest, 18.8 ppb, • and the highest,19.4 ppb, measurements.	The range is 18.8 – 19.4 ppb	18.8 – 19.4 ppb is called the **range** of the measurements.

Comparing NO_2 concentrations

The graph on the right shows the mean and range for the concentration of nitrogen dioxide in three different places.

- Compare London and York. The means are different but the ranges overlap.
- The range for London overlaps the range for York. So the true value for London could be the same as the true value for York. You cannot be confident that their NO_2 concentrations are different.
- Compare London and Harwell. The means are different and the ranges do not overlap.
- You can be very confident that there is a real difference between the NO_2 concentrations in London and Harwell.

When you compare data, do not just look at the means. To make sure that there is a **real difference**, check that the ranges do not overlap.

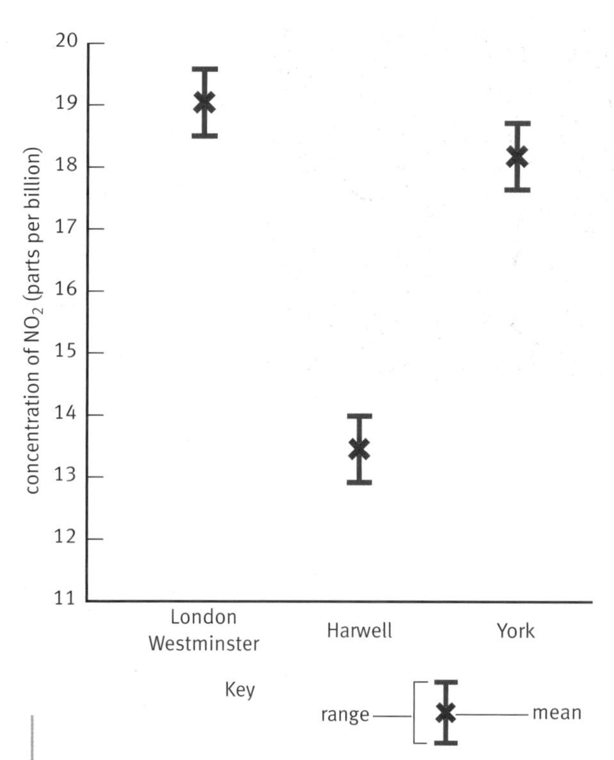

NO_2 concentrations in air from three places in England. All the measurements were made at the same time of day.

Questions

1 Jess measured the NO_2 concentration in the middle of a town. She took six readings: 22 ppb, 20 ppb, 18 ppb, 24 ppb, 21 ppb, 23 ppb. Jess used new equipment and was careful taking her measurements.
   a Calculate the mean value of the measurements.
   b Write down the best estimate and the range for the NO_2 concentration in this sample of air.

2 Look at the graph above. Does it show that there is a real difference in NO_2 levels between Harwell and York? Explain your answer.

3 Repeat measurements on an air sample produced these results for the NO_2 concentration:

Reading 1 – 39.4 ppb	Reading 2 – 45.8 ppb
Reading 3 – 42.3 ppb	Reading 4 – 38.7 ppb
Reading 5 – 39.7 ppb	Reading 6 – 32.7 ppb

There had been some problems with the equipment that day.
   a Plot these six readings on a number line.
   b Work out the mean NO_2 concentration and range for this sample.
   c Another sample was taken from a second place in the same town. The mean NO_2 concentration for this sample was found to be 44.1 ppb. Can you say with confidence that the second location had a higher NO_2 concentration than the first? Explain your answer.

4 A scientist took one measurement of NO_2 in an air sample. Explain why this would not give you much confidence in the accuracy of the result.

How are air pollutants formed?

Find out about

- what fuels are made from
- the products that result from burning fuels
- how burning fuels can make atmospheric pollutants

Many atmospheric pollutants are made by the burning of fossil fuels. This happens in power stations and in the engines of vehicles.

What happens when fuel burns in a power station?

Most power stations are fuelled by either coal or natural gas. Fuel and air go into the furnace and waste chemicals come out of the chimney. Use the diagram below to compare what goes in with what comes out of each type of power station.

Any change that forms a new chemical is called a **chemical change** or a **chemical reaction**.

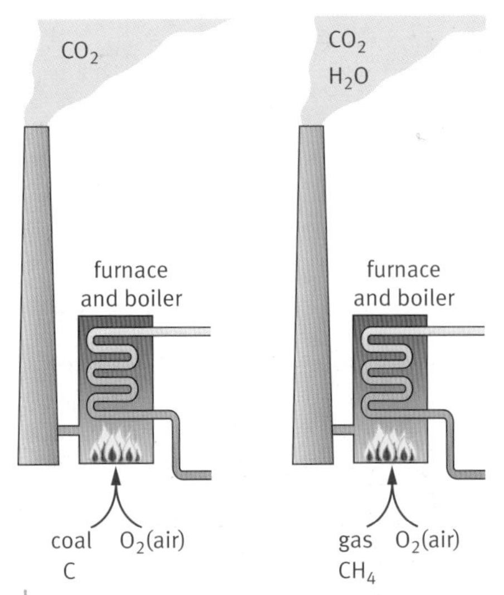

The chemicals going into and coming out of power station furnaces.

Coal-fired power station

Coal is mainly made up of carbon atoms. Oxygen molecules from the air are needed for coal to burn. The main product that comes out of the chimney of a coal-fired power station is carbon dioxide (CO_2). It must have been formed by the oxygen molecules (O_2) separating into oxygen **atoms**. These oxygen atoms then combine with the carbon atoms to make carbon dioxide. Reactions where a chemical joins with oxygen are called oxidation reactions.

Gas-fired power station

Natural gas is mainly methane (CH_4). Methane is a **hydrocarbon**. It is made of carbon and hydrogen atoms.

The main products from the burning of natural gas are CO_2 and H_2O.

These must have been formed by:

- carbon atoms and hydrogen atoms in CH_4 separating
- then carbon atoms combining with oxygen atoms to form CO_2
- and hydrogen atoms combining with oxygen atoms to form H_2O.

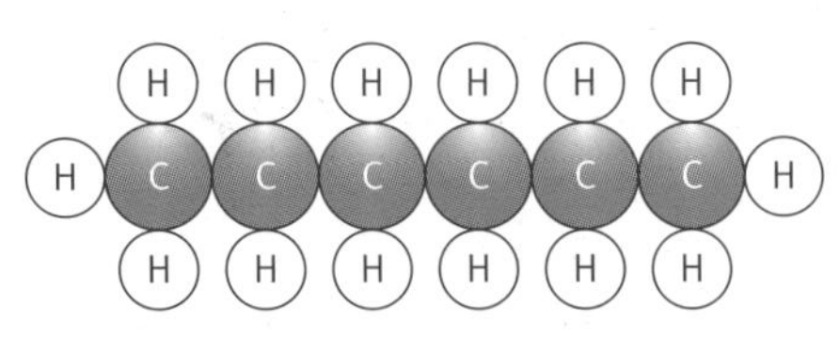

A hydrocarbon molecule. Natural gas, petrol, diesel, and fuel oil are all mainly made up of hydrocarbon molecules.

Other products from burning fuel

Burning coal and gas can also produce smaller amounts of these air pollutants:

- particulates – small pieces of unburned carbon
- carbon monoxide (CO) – formed when carbon burns in a limited supply of oxygen
- sulfur dioxide (SO_2) – formed if the fuel contains some sulfur atoms.
- nitrogen monoxide (NO) – formed when some of the nitrogen in the air reacts with oxygen at the high temperatures in the furnace

Nitrogen monoxide (NO) can then react with oxygen in the air to form nitrogen dioxide (NO_2). This is an oxidation reaction. Together, NO and NO_2 are referred to as nitrogen oxides.

Key words

- chemical change/reaction
- atom
- hydrocarbon

What happens when fuel burns in a car engine?

Vehicle engines burn petrol or diesel. These are also made up of hydrocarbon molecules. Use the diagram to compare what goes into a car engine with what comes out.

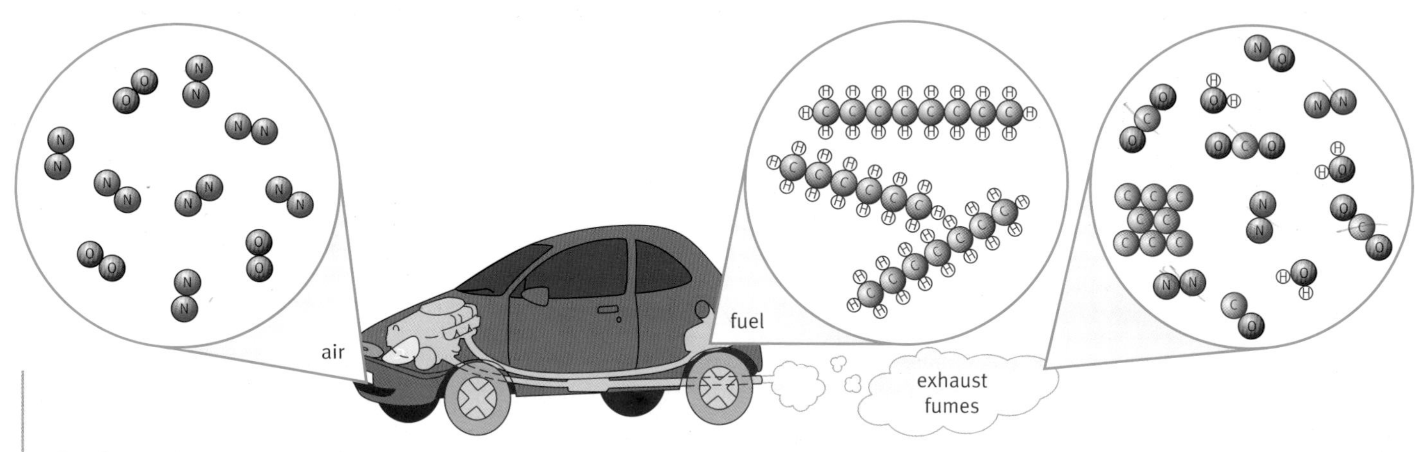

The chemicals going into and coming out of a car engine.

The overall change can be summarised as:

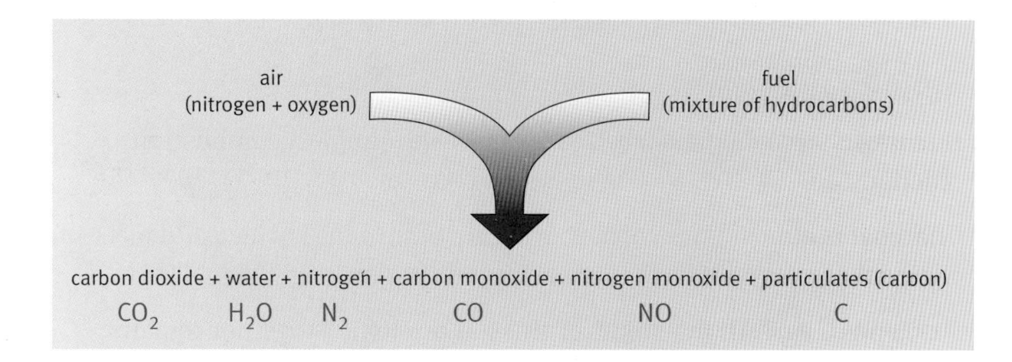

Check that you know which of the pictures in the three circles represents each of the chemicals mentioned in the summary. You can use these pictures to work out the chemical changes happening in the engine, and therefore how the pollutants are formed. For example, nitrogen monoxide (NO) must have been formed from nitrogen (N_2) and oxygen (O_2) in the air. These must have first split apart into atoms and then combined to make NO.

Questions

1. List the air pollutants that can be released from a coal-burning power station.
2. List the air pollutants released from a car engine when it burns fuel.
3. Use ideas about atoms separating and then joining together in different ways to explain how:
   - **a** H_2O forms when methane gas (CH_4) burns in a power station.
   - **b** CO forms when coal (C) burns in a power station.
   - **c** CO_2 forms when petrol burns in a car.

F What happens during combustion?

Find out about

- how atoms are rearranged during combustion reactions
- different ways of representing chemical changes

Reactions where a chemical joins with oxygen are called **oxidation** reactions.

Some chemicals can react rapidly with oxygen to release energy and possibly light. This type of oxidation reaction is called **combustion** or burning.

Fuel has escaped during this racing car crash. An uncontrolled combustion reaction is happening. The fuel and air mixture has been heated by either a spark or the hot engine.

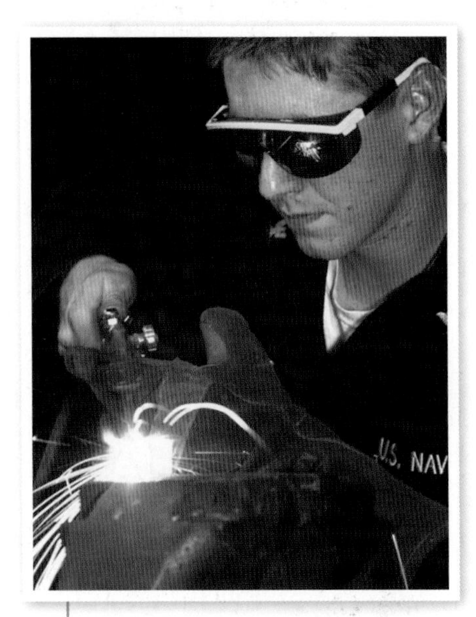

Fuel burns more rapidly in pure oxygen than in air. Oxygen obtained from the atmosphere is used in this oxy-fuel welding torch.

Burning charcoal

Burning charcoal on a barbeque is one of the simplest combustion reactions.

Charcoal is almost pure carbon. You can picture the surface of a piece of charcoal as a layer of carbon atoms tightly packed together.

Oxygen is a gas. All the atoms of this gas are joined together in pairs (O_2). These are molecules of oxygen.

During a combustion reaction, the atoms of carbon and oxygen are rearranged.

It will help you to understand this reaction if you can picture what happens to the atoms and molecules involved.

Air contains oxygen gas. One molecule of oxygen is two oxygen atoms joined together OO. Oxygen molecules split and react with carbon atoms in the charcoal. This forms carbon dioxide gas OCO.

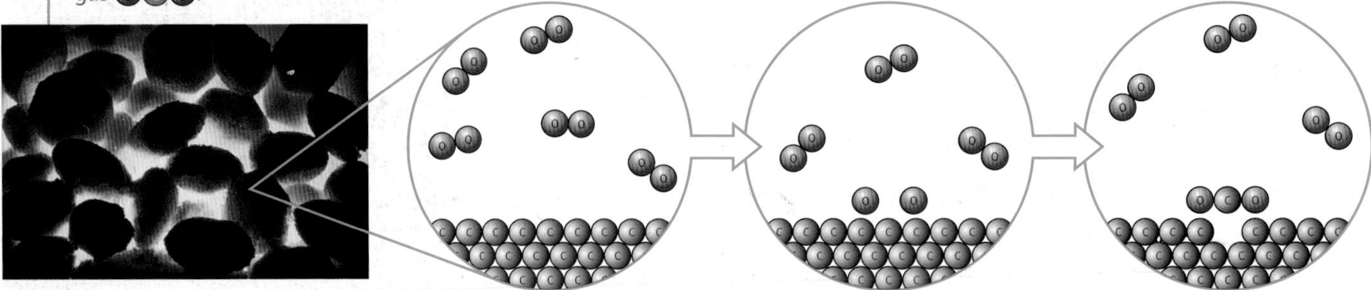

Describing combustion reactions

You can use pictures to describe the chemical change that happens when carbon dioxide burns.

+

The chemicals before the arrow are the ones that react together. We call them **reactants**.

The chemicals after the arrow are the new chemicals that are made. We call them **products**.

It would be time consuming if you always had to draw pictures to describe chemical reactions. So scientists use equations to summarise the pictures.

The combustion of charcoal can be summarised in this **word equation**:

carbon + oxygen ⟶ carbon dioxide

If you want more detail, you can write the **chemical equation** to show the atoms that make up each of the chemicals involved. This uses symbols for each chemical. These are called **chemical formulae**.

This is the chemical equation for the combustion of charcoal.

$$C + O_2 \longrightarrow CO_2$$

The chemical equation is a more useful description of the reaction than the word equation. It tells you how many atoms and molecules are involved and what happens to each atom.

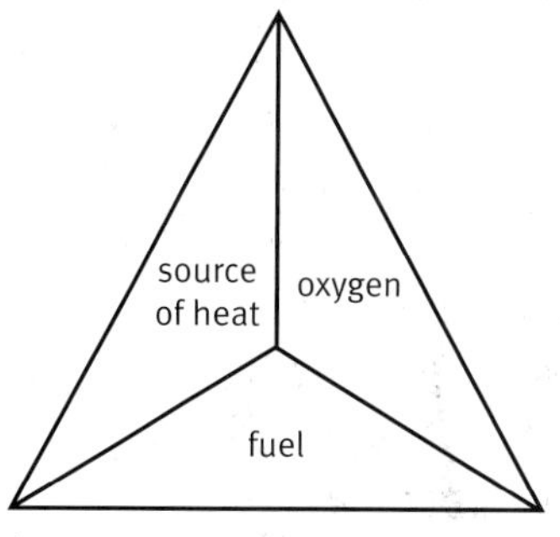

Three things are needed for a fire, or combustion reaction:
- a **fuel**, mixed with
- **oxygen** (air), and a
- source of **heat**, to raise the temperature of the mixture.

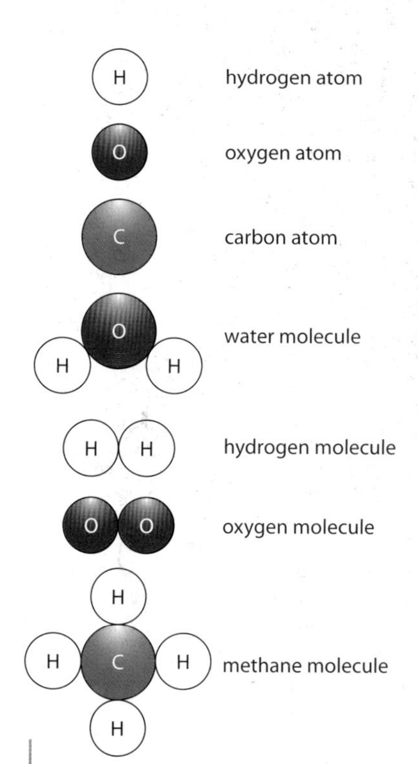

Atoms and molecules

Questions

1 How is a welding torch designed to produce a flame hot enough to melt and join metal?

2 What are the reactants and what are the products in each of the following chemical changes:
  a carbon combines with oxygen to form carbon dioxide
  b a hydrocarbon in petrol burns in oxygen to form carbon dioxide and water.

3 You and your cousin are having a barbecue. Your cousin asks you what happens to the charcoal when it burns. Write down what you would say. Include the words: atom, molecule, combustion, reactants, products, chemical change.

4 Draw pictures to represent these chemical changes:
  a hydrogen burning in oxygen to form water
  b methane burning in oxygen to form water and carbon dioxide.

Key words

- ✔ oxidation
- ✔ combustion
- ✔ reactants
- ✔ products
- ✔ chemical formula
- ✔ word equation
- ✔ chemical equation

Where do all the atoms go?

Find out about

- what happens to atoms during chemical reactions
- how the properties of reactants and products are different

Look at the picture below. How many atoms of hydrogen (H) are there before and after the reaction? Count the atoms of oxygen (O) before and after. What does this show?

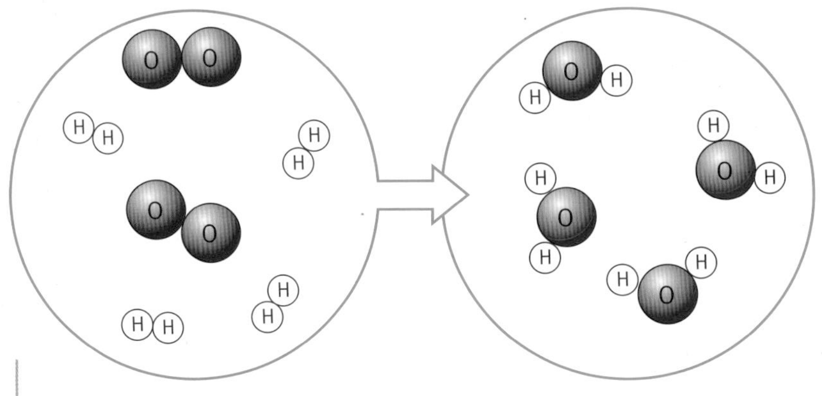

The reaction of hydrogen and oxygen to form water.

When you have had a bonfire, some of the atoms that made up the rubbish are in the ashes left on the ground. The others are in the products released into the air.

Conservation of atoms

All the atoms present at the beginning of a chemical reaction are still there at the end. No atoms are destroyed and no new atoms are formed. The atoms are conserved. They rearrange to form new chemicals but they are still there. This is called **conservation of atoms**.

For example, when a car engine burns fuel, the atoms in the petrol or diesel are not destroyed. They rearrange to form the new chemicals found in the exhaust gases.

Look again at the picture of hydrogen reacting with oxygen to form water. *Two* molecules of hydrogen react with just *one* molecule of oxygen. This produces *two* molecules of water. We can represent this change by:

Notice that there are the same numbers of each kind of atom on each side of the equation. All the atoms that are in the reactants end up in the products. The atoms are conserved.

All atoms have mass. Because the atoms are conserved, the mass of the reactants is the same as the mass of the products. This is called **conservation of mass**.

Properties of reactants and products

The **properties** of a chemical are what make it different from other chemicals.

For example, some chemicals are solids, some are liquids, and some are gases at normal temperatures. Some are coloured, some burn easily, some smell, some react with metals, some dissolve in water, and so on. Each chemical has its own set of properties.

The table compares the properties of the reactants and products of the reaction between sulfur and oxygen.

Chemical	Properties
sulfur (reactant)	yellow solid
oxygen (reactant)	colourless gas; no smell; supports life
sulfur dioxide (product)	colourless gas; sharp, choking smell; harmful to breathe; dissolves in water to form an acid

The atoms and molecules involved in the burning of sulfur.

In any chemical reaction, all the atoms you start with are still there at the end. But they are combined in a different way. So the properties of the products are different from the properties of the reactants.

This is very important for air quality. You can have a piece of coal that is a harmless black stone.

But the coal may contain a small amount of sulfur. When it burns the sulfur will react with oxygen to form the gas sulfur dioxide.

The sulfur dioxide escapes into the atmosphere. It will harm the quality of our air. It will react with water and oxygen in the air to form acid rain. This is harmful to plants, animals, and buildings. The harmless piece of coal has produced a harmful gas.

Key words

- conservation of atoms
- conservation of mass
- properties

Questions

1 You have learned that atoms are conserved during a chemical reaction. Work out how many molecules of CO_2 and H_2O will be produced when one molecule of methane (CH_4) is burnt. Draw a picture to show the atoms and molecules in the reaction. Work out how many molecules of O_2 will be used.

2 Water (H_2O) is made by reacting molecules of hydrogen and oxygen together. How are the properties of water different from the properties of the reactants it is made from?

3 Burning rubbish gets rid of it forever. Is this a true statement? Think about the atoms in the rubbish. Fully explain your answer.

H How does air quality affect our health?

Find out about

- how to look for links between air-quality data and the symptoms of an illness
- how pollen causes hayfever
- the link between asthma and air quality

Hayfever

Do you suffer from a runny nose, sneezing, and itchy eyes in the summer? This could be hayfever.

Hayfever got its name because people noticed that it happens in the summer. This is when grass is being cut to make hay. It is also the time when pollen from plants is at its highest.

To find out what causes hayfever it is important to first look at what **factors** are linked with hayfever. Factors are variables that may affect the outcome. In this case, hayfever is the **outcome.** Pollen is a factor that may affect hayfever.

Pollen traps collect pollen grains so that they can be counted using a microscope. This gives the 'pollen count'. Newspapers, radio, and television report the pollen count during the summer.

Pollen is released by plants and may travel many kilometres on the wind. Pollen grains are in the air that we breathe.

Is there a link between hayfever and pollen?

If an outcome increases (or decreases) as a particular factor increases, this is called a correlation. So, do more people suffer from hayfever when the pollen count increases?

Looking at thousands of people's medical records shows that hayfever is highest in the summer months. This is when most pollen is in the air. It is important to look at a randomly selected sample of medical records to collect data that is representative of the whole population.

This evidence shows that there is a correlation between pollen levels and hayfever symptoms. But does this mean that there is a causal link between pollen and hayfever – is pollen the **cause** of hayfever?

Pollen grains under the microscope. Different plants release different types of pollen. (Magnification approximately ×1360.)

Does pollen cause hayfever?

An increase in two things could be caused by a third factor that has not been measured. Or it could be a coincidence that the two things increase at the same time.

Think about ice cream. Most ice cream is sold in the summer months but nobody would say that ice cream causes hayfever. It may be just a coincidence. Or both increases may be caused by some other factor.

To claim that pollen causes hayfever you need some supporting evidence. You need to show that there is a correlation and you need to be able to explain *how* pollen causes hayfever.

Different types of pollen are released at different times of year. Some people have hayfever in months that correlate with particular types of pollen being released. This is strong extra evidence for the correlation between pollen and hayfever.

Skin-prick tests show that people who suffer from hayfever are allergic to pollen. Hayfever is an allergic reaction caused by pollen. So there is a **correlation** between hayfever and pollen, because pollen causes hayfever.

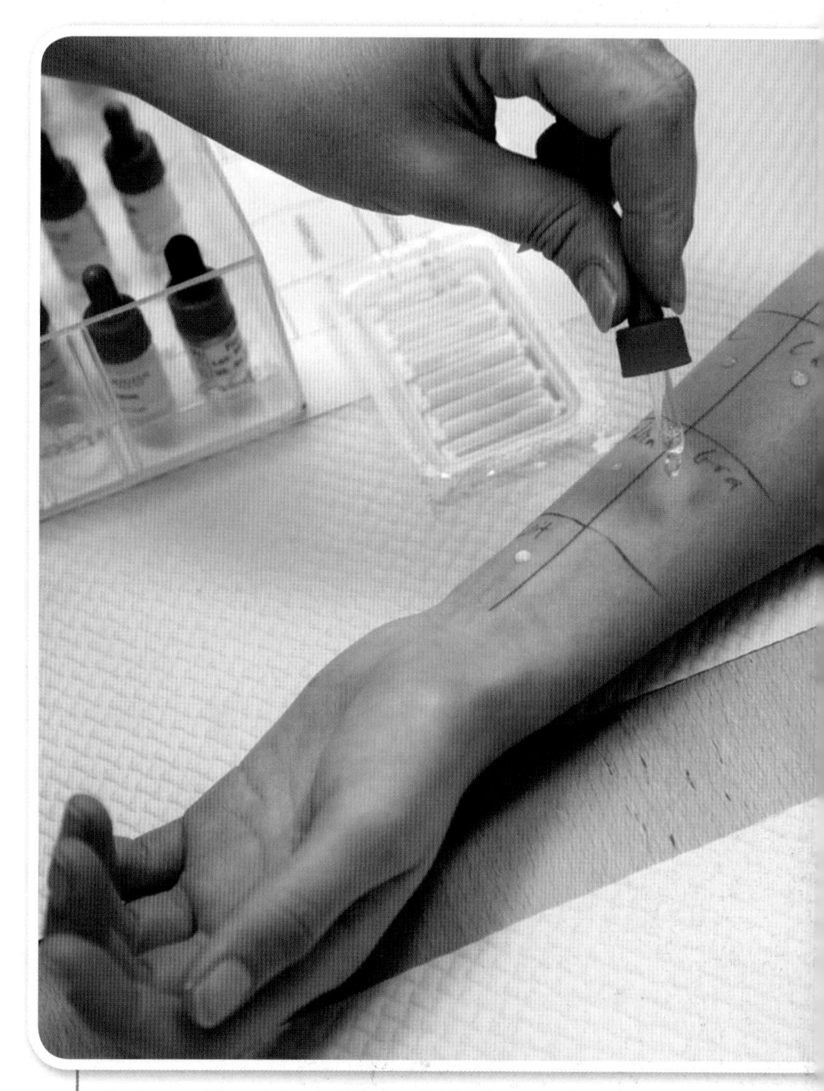

During a skin-prick test, drops of solution are placed on the skin. The skin beneath is pricked. If the patient is allergic to the substance in the solution (eg pollen), their skin will turn red and itchy.

Questions

1. Suggest why it is useful to report the levels of pollen in the air during the summer months.
2. Write a note to a friend explaining:
   - a what is meant by 'there is a correlation between pollen count and hayfever symptons'
   - b why you need to look at medical records of a large number of people to be sure there is a correlation
   - c why a correlation between ice cream sales and hayfever does not mean that ice cream causes hayfever.

Key words

- correlation
- cause
- factor
- outcome

Asthma and air quality

Asthma is a common problem. During an asthma attack, a person's chest feels very tight. They find it difficult to breathe.

Many things can trigger asthma attacks in people who have asthma. These include:

- tree or grass pollen
- animal skin flakes
- dustmite droppings
- air pollution
- nuts, shellfish
- food additives
- dusty materials
- strong perfumes
- getting emotional
- stress
- exercise (especially in cold weather)
- colds and flu.

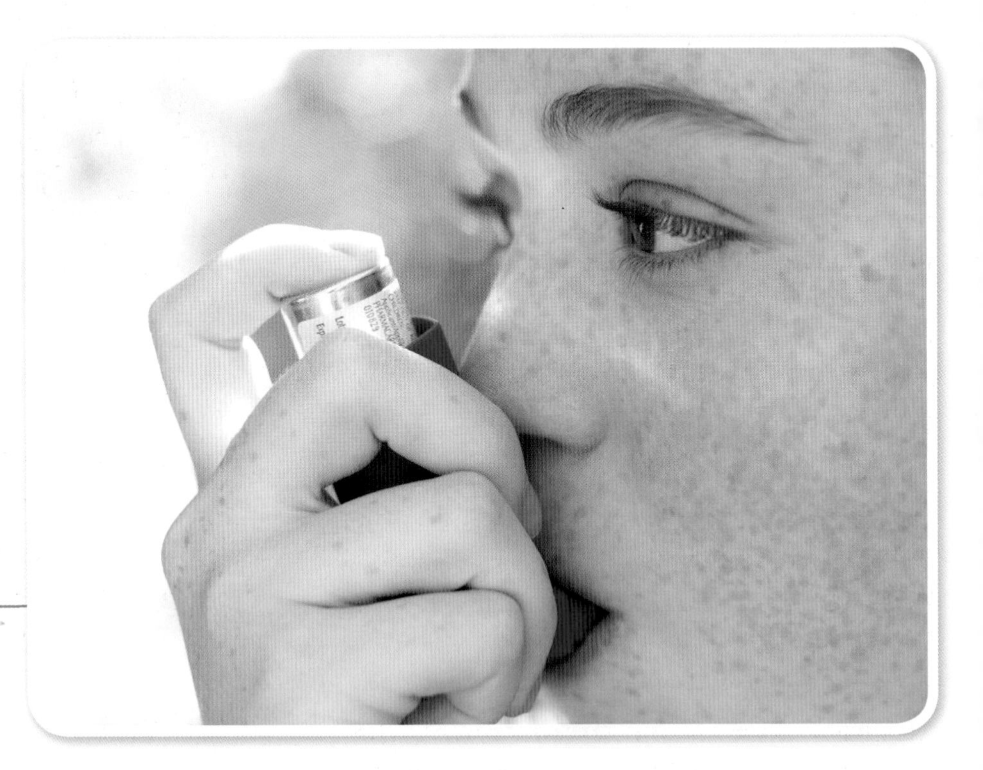

Inhalers are used to treat asthma attacks and to keep the condition under control. They contain medicines that help people's airways 'open up', allowing them to breathe more freely.

Causes of asthma

Medical evidence shows that asthma attacks are triggered by many different factors.

Nitrogen dioxide is an air pollutant that comes mainly from traffic exhausts. Large-scale studies have shown that if the concentration of nitrogen dioxide stays high for several days, there is an increased risk of people with asthma suffering from asthma attacks. This is a correlation.

People who have asthma have sensitive lungs. Air pollutants may irritate a person's lungs, particularly if their lungs are sensitive. If the level of nitrogen oxide pollution stays high for several days there is an increase in the number of asthma attacks. The data shows that there is a correlation between levels of pollution and asthma, but the data does not give clear evidence that nitrogen oxides cause asthma.

Exposure to nitrogen oxides increases the chance of an asthma attack but does not mean that all people with asthma will suffer an attack. Even so, the link between the factor (exposure to polluted air) and the outcome (an asthma attack) is still described as a correlation.

Exposure of children to high levels of traffic pollution may lead to an increased chance of their developing asthma.

Studying asthma

The causes of asthma are not fully understood. There is evidence to show that some people are more likely to have asthma attacks because of their genes. Other evidence indicates that environmental factors, such as air pollution, are also involved.

Scientists who study asthma publish their findings at conferences and in journals. This makes it possible for other scientists to evaluate their claims critically. Reviewers look carefully at the way that the scientists describe their methods of investigation, at their presentation of the data, and at the way they interpret that data and come to conclusions.

It has been suggested, for example, that stress could be a factor that increases the chance of an asthma attack. Supporters of the stress theory suggest that stress changes the way the lining of the lungs reacts to irritants.

The stress theory is just one of several explanations for the worldwide increase in the number of people affected by asthma. The data is complex and hard to interpret. This means that different scientists come up with different conclusions about the causes of asthma.

Questions

3 a Make a list of some of factors that might be a cause of asthma.

b Explain why scientists are finding it hard to work out why more and more people are being affected by asthma.

4 Why is it important that scientists publish their data and explanations?

5 Why do scientists look to see if published results can be replicated by other scientists before deciding to accept a scientific claim?

Find out about

- how laws and regulations can help improve air quality
- how new technology can reduce harmful emissions from cars and power stations
- what we can all do to reduce air pollution

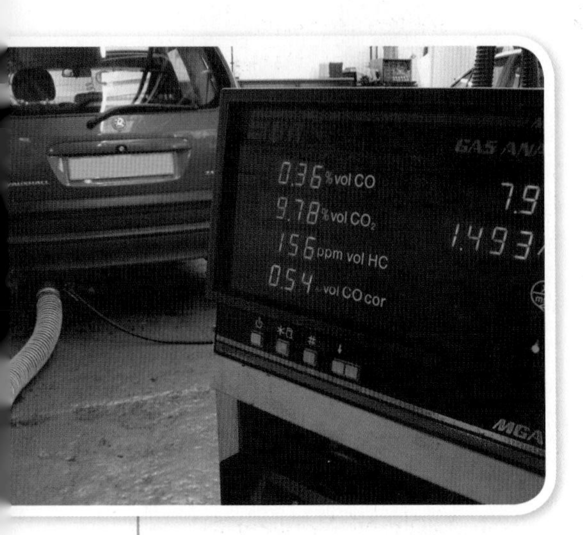

MOT exhaust emission analysis. Exhaust emissions are tested for carbon monoxide and unburnt fuel.

Reducing pollutants from cars

For individuals, the simplest way to improve air quality is to use the car less. A good alternative is to use public transport such as buses or trains, or walk or cycle. Fifty people travelling by bus use a lot less fuel than if they each travel in their own private car. Less fuel burnt means fewer pollutants released. By making it easier for people to use buses and trains, governments can help improve air quality.

Governments can also set legal limits on the amount of pollutants a vehicle is allowed to produce. In Britain, MOT tests include a vehicle emissions test, which means that cars with emissions above the legal limit are not allowed on the roads. **Regulations** state that vehicles over three years old must have an MOT test every year.

Efficient engines and catalytic converters

Engineers are continually working on improving the **efficiency** of car engines. A more efficient engine means that a car will burn less fuel to travel the same distance, reducing pollutants. This is good for car owners too because they do not need to buy so much fuel.

Even a very efficient engine will still produce air pollutants. This is why scientists have developed ways of removing the worst pollutants from the exhaust. All new cars have catalytic converters fitted to their exhaust systems. The waste gases pass through a metal honeycomb structure with a large surface area. The metal surface speeds up certain chemical reactions. A **catalytic converter** changes the pollutants carbon monoxide (CO) and nitrogen monoxide (NO) into less harmful gases.

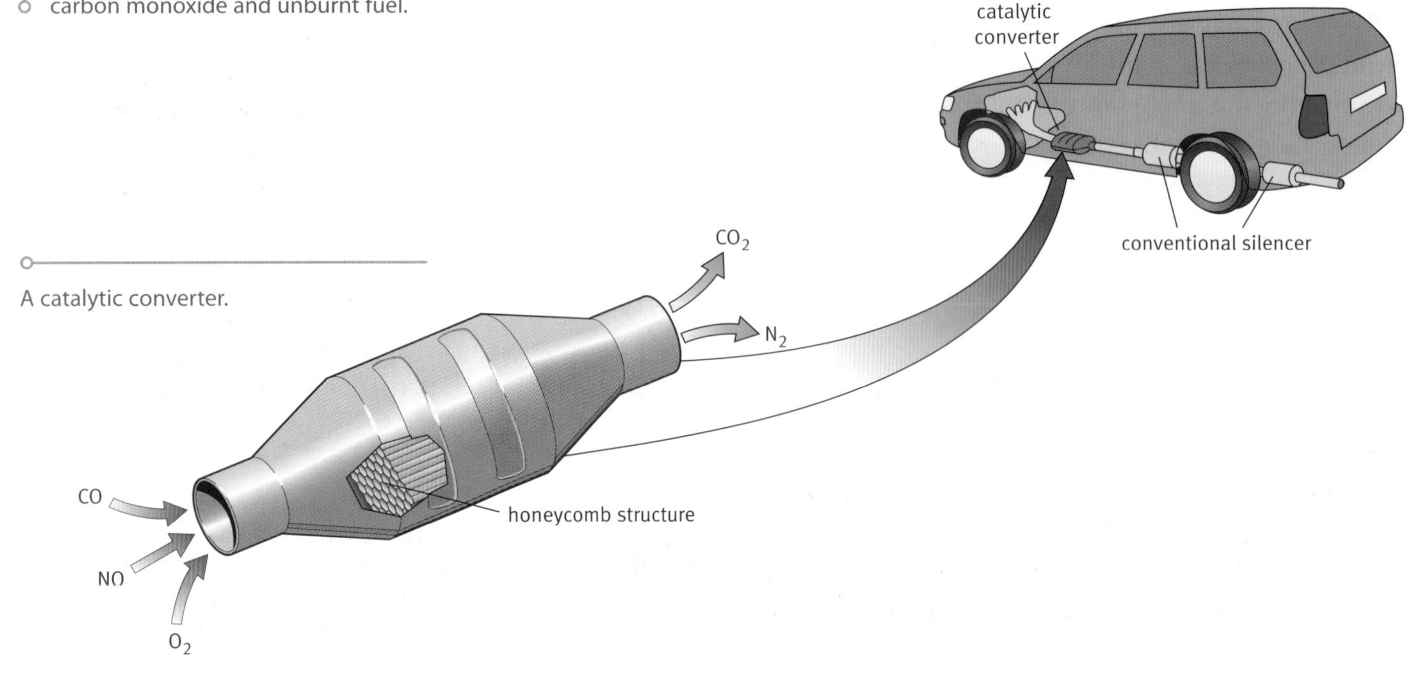

A catalytic converter.

The chemical reactions that occur in a catalytic converter are:

carbon monoxide + oxygen ⟶ carbon dioxide

nitrogen monoxide ⟶ nitrogen + oxygen

Oxygen is added to carbon monoxide. This is an oxidation reaction.

Oxygen is removed from nitrogen monoxide. This is a **reduction** reaction.

Even with a catalytic converter, carbon dioxide is still released. Carbon dioxide levels are a concern due to the link with global warming. The only way of producing less carbon dioxide is to burn less fossil fuel.

Cleaner transport

Transport can be made cleaner by improving existing fuels or using other sources of power.

Diesel fuel contains small amounts of sulfur compounds. Low-sulfur fuels have had these compounds removed. This reduces the amount of sulfur dioxide in exhaust emissions.

Biofuels are a renewable alternative to fossil fuels for motor vehicles. They are made from plants such as sugar cane, corn, and oilseed rape. The plants absorb carbon dioxide as they grow. When the fuels are burnt, this carbon dioxide returns to the atmosphere. Producing more biofuel would need more land given over to growing plants for fuel instead of food. A small amount (2–3%) of biofuel can be mixed into petrol or diesel without changing engines or fuel stations.

Some vehicles are now powered using electricity stored in batteries, rather than fuels such as petrol, diesel, or biofuels. Electric vehicles do not make any waste gases when they are used. However, the electricity comes from power stations, which in many cases burn fossil fuels. Using electric vehicles in congested cities means the pollutants are not released in the city but elsewhere. In comparison with petrol vehicles, electric vehicles have a short range before needing to be recharged. Recharging can take several hours.

This hybrid car runs on both electricity and petrol. The electricity comes from power stations through the national grid. By choosing a car that can use electricity, drivers can reduce the amount of petrol they use.

Key words

- efficiency
- catalytic converter
- reduction
- regulations

Questions

1 All new cars are fitted with a catalytic converter.
   a Which two pollutants are removed by catalytic converters?
   b For each pollutant, which less harmful gas is it changed into? What type of reaction is needed to do this?
   c Why is this not a perfect solution?

2 Hybrid cars use electricity, but also have a normal fuel engine for when it is needed.
   a Explain why using a hybrid car could reduce traffic pollution in a town.
   b Explain why using a hybrid car in electric mode can still result in pollutants reaching the atmosphere.

3 Write a letter to your local councillor, suggesting how people living in your area could reduce air pollution.

Trees killed by acid rain in the Czech Republic. Sulfur dioxide is a waste gas produced by power stations. It reacts with water and oxygen to form acid rain.

Reducing pollutants from power stations

Individuals, governments, and industries all have a role to play in reducing air pollution from power stations.

Each individual can contribute towards a reduction in pollutants released from power stations by using less electricity. However, even if people use less electricity, it is still good to reduce the air pollution produced in making the electricity we do use.

Most power stations burn coal, natural gas, or fuel oil. When these fossil fuels are burnt, waste gases and particulates are produced. The main product is carbon dioxide. Fossil fuels also contain impurities such as sulfur. When sulfur burns, the pollutant sulfur dioxide is produced. This can go on to produce acid rain.

Sulfur dioxide

One way of reducing the amount of sulfur dioxide released into the atmosphere is to remove sulfur from fuels before they are burnt.

Natural gas and fuel oil can be refined to remove sulfur, before they are burnt in a power station. This means that less sulfur dioxide is formed.

Scientists have also devised a way of removing sulfur dioxide from waste gases (or flue gases) before it can escape from the power-station chimney. This is called **wet scrubbing**.

Sulfur dioxide is an acid gas. Acids react with alkaline chemicals. Wet scrubbing is a process that uses an alkali to react with sulfur dioxide and remove it from flue gases.

Seawater is naturally slightly alkaline and can be used for wet scrubbing. The flue gases are sprayed with seawater droplets, which absorb and react with the sulfur dioxide. The droplets are collected and removed and the cleaned flue gases are released through the power-station chimney.

Other alkalis can be used for wet scrubbing. Powdered lime (calcium oxide) and water can be mixed together to form an alkaline slurry. The flue gases are mixed with air and sprayed with this alkaline slurry. The sulfur dioxide in the flue gases reacts and forms a new solid chemical called calcium sulfate. The solid is collected and removed, and the cleaned gases continue up the chimney. Calcium sulfate can be sold and used as building plaster.

cleaned gases to chimney
waste gases from furnace
calcium sulfate
air
lime and water

Removing sulfur dioxide from flue gases by wet scrubbing with an alkaline slurry.

Particulates

Particulates (tiny particles of carbon and ash) are also found in power-station flue gases. These are a problem because they can make surfaces of buildings dirty, and cause breathing problems. They can be removed by passing them through an electrostatic precipitator. This contains electrically charged plates. The particulates pick up a negative charge, are attracted to the positive plate, and are then collected and removed.

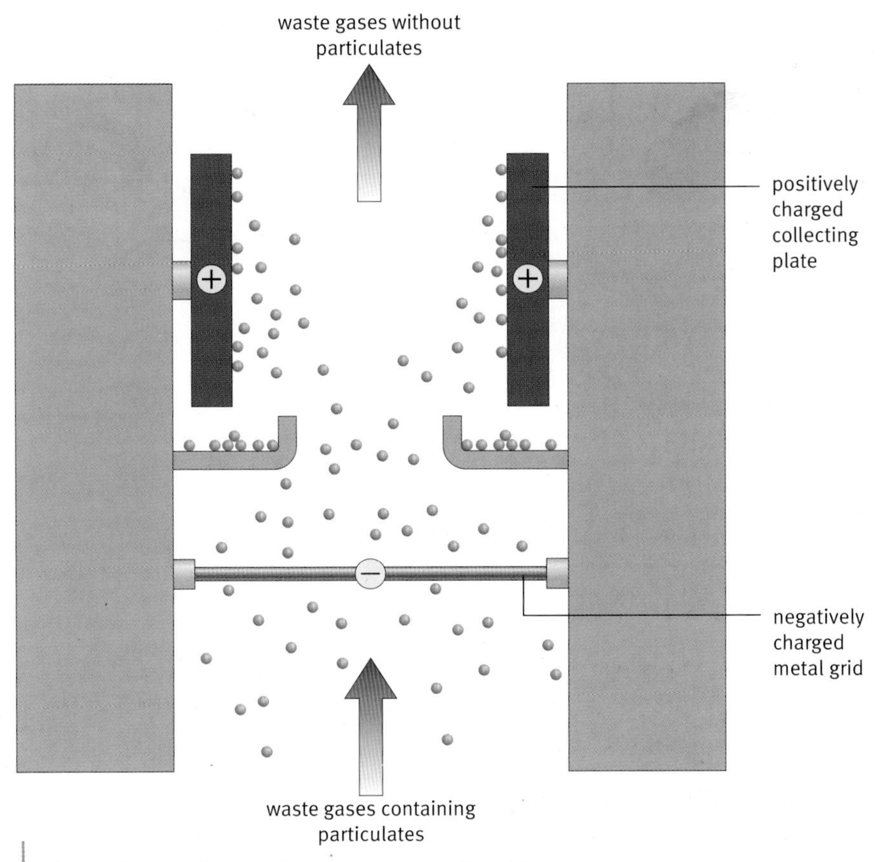

Removing particulates from a power-station chimney to prevent them being released into the atmosphere.

Key word

- wet scrubbing

Questions

4 Name three types of fossil fuel that are burned in power stations.

5 What problems are caused by sulfur dioxide in the atmosphere?

6 Describe one way of stopping sulfur dioxide being formed in power stations, and another way of removing it from waste gases if it is formed.

7 Describe how particulates are removed from power-station waste gases.

Science Explanations

Air pollution can affect people's health and the environment. In order to improve air quality it is important to understand where air pollutants come from and how they are made.

You should know:

- how the Earth's early atmosphere was formed and how it changed as a result of the evolution of photosynthesising organisms
- the importance of the oceans in removing carbon dioxide from the atmosphere leading to the formation of sedimentary rocks and fossil fuels
- which gases now make up the Earth's atmosphere, and how natural events and human activity continue to add gases and particulates to the air
- that burning fossil fuels changes the atmosphere by adding extra carbon dioxide (contributing to global warming) and smaller amounts of other pollutant gases as well as tiny particles of solids (such as particulate carbon)
- that the reactions when fuels burn are oxidation reactions
- that, when a hydrocarbon burns, the oxygen atoms from the air combine with the carbon atoms to form carbon dioxide, and with the hydrogen atoms to form water
- why some pollutants are directly harmful to humans and some are harmful to the environment
- that in any chemical reaction the atoms of the reactants separate and recombine to form different chemical products
- that the number of atoms of each kind is the same in the products as in the reactants
- that the properties of the reactants and products of chemical changes are different
- why the incomplete burning of fuels produces particulate carbon and a poisonous gas, carbon monoxide
- why some fuels produce sulfur dioxide gas when they burn, and why this gas gives rise to acid rain if released into the air
- why the waste gases from fuels burning inside a furnace or engine produce nitrogen oxide gas
- that, when released into the air, nitrogen oxide combines with more oxygen to make nitrogen dioxide gas, which can also contribute to acid rain
- that technological developments such as catalytic converters and wet scrubbing can reduce the amounts of pollutants released into the atmosphere.

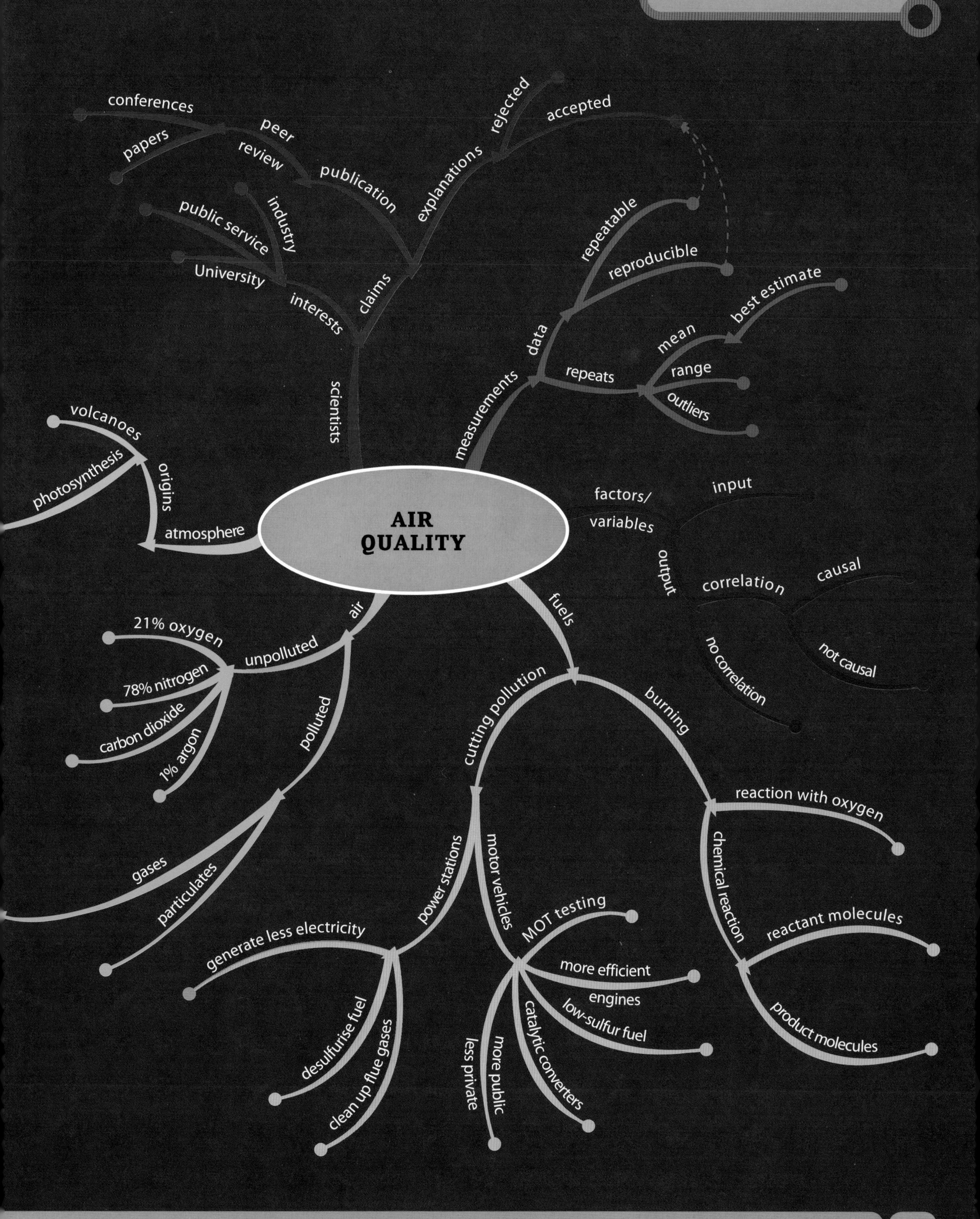
AIR QUALITY
scientists
interests
University
public service
industry
claims
publication
peer review
papers
conferences
explanations
rejected
accepted
measurements
data
repeatable
reproducible
repeats
mean
range
outliers
best estimate
factors/variables
input
output
correlation
causal
not causal
no correlation
atmosphere
origins
volcanoes
photosynthesis
air
unpolluted
21% oxygen
78% nitrogen
carbon dioxide
1% argon
polluted
gases
particulates
fuels
cutting pollution
power stations
generate less electricity
desulfurise fuel
clean up flue gases
motor vehicles
MOT testing
more efficient engines
low-sulfur fuel
catalytic converters
more public
less private
burning
reaction with oxygen
chemical reaction
reactant molecules
product molecules

Ideas about Science

Scientists use data, rather than opinions, to justify their explanations. They collect large amounts of data when they investigate the causes and effects of air pollutants. They can never be sure that a measurement tells them the true value of the quantity being measured.

If you take several measurements of the same quantity, the results are likely to vary. This may be because:

- you have to measure several individual examples, for example, exhaust gases from different cars of the same make
- the quantity you are measuring is varying, for example, the level of nitrogen oxides in exhaust gases
- the limitations of the measuring equipment or because of the way you use the equipment.

The best estimate of the true value of a quantity is the mean of several measurements. The true value lies in the spread of values in a set of repeat measurements.

- A measurement may be an outlier if it lies outside the range of the other values in a set of repeat measurements.
- When comparing information on air quality from different places a difference between their means is real if their ranges do not overlap.
- A correlation shows a link between a factor and an outcome, for example, as the level of particulates in the air goes up the number of people suffering from lung disease goes up.
- A correlation does not always mean that the factor causes the outcome.

Scientists publish their results so that their data and claims can be checked by others. Scientific claims are only accepted once they have been evaluated critically by other scientists.

- Reviewers check claims to make sure the scientists who did the work have checked their findings by repeating them.
- The scientific community generally does not accept new claims unless they have been reproduced by other scientists.
- Scientists may come to different conclusions about the same data.

Some applications of science, such as using fuels, can have a negative impact on the quality of life or the environment.

- The only way of producing less carbon dioxide is to burn less fossil fuel.
- Pollution caused by power stations that burn fossil fuels can be reduced by using less electricity, removing sulfur from natural gas and fuel oil, and by removing sulfur dioxide and particulates (carbon and ash) from the flue gases.
- Pollution caused by vehicles can be reduced by burning less fuel in more efficient engines, using low-sulfur fuels, and using catalytic converters.

Review Questions

1 The table shows the percentage of gases in dry air. Copy and complete the table by writing the correct gas next to each percentage.
Choose from these gases.

argon **hydrogen**
nitrogen **oxygen**

Gas	Percentage
	1
	21
	78

2 Ethanol can be used as a fuel instead of petrol. When pure ethanol burns completely, it reacts with oxygen (O_2) and produces carbon dioxide (CO_2) and water (H_2O) as the only products. Copy and complete this drawing to show the products of the reaction when pure ethanol is burned completely.

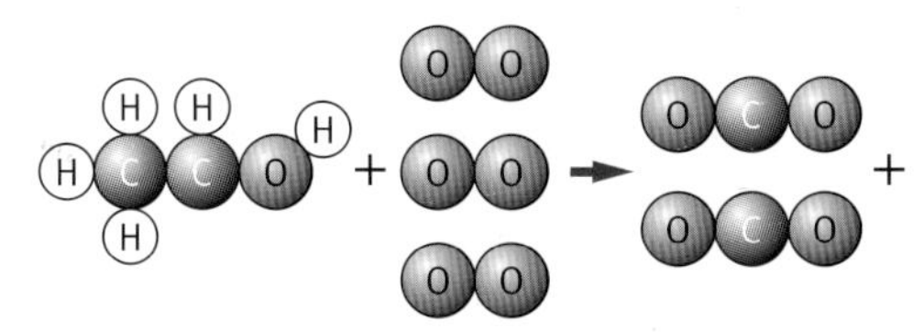

3 Students measured the pH of rain falling on their school playground. They collected and tested six samples of water on the same day. To get a best estimate, they worked out the mean pH from their results.
Their results are shown in the table.

a What is the range?
b What is the mean?

Sample	1	2	3	4	5	6
pH	5.8	5.6	5.8	4.2	5.7	5.6

4 The fumes that are released into the air from a car exhaust cause pollution.

Nitrogen monoxide is made as fuel burns in a car engine. Which chemicals react to make nitrogen monoxide?

- **ammonia**
- **hydrogen**
- **oxygen**
- **carbon dioxide**
- **nitrogen**
- **sulfur**

5 When nitrogen monoxide is released into the air, it reacts with oxygen to form nitrogen dioxide. Why is nitrogen dioxide not formed in the car engine?

6 Which of these measures could do the most to reduce the amount of nitrogen dioxide pollution in the air?

a Use low-sulfur fuels.
b Adjust the balance between public and private transport.
c Encourage people to use less electricity.

7 Several pollutants are released from car exhaust systems. These pollutants do not stay in the air. Over a period of time they are removed by a number of processes. Which pollutants are removed by each of the processes described in the table?

Process	Pollutant
reacts with water and oxygen to form acid rain	
deposited on surfaces, making them dirty	
used by plants and dissolves in sea water	

C2 Material choices

Why study material choices?

All the things we buy are made of 'stuff'. That stuff must come from somewhere. Before about 1900, virtually everything we used was made of naturally occurring materials that came from plants, animals, or rocks. Since then, people have discovered ways to change the properties of some materials that occur in nature – and also to make completely new materials. Materials are chosen for specific jobs because of their properties.

What you already know

- Chemicals can be elements or compounds.
- Chemicals can also be mixtures – two or more chemicals mixed together but not chemically combined.
- Different chemicals and materials have different properties.
- The properties of a material make it suitable for particular uses.
- Molecules are made of atoms.
- In a chemical reaction, atoms separate and recombine to form different chemicals.
- In a chemical reaction, the atoms are conserved.
- How to analyse the results of an experiment.

Find out about

- the testing and measurement that helps people to make good choices when buying products
- some of the explanations scientists use to design better materials
- the variety of polymers and plastics, and how they are used to meet our needs
- how nanotechnology is helping scientists to design new materials with a wide range of properties.

The Science

Scientists use their knowledge of molecules to explain why different materials behave in different ways. This gives them the ability to design new materials to meet a wide range of needs.

Ideas about Science

Scientists test products to check that they can do the job, are good value, and safe. You can use data from these tests when you buy a product. So you need to be able to judge whether or not the results can be trusted.

A Choosing the right stuff

Find out about

- materials and their properties
- natural and synthetic materials
- long-chain polymers

Key words

- material
- properties
- polymer
- natural
- synthetic
- ceramic
- metal
- mixture
- flexible

Latex is a natural polymer that can be tapped from rubber trees. After treatment, it is used in a wide variety of products, including the soles of shoes.

What the advertising agency didn't tell you

Of course chocolate shoes are a joke. Chocolate is not a good **material** for making shoes. Here are some reasons:

- chocolate would crack
- it would melt in warm weather
- dogs would follow you and lick your feet
- it would wear away too quickly
- it would leave a mess on the carpet.

Maybe not chocolate

Although chocolate does not have the right **properties**, the idea of moulded shoes is not new. South American Indians used to dip their feet in liquid latex straight from the rubber tree. They would sit in the sun to let the latex harden, forming the first, snugly fitting, wellies. Latex is more suitable than chocolate for making shoes. Let's see what properties it has that make it better.

Fantastic elastic

The most obvious difference between latex and chocolate is that latex is **flexible**. Any material chosen to make our shoes needs to be flexible so you can bend your feet.
It also needs to be:

- hard wearing because you will walk on it
- waterproof
- a solid at room temperature
- elastic so it keeps its shape
- tough so that it won't crack when it bends.

Latex has all these properties whereas chocolate does not. Latex is a polymer.

Products are made from a wide range of materials.

What are polymers?

All **polymers** have one thing in common. Their molecules are very long chains of atoms. This is true for **natural** polymers such as cotton, leather, and wool and for **synthetic** polymers such as polythene, nylon, and neoprene. Most of the materials used to make shoes are polymers.

Choosing materials

We make items from a huge range of materials. We use **ceramics** for mugs, plates, tiles, glass windows, bricks, and toilets. We use **metals** for an enormous number of products including aircraft, cars, pipes, wires, jewellery and sports equipment. We use polymers for making bags, clothes, window frames, and computers. All these materials are chemicals. Some of the metals are pure chemicals, not mixed with anything else; most of the materials we use are **mixtures** of chemicals.

When a designer is deciding how to make a product they can choose which material to use according to which properties they require the finished item to have.

What's in a name?

Sometimes words can have more than one meaning. The word 'material' can mean cloth or fabric, but to a scientist it means any sort of stuff you can use to make things from.

Questions

1. Look at the picture of young people in a car. Identify items that could be made from:
   a ceramics
   b metals
   c polymers.
2. Leather is a natural polymer. Suggest which properties of leather make it a good material for smart shoes.
3. Steel is a metal. Steel is sometimes used to make toecaps for work boots, but it is not used to make the whole boot. Suggest some properties that steel does *not* have that would be useful in a boot.

Using polymers

Find out about

- how natural and synthetic polymers meet our needs
- examples of polymers and their uses
- how and why natural polymers are being replaced with synthetic ones

All sorts of polymers meet people's most basic needs. These include:

- physical needs for shelter, warmth, and transport
- bodily needs for food, water, hygiene, and healthcare
- social and emotional needs for human contact, leisure, and entertainment
- needs of the mind to stimulate thinking and creativity.

Natural polymers

Before synthetic polymers were discovered, these needs were met using natural polymers as well as materials like metals, glass, and ceramics. The pictures below show some materials that are natural polymers.

Cotton fabric has been used for clothing and household textiles for many years. Cotton grows around the seeds of the cotton plant.

Silk looks and feels luxurious, and has long been popular for high-quality clothing. It is a protein fibre obtained from the cocoons of the silkworm.

Wool is good at regulating temperature, so is used in a variety of clothing and textile products. It is a protein fibre obtained from animals such as sheep.

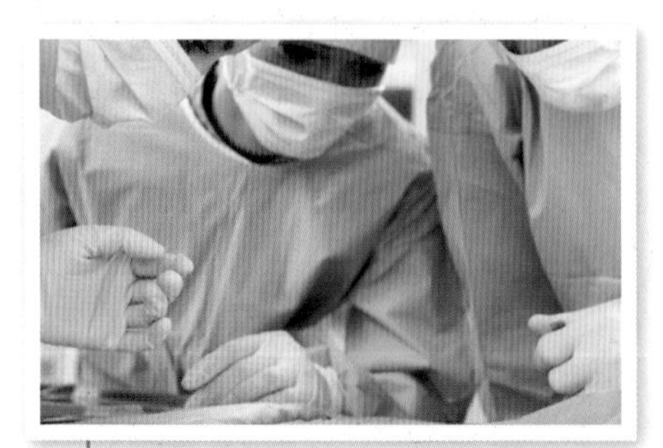

Doctors and other health workers wear gloves made of natural **rubber** (latex) for protection and to prevent infection. Latex is tapped from some plants, including the rubber tree.

Fur was one of the first materials used for clothing. Today its use is controversial – many view its use as cruel and unnecessary.

Paper has many uses including for decoration, for leisure, and for passing on information. It is made from wood pulp, which consists of cellulose fibres.

From natural to synthetic

The pictures below and on the following page show some synthetic polymers. Many items that used to be made from natural polymers are now made of synthetic ones. As more and more synthetic polymers are made, designers and engineers have more to choose from when they make new products. Each new polymer has different properties, which may be superior to those available in nature.

- Natural fibres for clothing are replaced with synthetic fibres, which may be easier to wash, hold their shape better, or be available in a wider range of colours.
- Fur and leather may be replaced with synthetic polymers, which avoids using animal products.
- Wood is often replaced with plastics, which are much lighter and do not rot or require painting.
- Paper bags are often replaced with plastic ones, which are lighter and also waterproof.

Key word

✓ rubber

Polythene bags help people to protect, store, and carry food.

This patient in Sri Lanka is fitting a new leg made of polypropylene.

The world's first inflatable church made from PVC.

PET is a polyester used to make soft-drinks bottles and other food containers.

Polyester is used to make hulls and sails.

This acrylic painting was on show in a shop in Zanzibar.

Manchester City stadium roof is made from polycarbonate 'glass'.

Kevlar helmets have saved many soldiers' lives.

A wet suit made from neoprene offers warmth and protection.

Questions

1. Create a chart, diagram, or table to show how polymers can meet our needs. Use the examples throughout Section B and any other examples that you know of.
2. Give at least three examples of objects that were once made of natural polymers but are now made of synthetic ones. In each case, suggest a reason for the change.
3. Name an object now made from plastic rather than metal and explain why.
4. Name an object now made from synthetic polymers rather than ceramics and explain why.
5. Give an example where replacing natural polymers with synthetic ones may be good for:
   a the environment
   b animal welfare.
6. Give an example where replacing natural polymers with synthetic polymers may be bad for the environment.

Getting the right material

Manufacturers and designers have to choose the right materials to make their products. They decide which materials to use based on their properties and cost. In many products, the materials include polymers.

Find out about

- words scientists use to describe materials
- testing materials to ensure quality and safety

Key words

- strong
- tension
- compression

Modern materials with special properties.

For example, the soles of shoes have to be flexible, hard wearing, and strong. Also, they must not crack when they bend – they have to be tough. A synthetic rubber is a good choice.

The case of a computer has to be very different from shoe soles. It needs to be stiff, strong, and tough. People want a case that resists scratches and keeps its appearance. So the polymer has to be hard.

Material words

When scientists describe the properties of materials, they use special words. Some of these, like 'strong', have everyday meanings that are similar to their technical meaning. However, some are a little different.

A material is **strong** if it takes a large force to break it. Some materials are strong when stretched. Examples are steel and nylon, which are strong in **tension**. Concrete tends to crack when in tension but it is very strong in **compression**. This makes it useful for pillars and foundations.

A suspension bridge must be strong. The cables are made of steel, a material that is strong in tension. The columns are made of concrete, a material that is strong in compression.

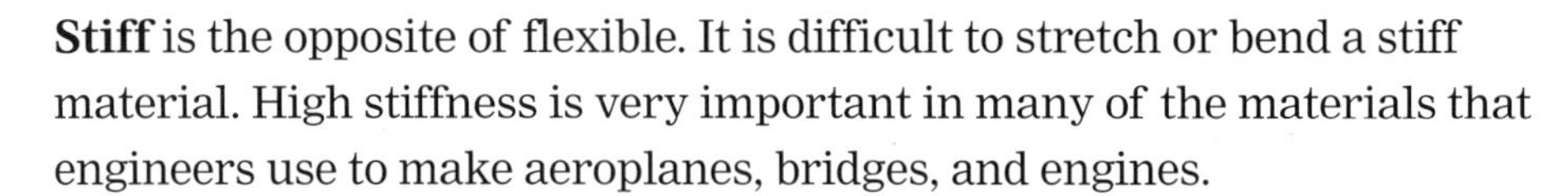

Stiff is the opposite of flexible. It is difficult to stretch or bend a stiff material. High stiffness is very important in many of the materials that engineers use to make aeroplanes, bridges, and engines.

Hard and **soft** are also opposites. The softer a material, the easier it is to scratch it. A harder material will always scratch a softer one.

In many applications it is also important to know how heavy a material is for its volume. Materials such as steel and concrete have a high **density**. Other materials are very light for their volume and have a low density. Examples are foam rubber and expanded polystyrene.

Key words

- stiff
- hard
- soft
- density

A testing machine for plastic packaging. Measuring the force needed to crush the container gives a value for the strength of the pack.

Measuring the words

Technical words help to describe materials. There are times when more than a description is needed. Accurate measurements of properties are necessary when it is important to compare materials and test their quality.

For example, a pole used for pole vaulting must be flexible, but not too flexible or it will not support the weight of the person using it. The landing pit must be soft but not too soft so that the pole vaulter lands safely. In situations like these the properties of the materials must be measured to make sure that they are just right.

Engineers test materials and products and measure their properties. The flexibility of a material can be found by measuring how much it bends under a given force. The force that breaks a material tells you its strength. The density of a material can be calculated from its mass and volume.

Questions

1 Look at the picture of rollerbladers. Identify items that they are wearing, or parts of the rollerblades, that are:
   a flexible b stiff c strong d hard

2 Look at the two pictures of material testing. Which is a test of strength in tension? Which is a test of strength in compression?

3 Suggest reasons for measuring the strength of packaging materials.

This machine measures the force needed to break sewing threads. Samples from every batch that leaves the factory are tested to make sure the threads are always the same.

Quality control

It is particularly important to take accurate measurements when someone's safety depends on it.

Abseiling ropes must not break when they are being used, so the ropes are tested to make sure that they are strong and safe. The ropes must be able to hold much more than the weight of one person.

Most ropes are produced in batches. A machine is used to test a sample of rope from each batch and find out the force that is needed to break it. Standard procedures are followed carefully to make sure the measurements are accurate. **Accuracy** is how close a measurement is to the true value.

Repeatable and reproducible data

Each time a test is carried out the same method is used. This means that if someone tested the same rope twice they would expect to get the same result – the measurement is **repeatable**. It also means that if anyone else carried out the test using this method they would expect to get the same results – the measurement is **reproducible**.

In reality there may be small differences in the results obtained each time a test is carried out. This is because there may be errors, for example, errors within the machine. These errors can be kept small by regularly calibrating the machine. Calibration is used to check that the readings given by the machine are accurate and adjusting the machine if they are not.

It is important to control all the factors that are not being tested that may affect the results, for example, temperature. The minimum breaking load is the force needed to break the rope at standard room temperature. If the rope was tested at a different temperature this might affect the strength of the rope and you could not be sure that the rope meets the requirements.

The best estimate of the true value of the breaking load is found by repeating measurements on at least three samples, then calculating the mean and finding the range.

Abseiling ropes must be strong in tension. Climbers want to be sure that their ropes have been tested.

Key words

- ✓ accuracy
- ✓ repeatable
- ✓ reproducible

Questions

4. Give one factor that is controlled when abseiling ropes are tested.
5. Explain why is it important to repeat measurements.
6. Samples of abseiling rope were tested for their strength in tension and the following results were collected:

   27 546 N   27 356 N   27 598 N   27 467 N.

   Calculate the mean and the range for this set of measurements.

D Zooming in

Find out about

- materials under the microscope
- molecules and atoms in materials
- models of molecules

A woollen jumper is very different from a silk shirt. The shirt is more formal and less stretchy than the jumper. They are both made from natural polymers but they are very different. Their properties depend on their make-up, from the large scale to the invisibly small:

- the visible weave of a fabric
- the microscopic shape and texture of the **fibres**
- the molecules that make up the polymer
- the atoms that make up the molecules.

Silk.

The visible weave

The fabric of a woven shirt is tightly woven but even so it is possible to see the criss-cross pattern of threads. The fabric is hard to stretch because the strong threads are held together so tightly.

On the other hand, a knitted jumper is soft and stretchy. The loose stitches allow the threads to move around.

The weave and the stitches are visible to the naked eye. They are **macroscopic** features. However, the properties of a fabric also depend on smaller structures.

Magnification: × 20. Visible: to naked eye. Width of circle: 4 millimetres.

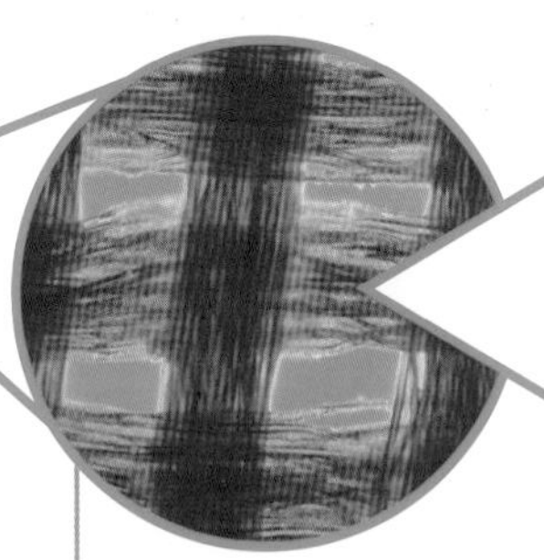

Magnification: × 1000. Visible: down a microscope. Width of circle: 80 micrometres.

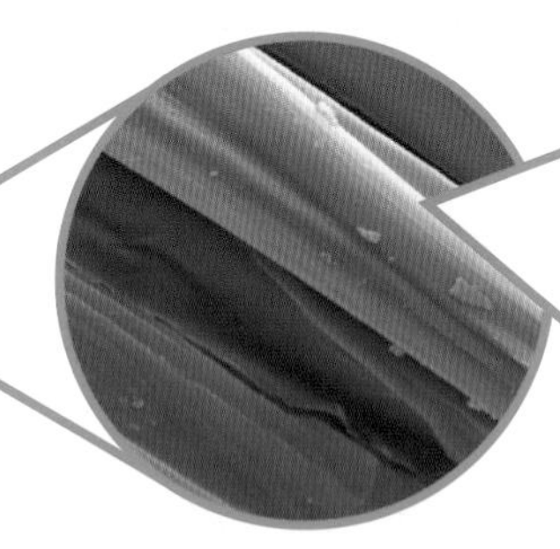

Magnification: × 50 million. Visible: not even in a microscope. Width of circle: 1.5 **nanometres**.

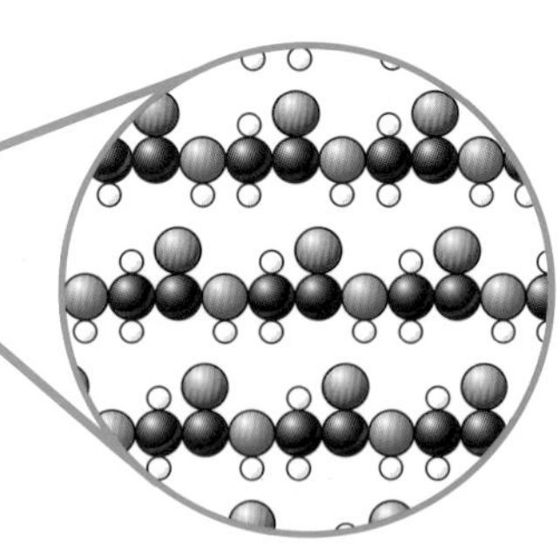

Levels of structure and detail. A millimetre is a thousandth of a metre. A micrometre is a thousandth of a millimetre. And a nanometre is a thousandth of a micrometre.

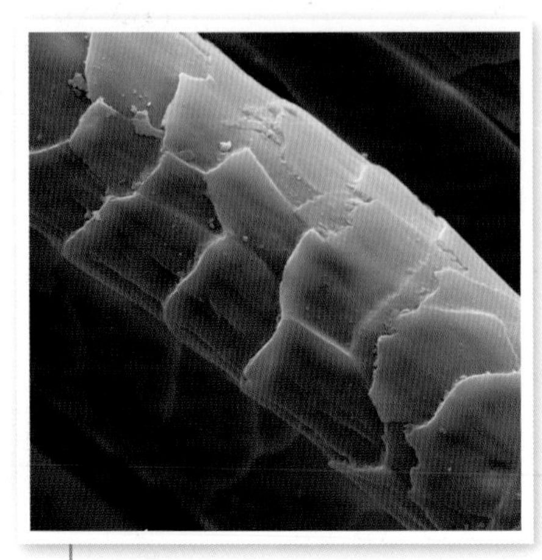

A wool fibre magnified 1250 times.

Taking a closer look

A microscope can show details of the individual fibres in a fabric. Silk, for example, has smooth, straight fibres that slide across each other.

Wool fibres have a rough surface that is covered in scales. The wool fibres tend to cling to each other in the thread and also make the threads cling together.

The invisible world of molecules

It is difficult to look much further into the structure of materials using microscopes of any kind. Scientists explain the differences between silk, wool, and other fibres by finding out about their molecules. Molecules are very small indeed, so small that it needs a giant leap of the imagination to think about them.

Scientists measure the sizes of atoms and molecules in nanometres (nm). There are 1 000 000 000 nanometres in a metre. Some molecules, such as the small molecules in air, are even smaller than one nanometre but many are bigger.

The molecules in fibres are big on the nanometre scale. They are very long – 1000 nanometres or more. The shape and size of the **long-chain molecules** in a fibre make the material what it is. Polymers have special properties because the molecules in them are so long.

Model molecules

Even the largest molecules and atoms are invisible. So in the nanoworld of molecules, scientists build models based on the results from their experiments.

Models of molecules can be compared with the map of the London tube system. The tube map does not look like an underground railway. But it has lots of useful information about the way the stations are connected. In a similar way, models of molecules do not look like real molecules. But they show what scientists have discovered about the atoms in the molecules and how they are joined together.

Questions

1 a Put the following in order of size, starting with the largest: fibre, fabric, atom, thread, molecule.
  b Use the words in part **a** to write four sentences that describe the decreasing structures. The first sentence might be: Fabrics are made by weaving together threads.

2 a How many chemical elements are there in silk?
  b A hydrocarbon contains hydrogen and carbon atoms only. Is silk a hydrocarbon?

3 A polymer molecule is about 1000 nanometres long. An atom is about 0.1 nanometres across.
  a Estimate how many atoms there are along the chain.
  b How many molecules would fit into a millimetre?

Key words

- fibre
- macroscopic
- nanometre (nm)
- long-chain molecules

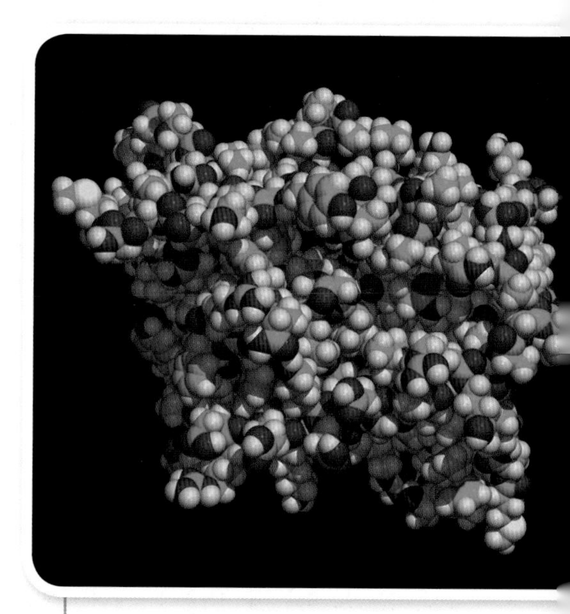

Computer model of a protein molecule. No-one knows what atoms and molecules look like. It helps to use models to understand what they do. In the real world the atoms are not coloured. In this computer image the atoms of each element are colour-coded: carbon (green), sulfur (yellow), nitrogen (blue), hydrogen (grey), and oxygen (red).

The big new idea

Find out about

- polymer discoveries
- polymers as long-chain molecules
- monomers that make up polymers

The 1930s was the decade of the first synthetic polymers. The world was a tense place and war was on its way. Governments were looking for scientific solutions to give them an advantage. This speeded up many scientific developments. Some of these used the big new idea: polymers. However, the first synthetic polymer was discovered by accident.

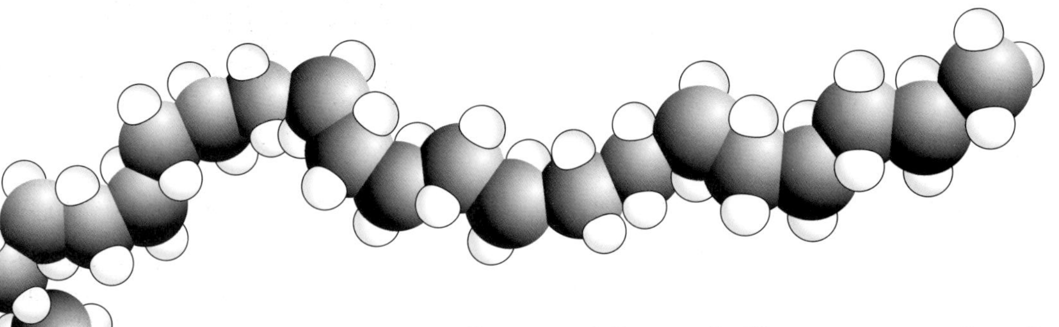

The accidental discovery of polythene

In 1933, two chemists made polythene thanks to a leaky container. Eric Fawcett and Reginald Gibson were working for ICI. Their job was to investigate the reactions of gases at very high pressures. They had put some ethene gas into the container and squashed it to 2000 times its normal pressure. However, some of the ethene escaped. When they added more ethene, they also let in some air.

Two days later, they found a white, waxy solid inside the apparatus. This was a surprise. They decided that the gas must have reacted with itself to form a solid. They realised that, in some way, the small molecules of ethene had joined with each other to make bigger molecules.

They worked out that the new molecules were like repeating chains. The chains were made from repeating links of ethene molecules.

Later they understood that oxygen in the air leaking into their apparatus had acted as a catalyst. The oxygen speeded up what would otherwise have been a very, very slow reaction to join the ethene molecules together.

What are polymers?

Polymers all have one thing in common: their molecules are long chains of repeating links. Each link in the chain is a smaller molecule. These small molecules are called **monomers**. They each connect to the next one to form the chain. This is true for natural polymers such as cotton, silk, and wool and for synthetic polymers such as polythene, nylon, and neoprene.

The common name for the polymer discovered by Fawcett and Gibson is polythene. This is short for the chemical name poly(ethene). The word poly means 'many'; a poly-ethene molecule is made from many ethene molecules joined together.

The original high-pressure container used by Fawcett and Gibson is on display at the Science Museum. The diagrams show what was happening to the small ethene molecules as they joined up in long chains to make polythene. This is polymerisation.

A polymer pioneer

Wallace Carothers was an American chemist who discovered neoprene and invented nylon. Neoprene was another accidental discovery. A worker in Carothers' laboratory left a mixture of chemicals in a jar for five weeks. When Carothers had a tidy up, he discovered a rubbery solid in the bottom of the jar. Carothers realised that this new stuff could be useful. He developed it into neoprene. This synthetic rubber first came on the market in 1931 and is still used today, to make wetsuits, for example. This discovery helped Carothers to work out a theory of how small molecules can **polymerise**.

The discovery of nylon

Japan and the USA were on bad terms in the years before World War II. Trade between them was difficult and the supply of silk was cut off. It became rare and expensive. Carothers started looking for a synthetic replacement. In 1934, his team came up with nylon. This is a polymer made from two different monomers. The different molecules join together as alternate links in the chain.

Sadly, Carothers died before he could see the effects of his discoveries. Nevertheless, they are both still in use today.

Key words

- monomer
- polymerise

Questions

1 What is a polymer, and what is a monomer?

2 a Write down the names of two polymers that were discovered by accident.

b Many scientists have made accidental discoveries. All of these words might be used to describe these scientists:

lucky, skilful, foresight, inventive, creative

Choose two of these words to describe the scientists. In each case, explain why you have chosen that word.

Molecules big and small

Find out about

- long and short polymer chains
- properties of polymers

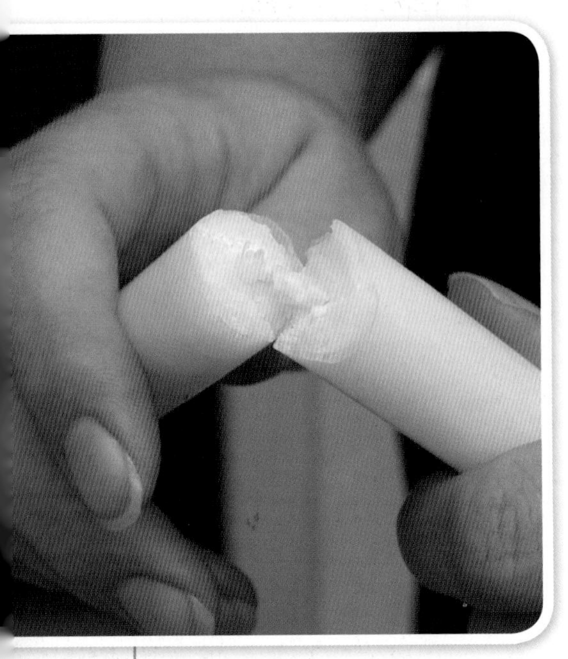

The molecules of candle wax are about 20 atoms long. Wax is weak and brittle.

The molecules of polythene are similar to those of candle wax. But they are about 5000 times longer. Polythene is much stronger and tougher than candle wax.

Long and short molecules

The properties of a polymer depend on the length of its molecules. The molecules in candle wax are very similar to those in polythene. However, wax is weaker and more brittle than polythene. Wax also melts at a lower temperature. This is because the wax molecules are much shorter. They contain only a few atoms; polythene molecules contain many thousands.

Two different bonds

The molecules are made of atoms. The bonds between *atoms* in the molecules are strong. So it is very hard to pull a molecule apart. The molecules do not break when materials are pulled apart.

But the forces between *molecules* are very weak. It is much easier to separate molecules from one another. They can slide past each other.

Breaking and melting wax and polythene

Stretch or bend a candle and it cracks. This is because separating the small molecules is not difficult. The forces between the molecules are very weak and they slip past each other quite easily.

Breaking a lump of polythene is much more difficult. Its long molecules are all jumbled up and tangled. It is harder to make them slide over each other. The long molecules make polythene stronger than wax.

Polythene also has a higher melting point than wax. This is because the forces between long polythene molecules are slightly stronger than the forces between short wax molecules. More energy is needed to separate the polythene molecules from each other so it melts at a higher temperature.

Question

1 Bowls of pasta can be used as an analogy to explain the difference between wax and polythene. One bowl contains cooked spaghetti. The other bowl contains cooked macaroni (or penne).
   - a In the analogy, what represents a molecule?
   - b Which kind of pasta represents wax and which represents polythene?
   - c Show how this analogy can help to explain why polythene is stronger than wax.

Hardening rubber

Natural rubber is a very flexible polymer, but it wears away easily. This makes it good at rubbing away pencil marks, but not much else.

Sometime around 1840, an American inventor called Charles Goodyear was experimenting with mixing sulfur and rubber. He was trying to improve the properties of the natural material. He accidentally dropped some of his mixture on top of a hot stove. He didn't bother to clean it off, and the next morning it had hardened.

It took two more years of research to find the best conditions for this new process, which Goodyear called **vulcanisation**. It made rubber into a stronger material that was more resistant to heat and wear. At that time no-one knew why this happened. They just knew that it worked and that it made rubber an excellent material for car tyres.

Goodyear started a business making tyres. He began with tyres for bicycles and prams. Now the business makes tyres for cars, motorcycles, and aeroplanes.

Find out about

- using science to change polymer properties
- cross-links to make polymers harder
- plasticisers to make polymers softer
- polymers that are crystalline

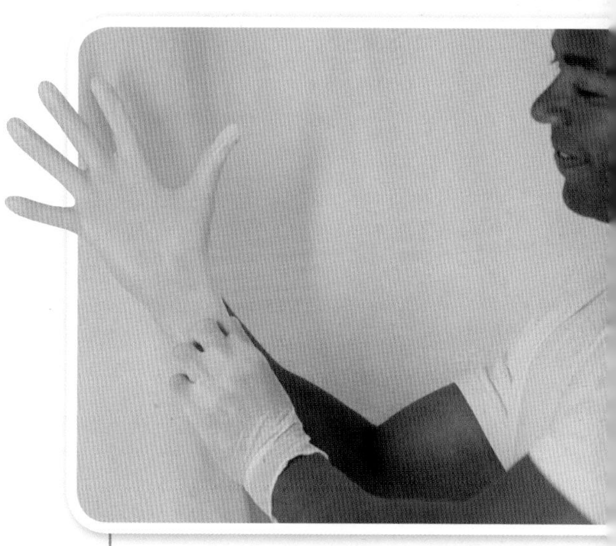

Vulcanising natural rubber produces gloves that are strong enough not to tear.

Cross-links

Goodyear was the first person to alter the properties of a polymer. He did not know why vulcanisation worked – only that it did. Now that we understand more about molecules, we know what's going on.

The sulfur makes **cross-links** between the long rubber molecules. The molecules are locked into a regular arrangement. This stops them from slipping over each other and makes the rubber less flexible, stronger, and harder. It also gives the rubber a higher melting point because more energy is needed for the molecules to separate and break out of the solid.

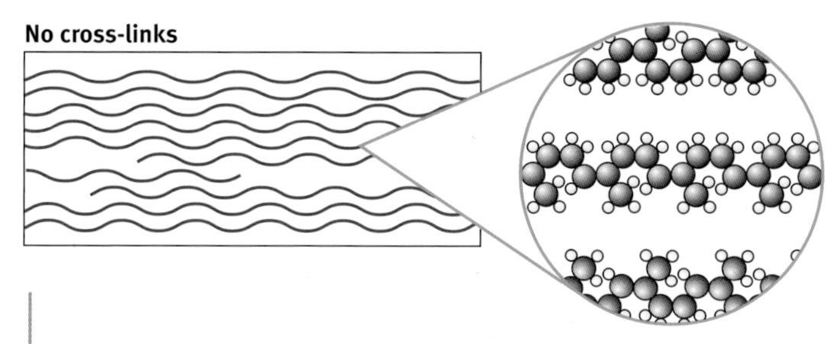

Each line represents a polymer molecule. Without cross-linking, the long chains can move easily, uncoil, and slide past each other.

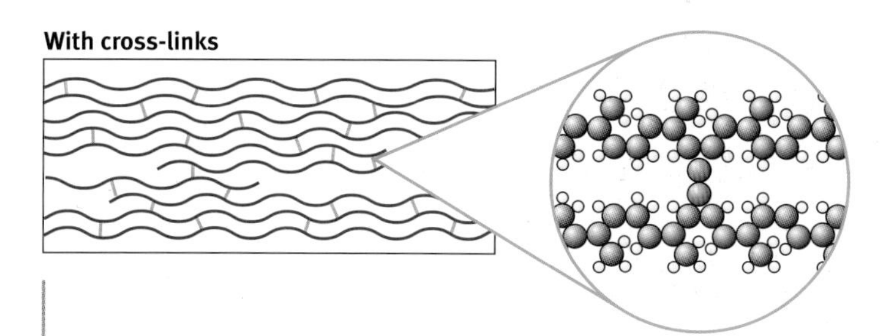

The sulfur atoms form cross-links across the polymer chains. This stops the rubber molecules uncoiling and sliding past each other.

Key words

- vulcanisation
- cross-links
- durable
- plasticiser

Softening up

PVC is a polymer often used for making window frames and guttering. These need to be **durable** and hard. PVC is also a good polymer for making clothing, but for this purpose it needs to be softer and more flexible.

To make PVC softer, the manufacturer adds a **plasticiser**. This is usually an oily liquid with small molecules. The small molecules sit between the polymer chains.

The polymer chains are now further apart. This weakens the forces between them, so less energy is needed to separate them and they slide over each other more easily. This means the polymer is softer and more flexible, and has a lower melting point.

This PVC is unplasticised. It is called uPVC.

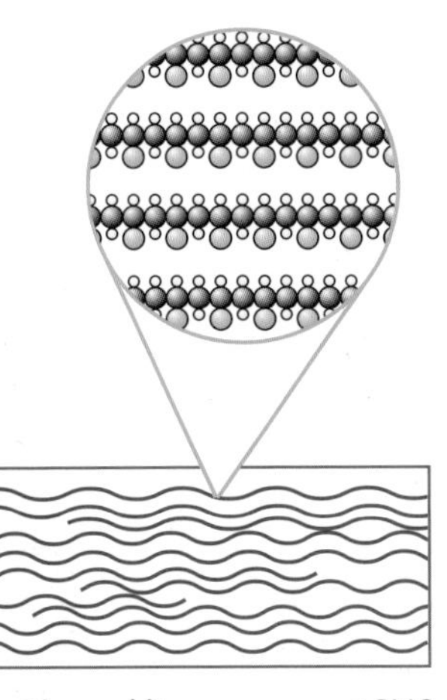

The red lines represent PVC molecules. These PVC molecules are long chains that lie close together. The closer they are, the stronger the forces between them.

This PVC has been plasticised to make it soft.

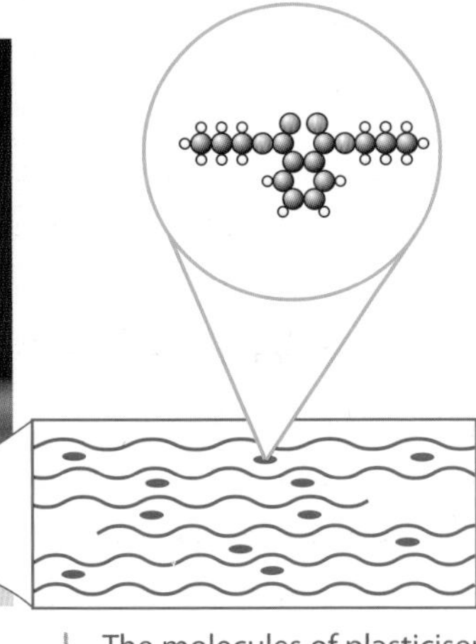

The molecules of plasticiser hold the PVC chains apart. This weakens their attraction and makes it easier for them to slide past each other.

Question

1 a Chemists can vary the extent of cross-linking between chains in rubber. How would you expect the properties of rubber to vary as the degree of cross-linking increases?

b What effect do plasticisers have on the properties of polymers?

Cling film

Cling film was first made from plasticised PVC. Unfortunately, the small plasticiser molecules were able to move through the polymer and into the food. Some people worried that the plasticiser might be bad for their health. The evidence that plasticisers are harmful is controversial and strongly challenged by the plastics industry (see C3, K: Benefits and risks of plasticisers).

There is stronger evidence that the regular use of plastic food wrap can cut down on food poisoning, which is a serious and growing risk to health.

Some cling film is now made using PVC and plasticisers that are much less likely to move from the polymer to food.

Original polythene

Chemists now know that in polythene made under high pressure the polymer chains have branches, which stop the molecules packing together neatly.

Compare a bonfire pile with a log pile. In a bonfire pile, the twigs and side branches stick out all over the place and the pile is full of holes. But a pile of logs is neatly stacked. It is the same with polymers but on a much smaller scale. If there are side chains, the structure is messy and full of holes. This is the case with polythene made under pressure, which has **branched chains**.

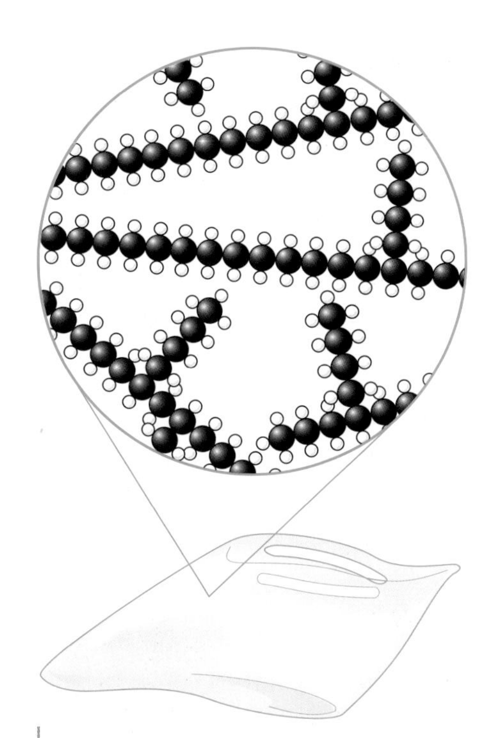

Polythene molecules made from ethene under pressure have side branches. This stops the polymer molecules lining up neatly. This type of polythene has a slightly lower density and is not crystalline.

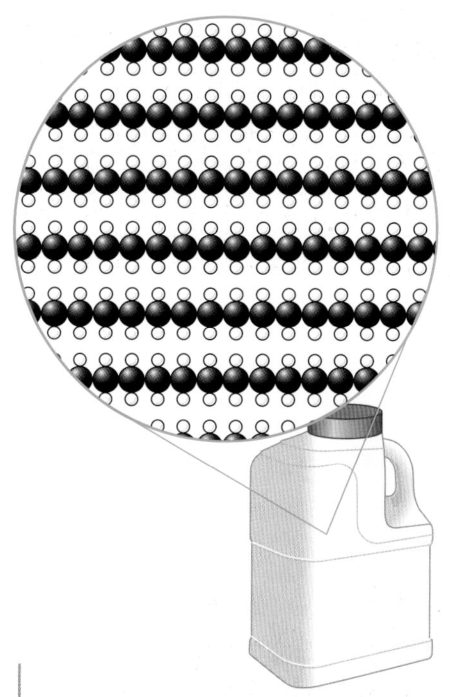

Polythene molecules made from ethene with a special catalyst do not have side branches. The polymer molecules line up neatly. This type of polythene has a slightly higher density and is crystalline.

A stronger, denser polythene

In some polymers the molecules are all jumbled up in a very irregular arrangement. In other polymers the molecules are arranged in neat lines; these are called **crystalline polymers**. There are also polymers that are partly irregular and partly crystalline. Scientists realised that increasing the crystallinity of polythene might make it stronger and denser. In the 1950s scientists found a way of making polythene molecules in neat piles of straight lines. It was an international effort by a German, Karl Ziegler, and an Italian, Giulio Natta.

They used special metal compounds as catalysts. These metal compounds act in a similar way to the oxygen in the high-pressure process. They speed up the rate at which the ethene molecules join together. The growing polymer chains latch onto the solid catalyst. The regular surface of the solid allows the molecules to build up more regularly.

In this new crystalline form of polythene, the molecules are more neatly packed together. The forces between the molecules are slightly stronger and more energy is needed to separate them. The new form is stronger and denser than the older type and softens at a higher temperature. Both types are still made – the old branched-molecule form is called low-density polythene (LDPE). The newer crystalline form is high-density polythene (HDPE).

Key words

- branched chains
- crystalline polymers

Questions

2 Why is HDPE slightly denser than LDPE? Suggest an explanation based on the structure and arrangements of molecules.

3 LDPE starts to soften at the temperature of boiling water. HDPE keeps its strength at 100°C. Suggest some products that would be better made of HDPE rather than LDPE and give your reasons.

Designer polymers

Find out about

- using science to design new polymer products

Gore-tex is waterproof and windproof, yet it allows the moisture from sweat to pass through.

Key word

- melting point

Ingenious layers: Gore-tex

Sometimes layers of different polymers with different properties are sandwiched together. An interesting example is the waterproof fabric Gore-tex, named after its inventor Bob Gore. He was working with a polymer called PTFE. This is the plastic coating for non-stick pans.

Gore discovered that if a sheet of PTFE is stretched, it develops very small holes and becomes porous. A single water molecule can pass through the small holes. But a whole water droplet is too large to get through. This got Gore thinking. His idea was that vapour evaporating from someone's skin would pass through the polymer sheet, but that rain drops would not.

Gore-tex has a layer of PTFE sandwiched between two layers of cloth. The wearer stays dry and comfortable no matter how energetic they are or what the weather is like. Sweat can always pass out through the fabric, but no water can get in.

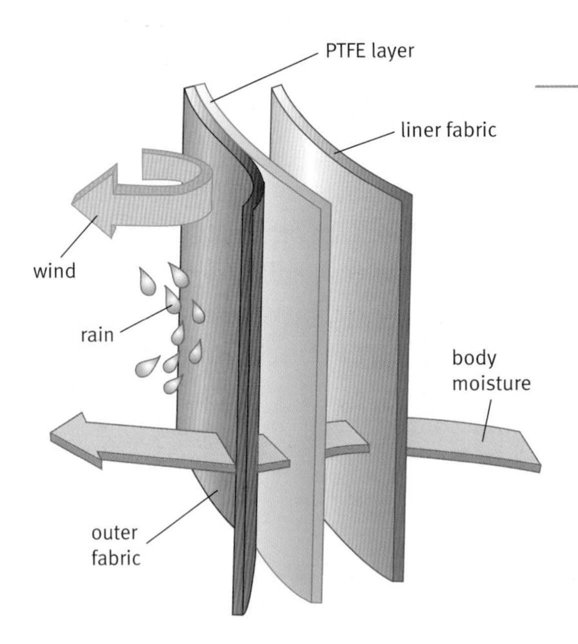

Gore-Tex membrane: there are billions of tiny holes in the film of PTFE. The holes are 20 000 times smaller than a raindrop but 700 times larger than a water molecule.

A very strong polymer: Kevlar

Nearly all the early synthetic polymers were discovered by accident. But once chemists started to understand how polymerisation works, they could predict how reactions might take place. This meant they could plan to make a polymer with certain properties.

Du Pont is a huge multinational company with a special interest in polymers. The company wanted to make a very strong but light-weight polymer with a high **melting point**. The chemists designed and made a polymer with very long molecules, linked together in sheets. These sheets were themselves tightly packed together in a circular pattern.

One of the scientists involved in the research was an American, Stephanie Kwolek. Her job was to make small quantities of the new polymer and turn it into a liquid. Once it was liquid, the polymer could be forced through a small hole to make fibres. The problem was that the polymer would not melt, nor would it dissolve in any of the usual solvents.

Stephanie Kwolek experimented with many solvents. She eventually found that the new polymer would dissolve in concentrated sulfuric acid. This is a highly dangerous chemical that can cause severe burns. But fibres of the new polymer were manufactured in this way. This was the origin of Kevlar, which is five times stronger than steel. It is used for bullet-proof vests and to reinforce tyres. A similar polymer called Nomex is used in protective clothing for racing drivers.

Stephanie Kwolek wearing protective gloves made of Kevlar. She discovered how to turn this polymer into fibres.

Copying nature: Velcro

The two surfaces of Velcro stick together with a strong bond but can be peeled apart. One surface is covered in hooks, the other in loops.

The inventor of Velcro, George de Mestral, was copying seed pods that he found stuck to his socks when he was out walking. The pods were covered with tiny hooks that attached themselves round threads in the socks.

De Mestral used nylon to make Velcro. He worked out how to weave the polymer thread in just the right way to produce hooks and loops.

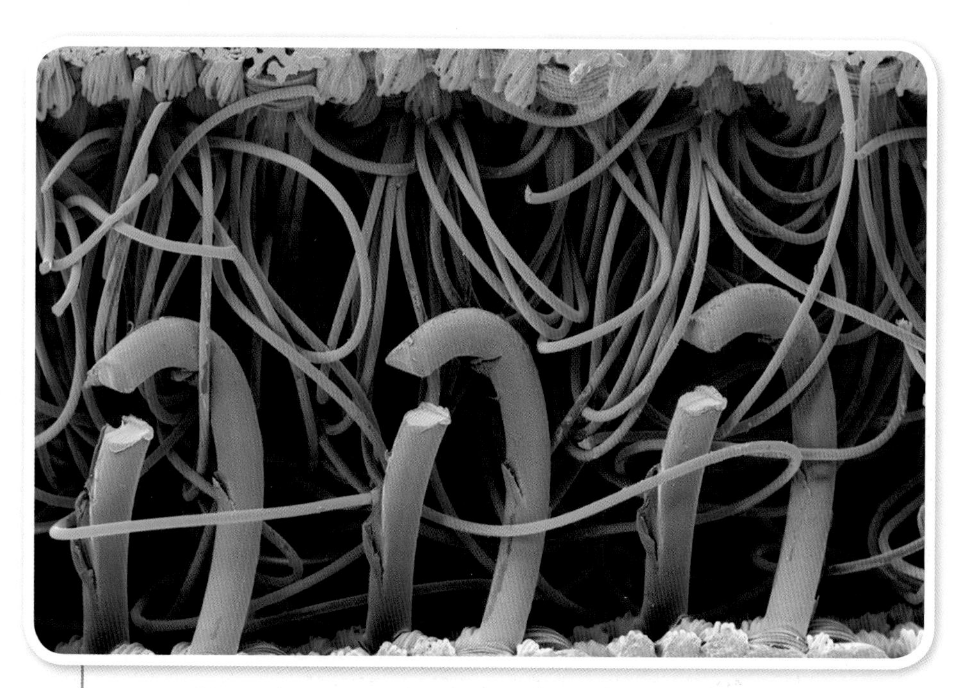

A magnified view of the nylon hooks and loops in Velcro material. This is a false-colour image taken with an electron microscope. The loops are loosely woven strands. The hooks are loops woven into the fabric and then cut. When the two surfaces are brought together they form a strong bond, which can be peeled apart. Magnification × 30.

Questions

1. Look through Sections E–H and identify two examples each of polymers, or polymer products:
   a. discovered by accident
   b. developed by design.
2. Explain how Gore-tex is waterproof, but still allows water to pass through it.
3. What words describe the properties of the polymer needed to make:
   a. the hooks in Velcro?
   b. the loops in Velcro?

Find out about

- ✔ crude oil and how it is made useful

Crude oil

Polymers are made from small molecules called monomers joined together in long chains. In most synthetic polymers, the small molecules originally come from **crude oil**.

Crude oil is a thick, sticky, dark-coloured liquid that formed over millions of years from the remains of tiny plants and animals called plankton. It is pumped out of the Earth's crust from wells under the ground or sea.

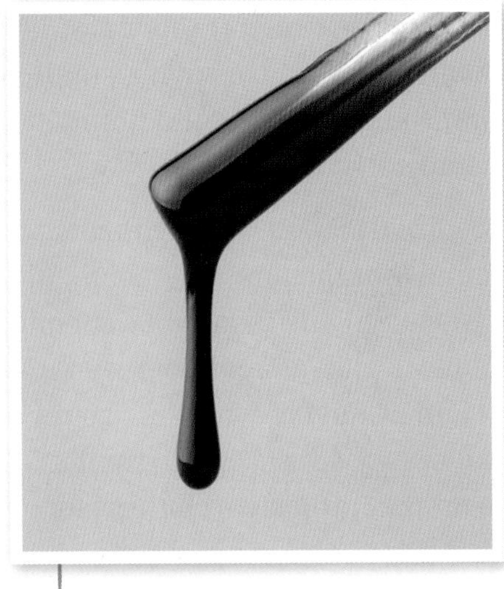

Crude oil is not very useful as it is.

Plant for processing chemicals from oil.

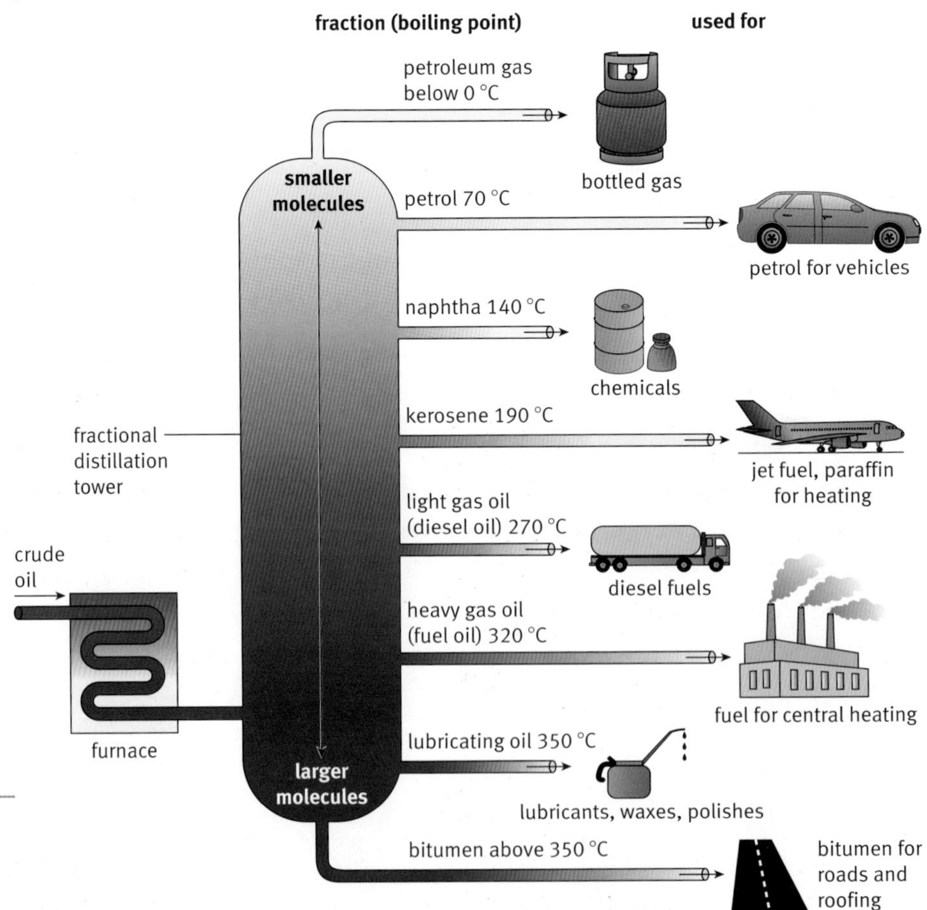

A fractionating tower, where crude oil is separated into fractions.

Crude oil is a mixture of **hydrocarbon** compounds. Because it is a mixture, crude oil is not very useful as it is. It needs to be separated into groups of molecules of similar size, called **fractions**. This is done by **fractional distillation**. When crude oil has been refined in this way it becomes very useful indeed, which is why it is so valuable.

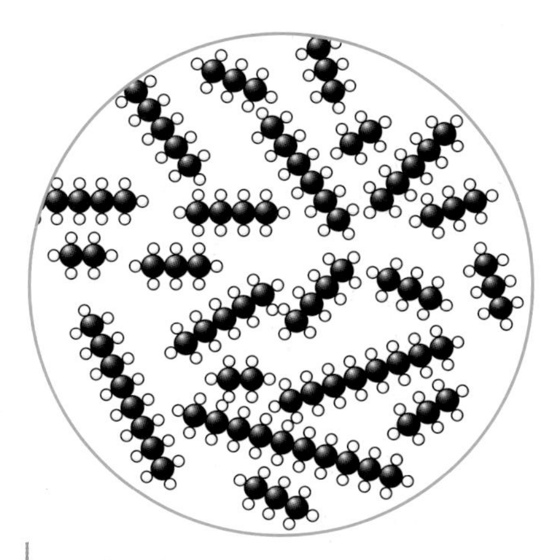

Crude oil is a mixture of hundreds of different hydrocarbons. A hydrocarbon contains hydrogen and carbon atoms only.

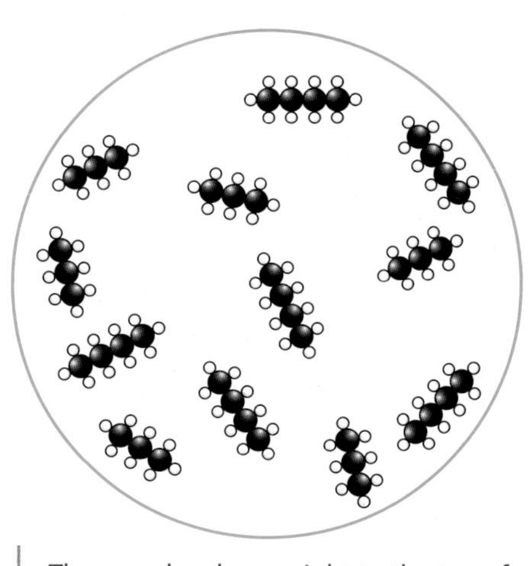

These molecules go right to the top of the fractionating column. This fraction contains some of the shortest hydrocarbons. There are weak forces between the molecules, giving them a low boiling point.

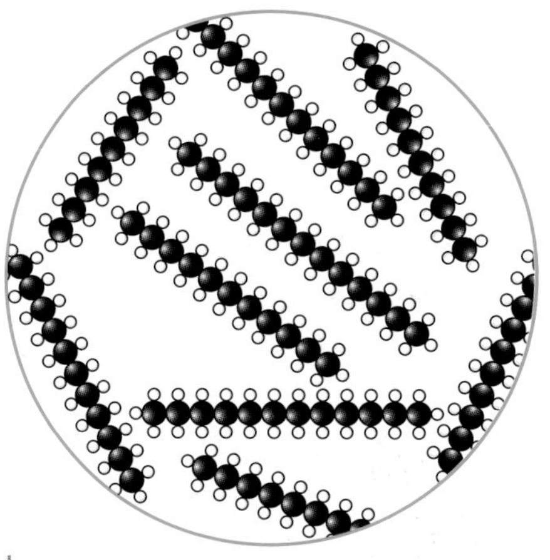

As you go down the tower the fractions have molecules with longer hydrocarbon chains. The forces between them get stronger, which increases the boiling points.

In the diagram on page 65, the crude oil is being heated in a furnace. The hydrocarbons in crude oil then go into the fractionating tower, which is hottest at the bottom and coolest at the top. The hydrocarbon molecules are separated by their boiling points.

The smallest molecules have the lowest boiling points and go to the top of the tower. This is because the forces between these molecules are very weak and only a little energy is needed for them to break out of the liquid and form a gas. The biggest molecules have the highest boiling points because the forces between the molecules are slightly stronger. These molecules stay at the bottom of the tower. Each fraction produced is still a mixture of molecules, but they are of similar size and boiling point.

Each fraction has different uses related to its properties. Most of the fractions from crude oil are used to provide energy for transport, homes, and industry. Some are used as lubricants. Some of the larger molecules are broken into smaller pieces, which can be used in **chemical synthesis** to manufacture new materials such as polymers. Only about 4% of crude oil is used in this way.

Key words

- crude oil
- hydrocarbon
- fraction
- fractional distillation
- chemical synthesis

Questions

1 Crude oil is made of hydrocarbons.
  a What is a hydrocarbon?
  b How do the hydrocarbons in crude oil vary?

2 Copy and complete:

  The fractionating tower separates the hydrocarbons into groups of molecules called __________. These groups of molecules are still mixtures, but they contain a __________ number of different hydrocarbons than the original crude oil. The hydrocarbon molecules in a fraction are similar in __________.

3 Explain why the fractional distillation of crude oil is so important.

Find out about

- the size of a nanometre
- the properties of nanoparticles
- new materials containing nanoparticles

Some bus companies use a nanoscale additive in diesel fuel. This reduces the amount of fuel used and the emissions from the vehicle, making the buses more efficient.

Geckos' feet have millions of nanometre-sized hairs that provide the 'stickiness' it needs to walk on ceilings.

Not just small – very small

Bus companies, holidaymakers, and people with serious injuries are just some of the groups benefiting from recent scientific and technological advances. New products have been developed that rely on the properties of materials at a very, very small scale. These technologies are called **nanotechnology**. 'Nano' comes from the Greek work *nanos*, which means dwarf. The particles used in nanotechnology are measured in nanometres (nm).

Nanotechnology is the use and control of structures called **nanoparticles**, which are very small. Some nanoparticles are made using specialist tools that build up new structures atom by atom. Other nanoparticles are made by chemical synthesis or by other techniques such as etching.

Each nanometre (nm) is a billionth of a metre or 0.000 000 001 m.
A nanometre is about:

- the width of a DNA molecule
- the distance your fingernails grow in a second
- 1/80 000 the thickness of the average human hair.

It is difficult to understand just how tiny a nanometre is.

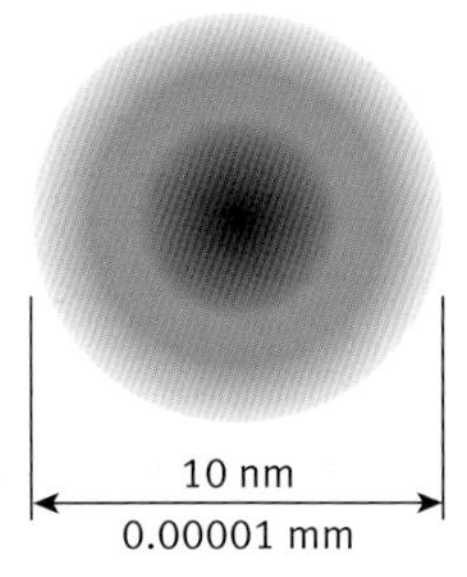

The diagram on the right shows a nanoparticle. It looks like a tiny piece of soot. It is 10 nm across, which is about 0.00001 mm. You could fit about 100 000 of them across 1 mm on your ruler. Let's think about it another way.

Imagine if a nanoparticle about 10 nm across was scaled up to the size of a football.

- An atom would become about the size of a 10p coin.
- A red blood cell would be the size of a football pitch.
- A cat would be about the same size as the Earth.

Properties of nanoparticles

Size of forces

If you wear a jumper and shirt both made of synthetic polymers, they sometimes seem to stick together. The force causing this 'stickiness' is similar to the force that allows a balloon rubbed against a jumper to stick to a wall. Usually we don't notice the forces holding surfaces together because they are very weak, but on the nanoscale they become very strong. These forces have the potential to be useful; they are what allow geckos to walk on ceilings. The geckos' feet have millions of tiny, nanometre-sized hairs, which give it a huge surface area and provide the stickiness required to hold the gecko to the ceiling. This has inspired scientists in the USA to design a robot that can walk up walls. They use a polymer that is similar to the bottom of the geckos' feet.

Size of surface

For nanoparticles, the **surface-area**-to-volume ratio is very large. In a solid 30-nm particle, about 5% of the atoms are on the surface. In a solid 3-nm particle, about half are. The atoms on the surface tend to be more reactive than those in the centre. This means materials containing nanoparticles are often highly reactive or have unusual properties.

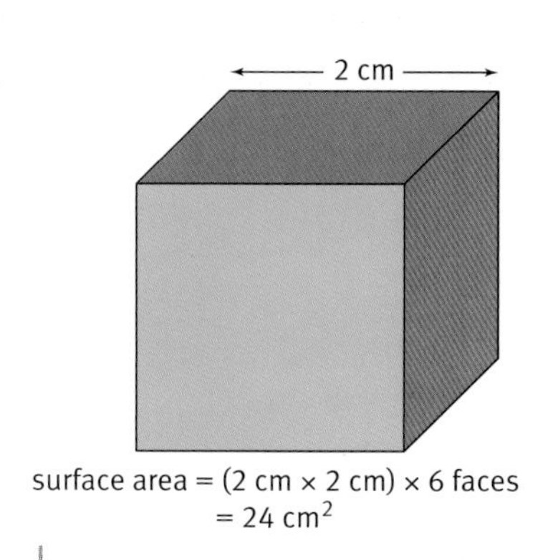

surface area = (2 cm × 2 cm) × 6 faces = 24 cm^2

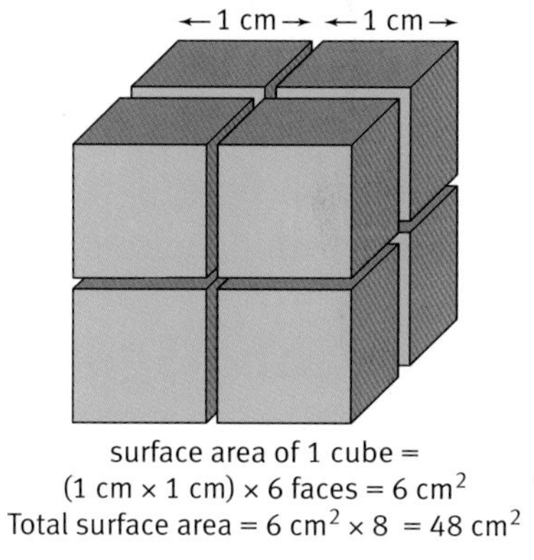

surface area of 1 cube = (1 cm × 1 cm) × 6 faces = 6 cm^2
Total surface area = 6 cm^2 × 8 = 48 cm^2

Total surface area is larger for smaller particles.

Some sunscreens contain nanoparticles.

Nanotechnology in medicine

Silver is best known for use in jewellery because it is shiny and unreactive. For a long time, it has been known that silver also has antibacterial properties. In 1999, a new bandage for serious injuries was launched. It is called Acticoat and contains silver nanoparticles. It is particularly important that the bacteria in a serious wound are killed before they can cause infections. The tiny nanoparticles of silver dissolve very quickly once they are moistened (for example, by blood from the wound) and the silver can get to work straight away.

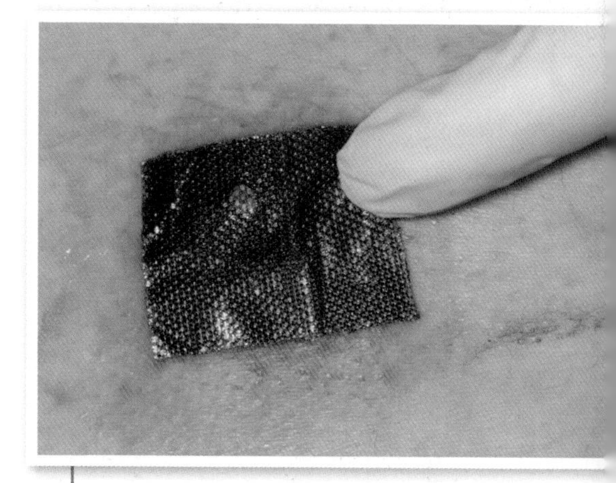

Acticoat is a type of bandage that contains nanometre-sized silver particles with antibacterial properties.

Questions

1 Write a description of what a nanometre is, for a student in Year 7.

2 Silver has many different uses.
   a Which properties of silver make it good for making jewellery?
   b How are the particles of silver in Acticoat different from the silver particles in jewellery?
   c Why will nanoparticles of silver be more effective in a wound dressing than ordinary silver?

Key words

- nanotechnology
- nanoparticles
- surface area

Find out about

- nanotechnology in nature
- uses of nanotechnology
- risks of using nanotechnology

Nanoparticles of salt are formed above the sea.

Nanotechnology in nature

Nanotechnology sounds very strange and new – but there are nanoparticles in nature. There are even nanoscale structures in living cells that can move and turn in a controlled way and carry out complex jobs. They are far more advanced than any of the synthetic nanotechnologies currently in use.

All these are on the nanoscale:

- tiny salt particles in the atmosphere, formed by ocean waves in windy conditions, which help in forming rain and snow
- proteins that control biological systems very precisely
- the enamel in your teeth, which is partly made of nanoparticles.

Humans have been making nanoparticles for years by accident, without knowing it. Some fires, particularly those burning solid fuels, produce nanoparticles (along with other waste).

Uses of nanotechnology

There are products already available that use nanotechnology. These include healthcare products, sports gear, and clothing.

Sunscreen

Many sunscreens contain particles of either zinc oxide or titanium oxide. These are white solids. In older formulations, the particles are relatively large and leave the skin looking white. More modern formulations use nanoparticles instead, which can be rubbed in and have a more natural appearance.

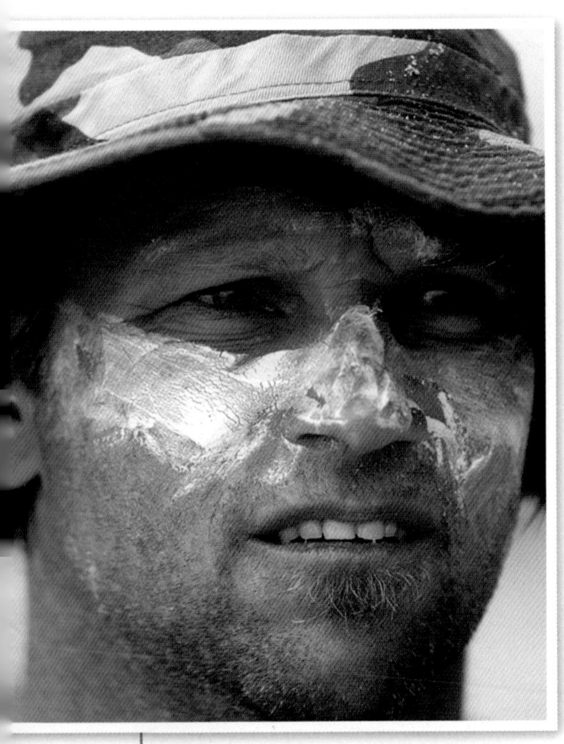

Older sunscreens could be seen on the skin.

Tennis balls and rackets

Nanotechnology was first used in a professional tennis match in 2002. The 'double-core' balls have an extra layer inside them made of 1-nm-sized clay particles mixed with rubber. This helps to slow down air escaping from the balls, keeping them inflated for longer.

Nanotechnology is improving the performance of sports equipment.

Nanoparticles are also added to materials such as the carbon fibre used to make tennis rackets. The resulting materials are lighter and stronger.

Electronic paper

Engineers at Bridgestone have developed a flexible display using nanotechnology. A substance they call 'liquid powder' is placed between two sheets of glass or plastic, creating a light-weight, flexible display. Electricity is needed to change the display, but when it is switched off it retains its image. These displays could replace paper posters, saving paper – and electric signs, saving energy.

Clothing

Scientists have developed clothes that contain zinc oxide nanoparticles. These are the same particles as those used in sunscreen. Clothes with these particles offer better UV protection. Stain-resistant clothes have also been produced. These have tiny nanoscale hairs that help repel water and other materials. This has the potential to reduce the amount of water and energy used in washing clothes.

Socks have been made that contain nanoparticles of silver or other chemicals. This gives the socks antibacterial properties to help prevent feet from smelling.

Coatings containing nanoparticles can repel water and other chemicals that might stain.

Self-cleaning windows

A company called Pilkington offers a product called Activ Glass, which is coated in nanoparticles. When light hits these particles, they break down any dirt on the glass. The surface is also hydrophilic (*hydro* - water, *philic* - loving), which means that water falling on it spreads over the surface, helping to wash it.

Nanotechnology and risk

Different properties, different risks

Nanoparticles have different properties compared to larger particles of the same material. This may mean that the nanoparticles have different effects on plants, animals, and the environment. It may also mean that they are more toxic to people. Some doctors are concerned that nanoparticles are so small they may be able to enter the brain from the bloodstream. If this is true, it could mean some chemicals that are normally harmless become highly toxic at the nanoscale.

Exactly how all the various nanoscale substances differ from larger particles of the same material is not fully understood. At present, there are no requirements for health and safety studies for nanoparticles to be different from those for larger particles. But some groups and organisations think that there should be.

Questions

1. Using Sections J and K, give an example of a sports product, a healthcare product, a lifestyle product, a building material, and clothing that may now contain nanotechnology.
2. List three ways in which nanotechnology products may be beneficial to the environment.
3. Explain why some people are concerned about the possible effects of nanotechnology on the environment.

Science Explanations

Scientists use their knowledge of molecules to develop new materials with useful properties. A wide range of different synthetic polymers can be made from hydrocarbons obtained from crude oil.

You should know:

- that one way of comparing materials is to measure their properties, such as melting point, strength, stiffness, hardness, and density
- why it helps to have an accurate knowledge of the properties of materials when choosing a material for a particular purpose
- that polymers such as plastics, rubbers, and fibres are made up of long-chain molecules
- why modern materials made of synthetic polymers have often replaced materials used in the past such as wood, iron, and glass
- that crude oil is one of the raw materials from the Earth's crust, which is used to make synthetic polymers
- that crude oil consists mainly of hydrocarbons, which are chain molecules of varying lengths
- that hydrocarbons are made from carbon and hydrogen atoms only
- why the boiling temperature of a hydrocarbon depends on its chain length
- that the petrochemical industry makes useful products by refining crude oil
- that fractional distillation separates the hydrocarbons in crude oil into fractions according to their chain length
- that some of the small molecules from refining crude oil are used to make new chemicals
- that polymerisation is a chemical reaction that joins up small monomer molecules into long chains
- how the properties of polymeric materials depend on the way in which the long molecules are arranged and held together
- how it is possible to modify polymer properties in various ways such as increasing the length of the chains, cross-linking the molecules, adding plasticisers, and changing the degree of crystallinity
- that nanotechnology is the use and control of structures that are very small (1–100 nm)
- that nanoparticles can occur naturally, by accident, and by design
- why nanoparticles of a material show different properties to larger particles of the same material
- how nanoparticles can be used to modify the properties of materials.

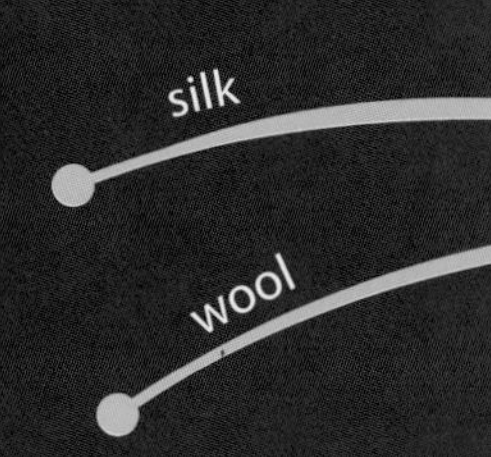

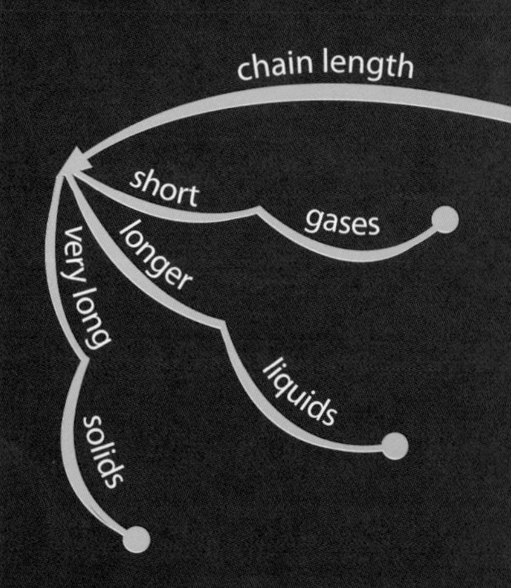

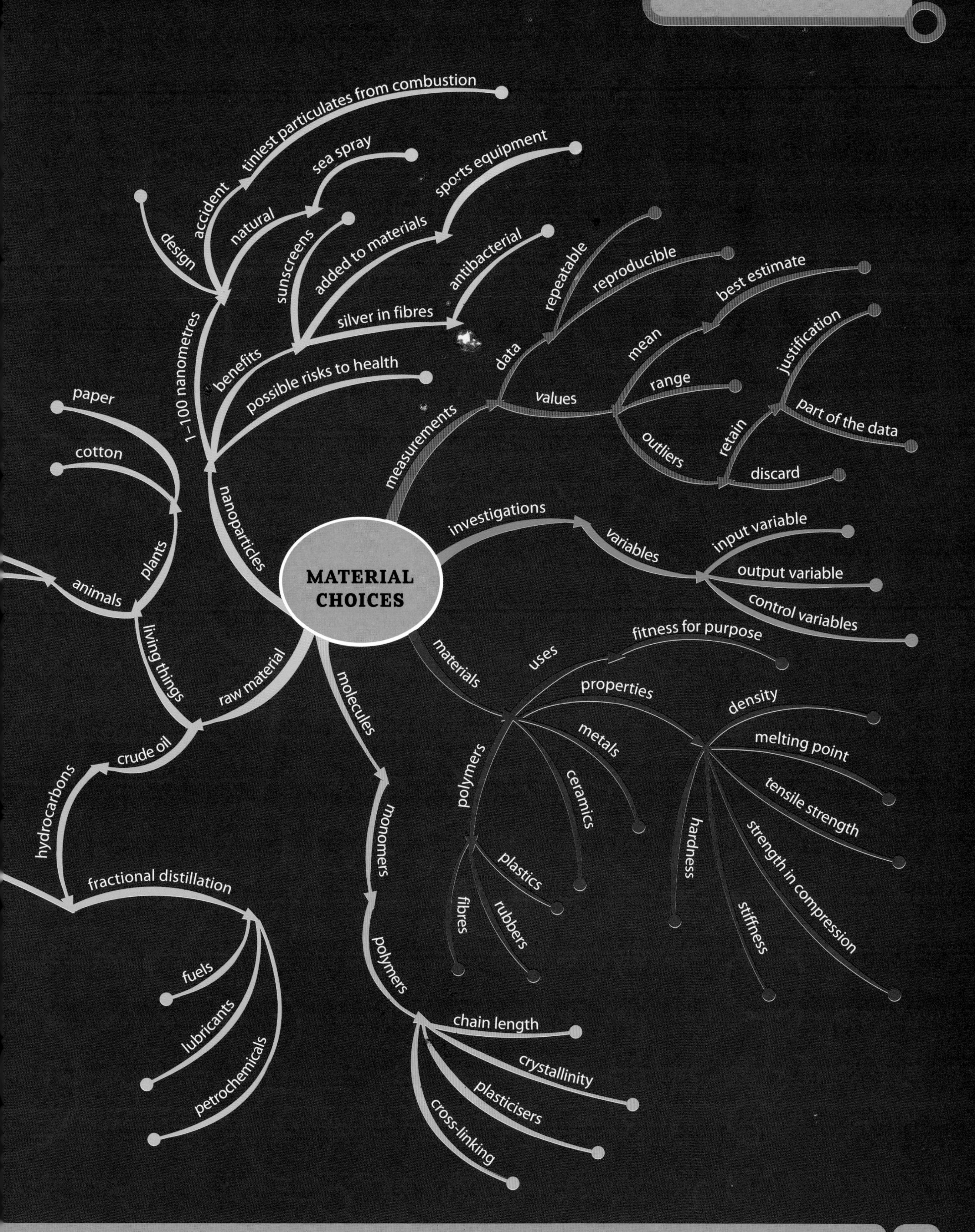

MATERIAL CHOICES
nanoparticles
1–100 nanometres
design
accident
tiniest particulates from combustion
natural
sea spray
benefits
sunscreens
added to materials
sports equipment
silver in fibres
antibacterial
possible risks to health
measurements
data
repeatable
reproducible
values
mean
best estimate
range
outliers
retain
justification
part of the data
discard
investigations
variables
input variable
output variable
control variables
materials
uses
fitness for purpose
properties
density
melting point
tensile strength
strength in compression
stiffness
hardness
metals
ceramics
polymers
plastics
rubbers
fibres
molecules
monomers
polymers
chain length
crystallinity
plasticisers
cross-linking
raw material
living things
plants
paper
cotton
animals
crude oil
hydrocarbons
fractional distillation
fuels
lubricants
petrochemicals

Ideas about Science

Scientists measure the properties of materials to decide what jobs they can be used for. Scientists use data rather than opinion to justify the choice of a material for a purpose.

Scientists can never be sure that a measurement tells them the true value of the quantity being measured. Data is more reliable if it can be repeated. When making several measurements of the same quantity, the results are likely to vary. This may be because:

- you have to measure several individual examples, for example, several samples of the same material
- the quantity you are measuring is varying, for example, different batches of a polymer made at different times
- the limitations of the measuring equipment or because of the way you use the equipment.

Usually the best estimate of the true value of a quantity is the mean (or average) of several repeat measurements. The spread of values in a set of repeat measurements, the lowest to the highest, gives a rough estimate of the range within which the true value probably lies. You should:

- be able to calculate the mean from a set of repeat measurements
- know that a measurement may be an outlier if it lies well outside the range of the other values in a set of repeat measurements
- be able to decide whether or not an outlier should be retained as part of the data or rejected when calculating the mean.

When comparing data on the properties of different samples of a material you should know that:

- a difference between their means is real if their ranges do not overlap.

To investigate the relationship between a factor and an outcome, it is important to control all the other factors that you think might affect the outcome. In a plan for an investigation into the properties of a material, you should be able to:

- identify the effect of a factor on an outcome
- explain why it is necessary to control all the factors that might affect the outcome other than the one being investigated
- recognise that the control of other factors is a positive design feature or that it is a design flaw if they are not controlled.

Some applications of science, such as using nanoparticles, can have unintended and undesirable impacts on the quality of life or the environment. Benefits need to be weighed against costs. You should know that:

- some nanoparticles may have harmful effects on health and that there is concern that products containing nanoparticles are being introduced before these effects have been fully investigated.

Review Questions

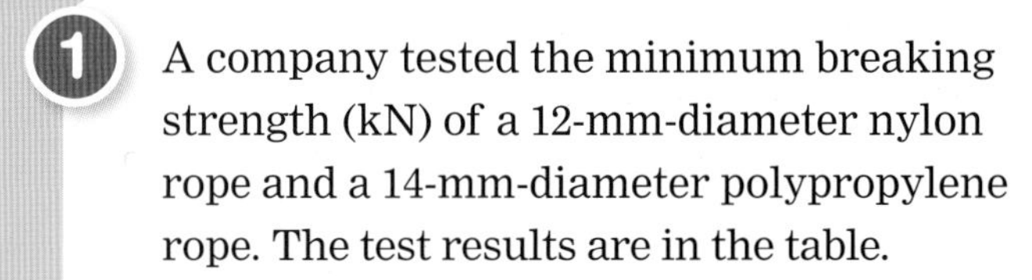

1 A company tested the minimum breaking strength (kN) of a 12-mm-diameter nylon rope and a 14-mm-diameter polypropylene rope. The test results are in the table.

Test number	1	2	3	4	5
Nylon rope	25.7	25.1	24.8	25.2	25.2
Polypropylene rope	19.9	20.8	20.1	20.4	20.8

a Suggest why the company repeated the test on each rope five times.

b For each rope, calculate the best estimate of the true value of the strength of the rope.

c Is there a real difference between the strengths of the two ropes? Use the test results to justify your answer.

2 **a** The table gives data about three hydrocarbons. The higher the number of carbon atoms in a molecule, the bigger the molecule. Describe the trend in boiling point as molecule size increases.

Hydrocarbon name	Boiling point (°C)	Number of carbon atoms in one molecule
Methane	–162	1
Hexane	69	6
Hexadecane	287	16

b Explain the trend in boiling points in terms of the strength of the forces between the molecules.

3 **a** Identify the property changes that occur when cross-links are made between long rubber molecules. Choose from the list below.

- flexibility increases
- flexibility decreases
- hardness increases
- hardness decreases
- melting point increases
- melting point decreases

b Identify two property changes that occur when a plasticiser is added to a polymer.

4 **a** Titanium dioxide nanoparticles are used in some sunblock creams. Because they are so small they do not reflect visible light.

i Within what range is the diameter of a nanoparticle?

ii Suggest one advantage of using nanoparticles in sunblock creams.

iii Explain why some people are concerned about the safety of using sunblock creams that contain nanoparticles.

b Give examples of two more uses of nanoparticles.

C3 Chemicals in our lives: Risks and benefits

Why study chemicals in our lives?

You are made up of chemicals, and so is everything around you. Many of the chemicals in the things you buy are natural, others are synthetic. Some chemicals are very good for you, others may be harmful. This module will help you to understand where some of these chemicals come from and why they are so useful.

What you already know

- Clues in the rocks help scientists discover how the Earth has changed.
- The movements of tectonic plates lead to changes in the Earth's surface.
- Chemical reactions rearrange atoms to give products with new properties, which may be either helpful or harmful.
- Alkalis neutralise acids to form salts.
- Crude oil is a valuable source of hydrocarbons.
- Polymerisation produces a wide range of plastics, rubbers and fibres.
- Plasticisers can be used to modify the properties of polymers.
- There are ways to weigh up the risks and benefits of scientific discoveries.

Find out about

- the geological history of Britain, which explains why it is rich in natural resources
- methods chemists use to turn raw materials, such as salt, into many valuable products
- ways to balance the benefits and risks of using chemicals
- the choices people make to ensure they use chemicals safely and sustainably.

The Science

Science can help to explain why Britain has deposits of valuable natural resources including salt and limestone as well as coal, gas and oil. These raw materials have been the basis of a chemical industry that has added to the wealth of the country for over 200 years.

Ideas about Science

Manufactured chemicals bring many benefits – but there are also risks. People are worried that many chemicals have never been thoroughly tested, so the risks are not fully understood. Choices about using chemicals should be based on evidence. The evidence comes from studying all the stages of the life of products.

A journey through geological time

Find out about

- how Britain came into existence as continents moved
- the different climates Britain has experienced
- magnetic clues that geologists use to track continents

A story of change

The Earth's outer layers (the crust and upper mantle) are divided into a number of **tectonic plates**. Each plate contains dense oceanic crust, often carrying some lighter continental crust on top of it. The plates move because of very slow **convection** currents in the underlying solid mantle.

Movements of the tectonic plates cause oceans to open up slowly between continents in some parts of the world. In other parts of the world, plate movements bring continents together with great force, creating mountain ranges. Most major volcanic eruptions and earthquakes happen at plate boundaries.

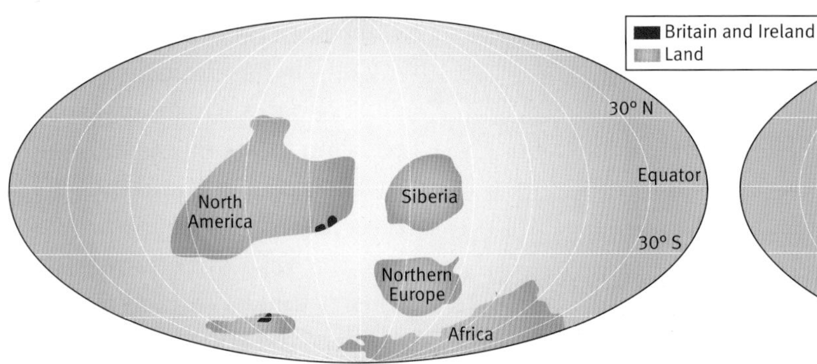

About 450 million years ago, in the Ordovician period, the two parts of the Earth's crust that would one day make up Britain were both south of the Equator. The northern and southern parts were separated by an ocean.

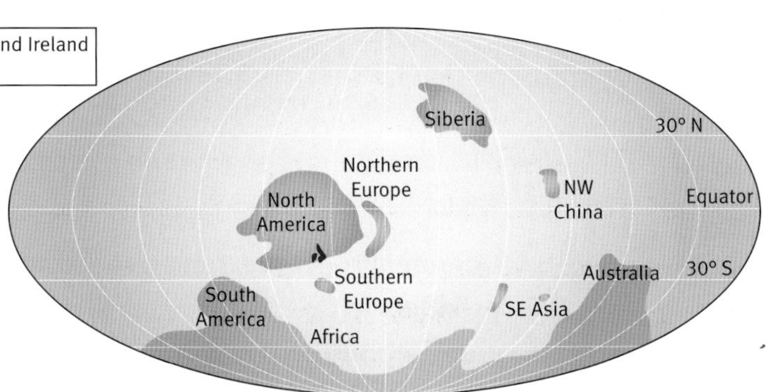

By about 360 million years ago, at the end of the Devonian period, the two parts of Britain collided. The collision between continents created a chain of mountains. The land that would become Britain was at the edge of this chain of mountains in a dry continent.

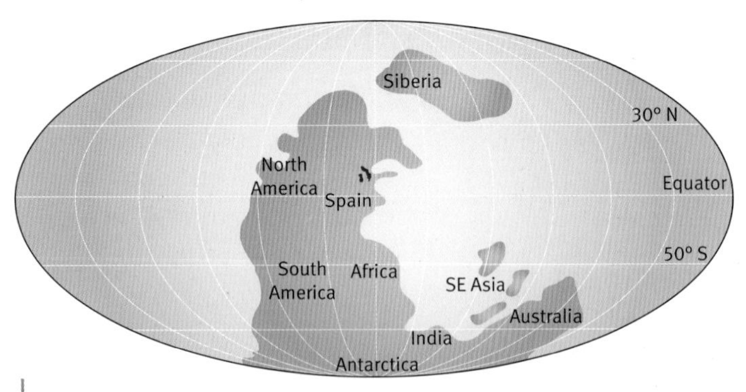

About 280 million years ago, at the beginning of the Permian period, Britain was just north of the Equator and had desert-like conditions.

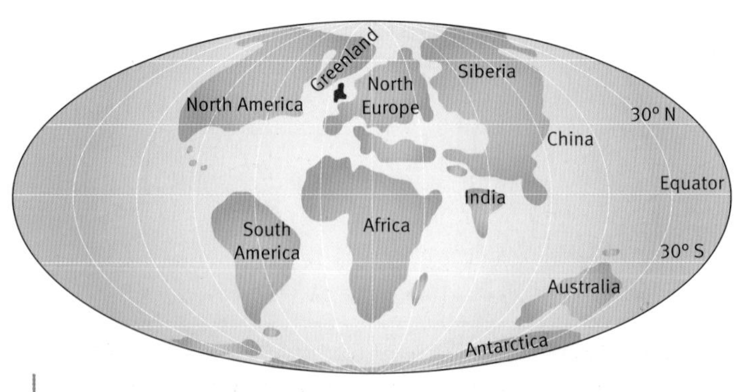

About 65 million years ago, dinosaurs (and many other groups of organisms) became extinct. Britain was on the edge of the North Atlantic ocean, just south of where it is today. The Atlantic Ocean was opening up as North America and Europe very slowly moved apart.

Magnetic clues to the past

In the 1950s, the idea that the continents were slowly moving across the Earth's surface was still controversial. Many geologists felt there was not enough evidence to support the idea. At around this time, a group of scientists at Imperial College London, showed that it is possible to track the position north or south of the Equator of a slowly drifting country by studying **magnetic** particles in the rocks.

Many volcanic lavas, and some sediments, contain the mineral magnetite. This mineral gets its name from the magnetic properties of its crystals. Magnetite in lava can be magnetised in a fixed direction once the rock has cooled enough. The magnetisation lines up in the direction of the Earth's magnetic field at that time, rather like iron filings around a bar magnet. The magnetisation of crystals in sediments can line up in a similar way.

Near the Equator, the magnetisation lies horizontally. Nearer to the Poles, the magnetisation is at an angle to the horizontal. So by measuring the angle at which crystals are magnetised in rocks, scientists can work out the **latitude** at which the rock was originally formed.

The measurements were combined with other clues to show that the rocks in Britain had drifted north from a position south of the Equator, over a period of millions of years. The evidence supported the theory of continental drift, and contributed to the development of the theory of plate tectonics. This movement means that Britain has experienced many different climates during its long history. Evidence for these various climates can be found in the different rocks that now make up the country.

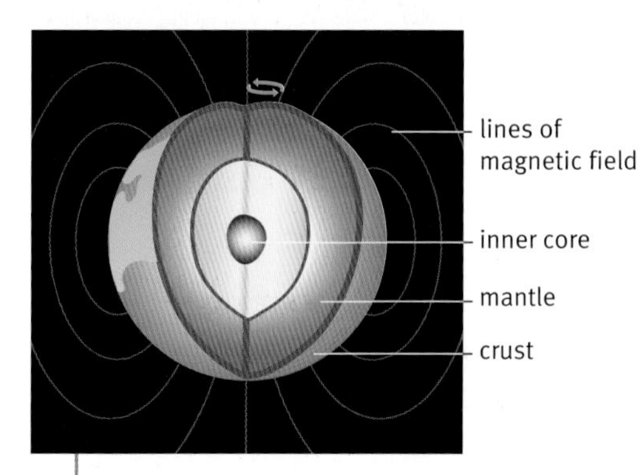

A cutaway diagram showing the Earth's magnetic field. The field is parallel to the ground at the Equator, but is more steeply angled nearer the Poles.

Questions

1. What causes continents to move over the surface of the Earth?
2. Do the observations of magnetic minerals made by the scientists at Imperial College support or conflict with the theory of plate tectonics?
3. Suggest evidence that geologists might look for to test the theory that the northern and southern parts of Britain were once on different continents.
4. Movements of the Earth's crust can cause layers of rock to bend and fold. Why might the folding of rocks make it very difficult to interpret the results from measuring the direction of magnetisation in rock samples?

Key words

- ✓ **tectonic plates**
- ✓ **magnetism**

B Mineral wealth in Britain

Find out about

- what geologists can learn by studying rocks
- the origins of some of the rocks in Britain

Sand dunes in the Namib desert, Namibia. Studying the ripples in today's sand dunes helps to explain the distribution of grain sizes and ripples seen in sandstone rocks.

Fossilised ripples in sandstone on the Maumturk Mountains, County Galway, Ireland.

Key words

- sedimentary rocks
- grains
- fossils
- erosion
- evaporation

Clues in the rocks

Geologists explain the history of the Earth's surface in terms of processes that can be observed today. For example, you can find out about the history of a **sedimentary rock** such as sandstone by looking at the shape and size of the sand **grains** in the rock. The sandstone may have formed from desert sand or river sediments. By comparing the sand grains in the rock with sand grains found in deserts and rivers today, geologists can find out about the conditions when the rock formed. Other clues come from the presence and shape of fossilised ripples in rocks, which may have been produced by the wind or water.

Some sedimentary rocks are rich in **fossils** of plants and animals. Geologists use fossils to put rock layers in order of their ages. This is possible because rocks may contain distinctive fossilised plants and animals from different periods of geological time. Comparing fossils with today's living organisms gives clues about the past environment where the fossilised plants and animals lived.

Different rocks in different climates

There is a rich variety of rocks in Britain, and some are very important economically. A chemical industry based on chlorine grew up by the River Mersey in north-west England because underground salt deposits, coal mines, and limestone quarries were nearby. These provided the raw materials for making chlorine. The salt, coal and limestone formed at different times and in different climates during Britain's long geological history.

Questions

1. Give an example to show how studying a natural process today can tell scientists that processes such as rock formation and mountain building are very slow and take place over millions of years.
2. The chemical industry uses limestone quarried in the Peak District National Park because it is very pure. How do geologists account for the purity of the limestone?
3. Why are fossils mainly found in sedimentary rocks, less commonly in metamorphic rocks, and not at all in igneous rocks?

Limestone from a cavern in the Peak District. It contains fossils of crinoid sea lilies. This limestone formed 350 million years ago. When it formed, the land that would become the Peak District lay below a shallow, warm sea, which was then just south of the Equator. At the time the water was very clear because rivers were bringing in very little sediment. The sea was full of living things. As the plants and animals died they sank to the bottom and formed fossils in the thickening mass of pure limestone.

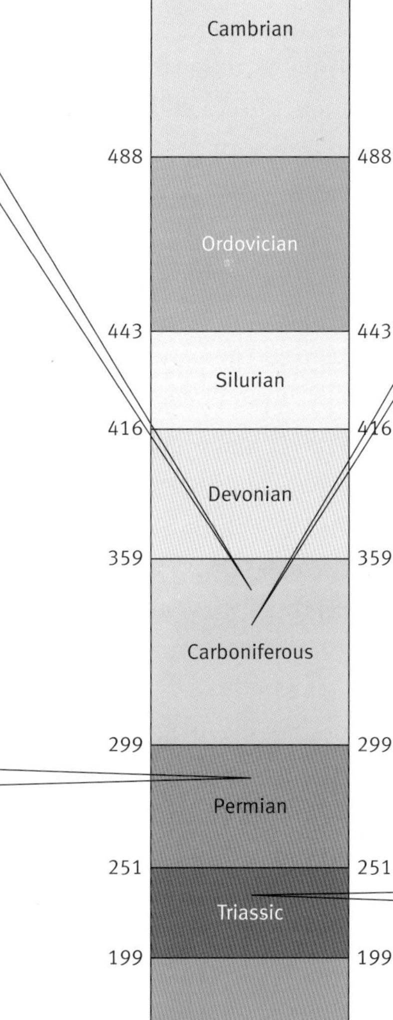

Sandstone in the Peak District. About 310 million years ago, the mountains to the north and east of the Peak District were **eroded** by fast-flowing rivers carrying sediments. Sand and small pebbles were deposited in layers, which were then compacted to form coarse sandstones. In the past, this rock was used to make millstones for grinding corn – it is called millstone grit.

Coal shale containing a fossil fern. About 280 million years ago, the river deltas in the area that is now Britain got bigger and created swamp land. Tree ferns grew in the swamps. As the plants died, they formed a layer of peat, which was covered by sediment, compressed, heated, and turned to coal. A period of mountain building followed and the rocks in the Peak District were pushed up towards the surface.

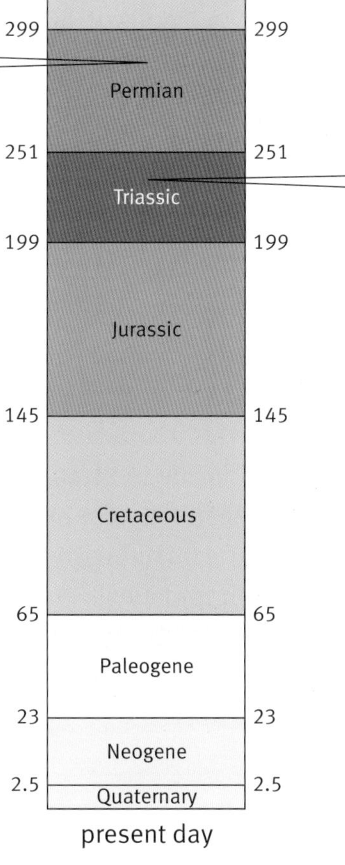

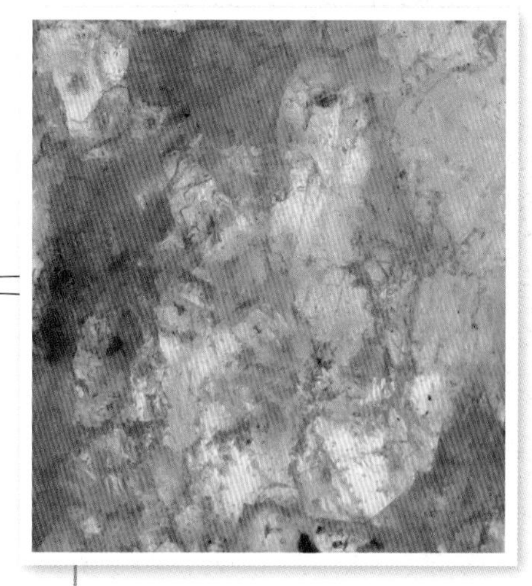

Rock salt mined in Cheshire, consisting mainly of the mineral halite. About 220 million years ago, seawater moved inland and created a chain of shallow salt marshes across land that is now part of Cheshire. Deposits of rock salt formed as the water in the marshes **evaporated**. This rock salt has a red-yellow coloration. The colour is from sand that blew into the salt marshes from surrounding deserts.

C Salt: sources and uses

Find out about

- the uses of salt
- where salt comes from
- the methods used to obtain salt

Salt cod in a fish market in Barcelona. In the days before refrigeration, cod caught in the seas off Newfoundland, Iceland, or Norway was salted and dried before being taken by ship to countries such as Portugal and Spain.

Carrying a basket of salt from salt pans near Mahabalipuram in India. Sea water is run into the pans and allowed to evaporate in the hot sunshine. The crystals contain mainly sodium chloride.

The importance of salt

Salt has played a vital part in human civilisation for thousands of years. Before there were modern ways of keeping food (such as canning, or freezing), salt was the only way to **preserve** meat and fish. After salting, these foods could be kept for a long time, for example, on long sea voyages. Today, salt is still used by the food industry to process and preserve food, and to add flavour. It is also used to treat icy roads in winter and as a source of chemicals such as chlorine.

Sea salt

Salt has been extracted from the sea off the east coast of Essex for over 2000 years. The rainfall in this part of Britain is lower than elsewhere. This means that the concentration of salt in the estuaries and rivers is higher, so less fuel is needed to **evaporate** the water and separate the salt. Small quantities of salt are still obtained in this way for home use.

Large-scale extraction of salt from the sea is only economical on coasts with hot and dry climates. In these places there is no need to burn fuel to separate the salt, because the energy needed to evaporate the water comes from the Sun.

Rock salt

There are two underground salt mines in England – one in North Yorkshire, the other at Winsford in Cheshire. This salt is mainly used to spread on the roads during freezing weather. Adding salt means that ice and snow melt. This is because salty water has a lower freezing point than pure water.

Miners use giant machines to extract rock salt. The rock contains about 90% sodium chloride in the form of the mineral halite. The halite is mixed with insoluble impurities, mainly a reddish clay. The salt used on roads does not need to be pure.

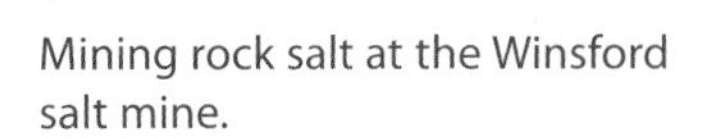

Mining rock salt at the Winsford salt mine.

Solution mining

The salt used for the chemical industry in Britain is not mined, it is extracted by pumping water down into the rock. The salt **dissolves** and is carried to the surface in **solution**. The impurities, such as clay, do not dissolve and so stay underground. The solution of salt in water is called **brine**.

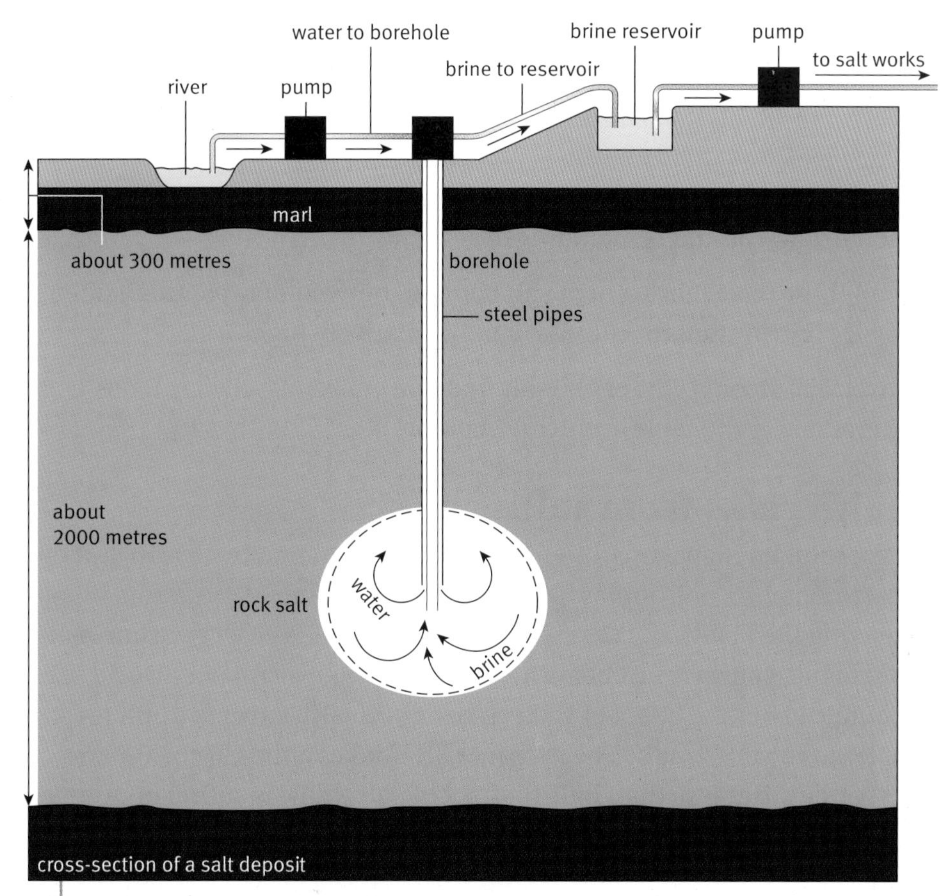

Using water to extract brine from an underground salt deposit.

Salt crystals are recovered from brine by evaporating the water. On such a large scale, the evaporators have to be as efficient as possible to minimise the energy needed to turn water into steam. The salt **crystallises** as the water evaporates. The crystals are separated from the remaining brine by **filtering** or using a **centrifuge**.

Sudden subsidence

Large-scale pumping to extract salt as brine started in Cheshire in about 1870. Uncontrolled pumping for about 60 years created very large underground holes. This led to widespread **subsidence** and flooding of land. From time to time there were disastrous collapses that destroyed buildings. Nowadays, pumping is planned so that the holes in the rock are spaced out and separated by pillars of rock. The rock that is left behind helps to prevent subsidence.

Key words

- preserve
- dissolve
- solution
- brine
- crystallise
- filter
- centrifuge
- subsidence

Subsidence caused by salt mining in 1891. The rear of Castle Chambers in Northwich suddenly sank into the ground. The building did not collapse because of its timber frame.

Questions

1. Explain why the east coast of Essex is a good place to extract sea salt.
2. Why is salt for treating roads extracted in a different way from salt used for food and salt used in the chemical industry?
3. Name the solvent and the solute in brine.
4. Outline how you would make some pure salt from rock salt on a small scale in a laboratory.

Find out about

- why we need salt in our diet
- the possible risks from eating too much salt
- the debate about the evidence for the claims made about salt and health

The sodium in salt is an essential part of your diet. It is found in your blood, tears, and nerves. Nerves conduct electrical signals thanks to the presence of salts.

Salt in food

Salt is sodium chloride. The sodium in salt is an essential part of a healthy diet – but you only need a small amount. Salt is used as a **flavouring** and also enhances other flavours present in food. Because humans need sodium in their diet, they naturally seek out salty-tasting food. The food industry also uses salt to preserve and process food.

The main sources of salt in the diet are:

- cereal products such as bread, chapattis, breakfast cereals, biscuits, and cakes
- processed meat and fish products
- some dairy products including cheese.

Not all these foods have a high salt content, but you may eat them often, so their contribution to your diet can be relatively high.

On average, around 75% of the salt that you eat is in everyday foods and 25% is added at the table or during cooking.

Health risks from salt

UK government departments such as the Department for Health and the Department for Environment, Food, and Rural Affairs have a role in protecting the health of the public. This role includes carrying out risk assessments concerning chemicals in food. The government also advises the public about the effect of food on health. Health experts think most people eat too much salt. The average salt intake in the UK in 2008 was 8.6 g per day. But less than half that, 4 g of salt a day, is sufficient for nearly everyone. Government agencies have been working with health experts, consumers and industry to reduce salt intake to a target of 6 g per day for adults, and less for children.

UK government agencies say that eating too much salt can raise people's blood pressure. This can increase the risk of developing heart disease or having a stroke. People can lower their blood pressure in as little as four weeks by cutting down on salt. Around a third of the population in the UK currently have high blood pressure. When blood pressure goes down, the risk of developing heart disease and stroke goes down too, at any age.

The Scientific Committee on Nutrition is an independent body of experts set up to advise the government. The committee has reviewed over 200 scientific papers from all over the world on the evidence on salt and health. It concluded that 'A reduction in the dietary salt intake of the population would lower the blood pressure risk for the whole population.'

The committee has also concluded that the risk from heart disease associated with high salt consumption is not only confined to those who already have high blood pressure. A large number of people with blood pressure in the normal range are also at risk.

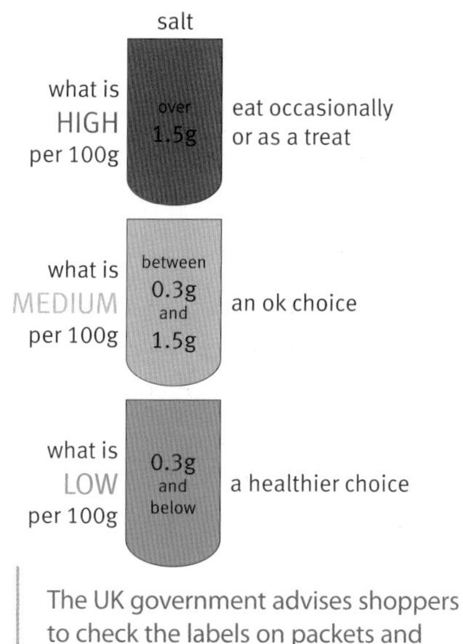

The UK government advises shoppers to check the labels on packets and choose foods with less salt.

Challenging the salt theory

The European Salt Producers' Association is an industry body representing salt producers across Europe. They have produced a report challenging the **theory** that reducing salt intake brings health benefits for everyone in the population. It suggests that there is no scientific proof for this theory. The report also suggests that a low-sodium diet could be harmful in some cases. To support this point of view, the association refers to two independent reviews of research carried out between 1966 and 2001.

Other scientists also argue that the evidence does not support an approach aimed at reducing salt intake in the whole population, as people with normal to low blood pressure might not benefit.

Key words

- ✓ flavouring
- ✓ theory

Questions

1. What are the main sources of salt in the diet?
2. A 25 g packet of crisps contains 0.6 g of salt. Do these crisps have a high, medium, or low salt content?
3. Suggest a reason why it is difficult to investigate scientifically the health risks of different levels of salt in the diet.
4. Suggest reasons why UK government agencies and the European Salt Association might come to different conclusions about the effects of salt levels in the diet on health.
5. Why might a member of the public, who is not an expert, ignore the advice of the government and eat more than the recommended amount of salt?

Alkalis and their uses

Find out about

- uses of alkalis
- where alkalis used to come from
- neutralisation of acids with alkalis

The old way to make soap. Soap was made by boiling up animal fat with the alkali potash. The potash (potassium carbonate) came from the ashes of burnt wood.

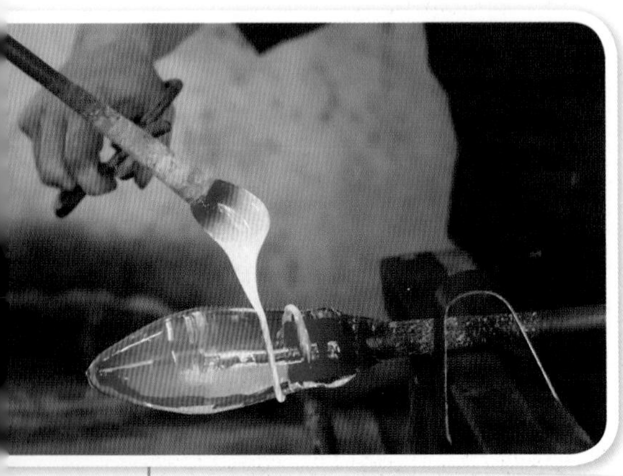
Glass is made by melting pure sand (silicon oxide) with lime (calcium oxide) and soda ash (sodium carbonate).

Traditional alkalis

Even before large-scale industrialisation, **alkalis** were needed:

- to neutralise acid soils
- to convert fats and oils into soap
- to make glass
- to make chemicals that bind natural dyes to cloth.

Alkalis for making alum

One of the first pure chemicals made in Britain was alum. The biggest use of alum was for dyeing cloth. Alum was needed to help natural dyes cling fast to cloth so that the colours did not fade too quickly during washing.

Alum was made on the north-east coast of Britain, where rock from the cliffs is unusually rich in aluminium compounds. Workers used to roast this rock in open-air fires for many months. Then they tipped the burnt rock into pits of water and stirred the mixture with long wooden poles.

After allowing the waste rock to settle, they ran the solution of soluble chemicals into lead pans. There they boiled the liquid to get rid of much of the water and added an alkali to neutralise acids in the solution. Finally they allowed the solution to cool in wooden casks. When they broke open the casks, found crystals of alum, which they could crush and put into bags for sale.

Some of the alkali used in this process was potash, from the ash of burnt wood. The rest was ammonia, from stale urine. Local people stored urine in wooden pails and this was collected in large barrels on horse-drawn carts. So much urine was needed that it was also brought in by sea from London. On the return journey, the ships delivered the bags of alum to dyers in the south of England.

Alkalis and their reactions

All alkalis are soluble in water – at least to some extent. When they dissolve, they raise the pH of water above 7. Alkalis are important because they **neutralise** acids.

Two very corrosive alkalis are sodium hydroxide and potassium hydroxide. When sodium hydroxide neutralises hydrochloric acid, there is a chemical change that produces sodium chloride. Chemists sometimes call this 'common salt', because it is just one of many different **salts** produced when acids and alkalis react. This reaction is shown by the **word equation**:

sodium hydroxide + hydrochloric acid ⟶ sodium chloride + water

There is a pattern to these reactions:

alkaline hydroxide + acid ⟶ salt + water

If the acid used is hydrochloric acid, the salt will be a chloride. If the acid used is sulfuric acid, the salt will be a sulfate. If the acid used is nitric acid, the salt will be a nitrate.

Sodium carbonate and potassium carbonate dissolve in water to form a solution with a pH above 7. They are also alkalis. They fizz when they are mixed with an acid because the reaction produces carbon dioxide as well as a salt and water.

potassium carbonate + sulfuric acid ⟶ potassium sulfate + carbon dioxide + water

Again, there is a pattern to the reactions:

alkaline carbonate + acid ⟶ salt + carbon dioxide + water

Questions

1 At which stages of the manufacture of alum were the following processes involved? Which of these processes involved chemical reactions to make new chemicals?
  a oxidation
  b dissolving
  c evaporation
  d neutralisation
  e crystallisation.

2 Stale urine contains 2 g of ammonia in 100 cm^3 of the liquid. The daily output of a person is about 1500 cm^3 of urine.
  a Estimate the mass of ammonia, in tonnes, that could be obtained per person per year (1 tonne = 1000 kg, 1 kg = 1000 g).
  b 3.75 tonnes of ammonia is needed to make 100 tonnes of alum. Estimate the number of people needed to supply the urine for an alum works producing 100 tonnes of alum per year.

3 Name the products of the reactions of:
  a calcium hydroxide with hydrochloric acid
  b potassium hydroxide with sulfuric acid
  c sodium carbonate with nitric acid.

Key words

- **alkali**
- **neutralise**
- **salt**
- **word equation**

Chalk and limestone contain calcium carbonate. They are insoluble in water. Heating them in a lime kiln, like the traditional one shown here, breaks down the calcium carbonate into calcium oxide. Calcium oxide reacts with water to make calcium hydroxide, which is slightly soluble in water, and can **neutralise** acids, including acids in soils.

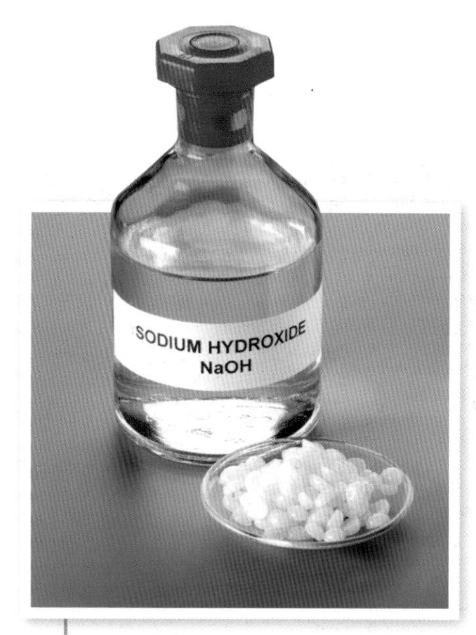

Pure sodium hydroxide is a white solid. It is soluble in water and used in solution as an alkali. Pure sodium hydroxide is very corrosive.

Chemicals from salt – the foul way

Find out about

- how alkalis were first manufactured on a large scale
- why this was such a polluting process
- how Parliament began to regulate the chemical industry

Air pollution from industry in Widnes in 1895.

Making alkali on a large scale

During the industrial revolution in the 1700s, natural sources of alkali became too scarce to meet demand. The shortage of alkali was particularly serious in France, where large amounts of alkali were needed for use in the glass industry.

In 1791, Nicolas Leblanc invented a new process that used chalk or limestone (calcium carbonate), salt (sodium chloride), and coal to make the alkali sodium carbonate. He was awarded a patent by the King of France and given enough money by the Duc d'Orléans to build a plant to operate the process.

This was the time of the French Revolution, and in 1794, the Duc d'Orléans was guillotined. As a result, Leblanc's factory was seized and the patent for his process became public property.

Now others could benefit from Leblanc's invention. A chemical industry began in England based on his process and continued for over 100 years. The industry grew rapidly. The rural areas of Widnes and Runcorn on the banks of the River Mersey were transformed into international centres for new industries based on salt.

The **Leblanc process** was highly polluting. For every tonne of the product sodium carbonate, the process created two tonnes of solid waste. It also released almost a tonne of **hydrogen chloride gas** into the air. This acid gas devastated all the land around. The solid waste was dumped in vast heaps outside the factory, where it slowly gave off a steady stream of toxic **hydrogen sulfide gas**. This gas has a sickening smell of bad eggs. Estimates suggest that by 1891, about 200 hectares of land around Widnes had been buried under an average depth of 3–4 metres of waste. Living and working conditions in the area were appalling.

First steps towards regulating the chemical industry

As industrialisation increased in the 1800s, the British public began to demand action from the government to control pollution. At that time the government was anxious not to restrict the chemical industry because it brought money to the economy and provided jobs. But Parliament did begin to pass laws to regulate working conditions and control pollution from railway engines and factory smoke.

Pollution by the chemical industry became so bad that in 1863, Parliament passed the first of the **Alkali Acts**. This Act set up an Alkali Inspectorate. Inspectors travelled the country to check that at least 95% of acid fumes were removed from the chimneys of chemical factories. The inspectors were scientists. Dressed in Victorian frock coats and top hats, they carried their measuring equipment up long ladders to the top of factory chimneys. There they sampled the smoke and fumes in all weathers.

Tackling the pollution problem

The first response of the Leblanc industry to the Alkali Acts was to dissolve the hydrogen chloride in water. They had no use for the hydrochloric acid that formed, so to begin with they just let it flow through sewers into the local rivers, where it killed all the life in the water.

In 1874, Henry Deacon invented a better way to use the acid gas from the Leblanc process. He found that it was possible to oxidise hydrogen chloride to **chlorine**. Chlorine is one of the elements that make up hydrogen chloride, but it has very different properties to the compound. Hydrogen chloride is corrosive and acidic, while chlorine can be used as a **bleach**.

In Henry Deacon's process, he mixed hydrogen chloride with oxygen and let the two gases flow over a hot catalyst. The products were chlorine and steam. The chlorine could be used to bleach paper and textiles.

The problems of the Leblanc process were eventually solved towards the end of the 1800s, not by government controls, but by developing new methods for manufacturing alkalis. The new processes are still in use today (see H: Chemicals from salt – a better way).

"...they were ruined when they were required to send labouring children to school; they were ruined when inspectors were appointed to look into their works; they were ruined when such inspectors considered it doubtful whether they were quite justified in chopping people up in their machinery; they were utterly undone when it was hinted that perhaps they need not make so much smoke."

Dickens mocked industrialists' attitudes to new controls when he wrote *Hard Times* in 1854.

Questions

1. Show how hydrogen chloride illustrates the fact that the properties of a chemical compound are very different from the properties of its elements.
2. Draw a diagram to show four molecules of hydrogen chloride, HCl, reacting with a molecule of oxygen to form two molecules of chlorine, Cl_2, and two molecules of steam (water).
3. Why was turning waste hydrogen chloride into chlorine better than dissolving it in water?
4. Suggest reasons why Parliament was slow to bring in laws to control the new chemical industry, despite the serious risks to health and unpleasantness for workers and for people living nearby.

Key words

- ✓ **hydrogen chloride gas**
- ✓ **hydrogen sulfide gas**
- ✓ **chlorine**
- ✓ **bleach**
- ✓ **Leblanc process**
- ✓ **Alkali Acts**

Benefits and risks of water treatment

Find out about

- the risks of waterborne diseases
- the benefits of chlorinating drinking water
- worries that some people have about chlorinated water

The threat of waterborne disease

Water that is contaminated by sewage can sometimes carry fatal diseases, such as cholera, typhoid, dysentery, and gastroenteritis. The World Health Organisation (WHO) reports that water quality is still a very serious threat to human health. According to the WHO, more than three million people still die each year as a direct result of drinking unsafe water. Waterborne infections cause about 1.7 million of these deaths. Those who die are mainly children in developing countries.

A polluted river in Arusha, Tanzania. In many parts of the world, people can't get clean drinking water. Much untreated sewage flows into streams and rivers.

Water treatment with chlorine

Prince Albert, the husband of Queen Victoria, died of typhoid fever in 1861. The poor state of the drains in Windsor Castle may have been to blame. It was only a few years later that it became normal to filter drinking water and then treat it with chlorine to kill **microorganisms**. This process might have saved Prince Albert's life.

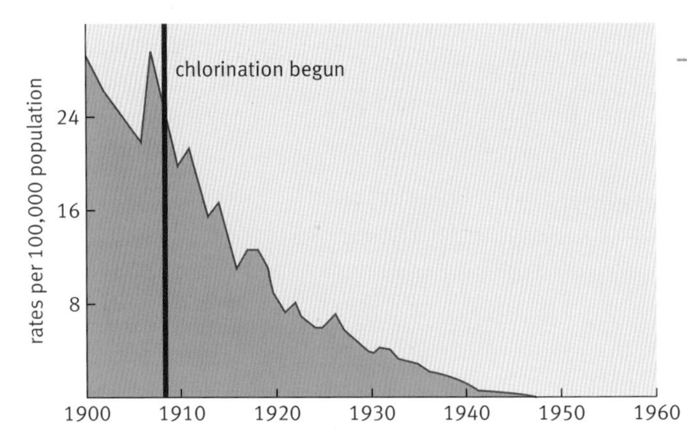

Death rate from typhoid fever in the USA, 1900–1960 (first published in the US Center for Disease Control and Prevention's Summary of Notifiable Diseases 1997).

Chlorination of drinking water in Britain became increasingly common in the early twentieth century. This led to a steep decline in deaths from typhoid. An advantage of using chlorine to treat water is that some of the chemical stays in the water. This means that it can protect against possible contamination in the pipes that carry it from the treatment works to the consumer.

As well as being part of the water purification process, chlorine removes unpleasant tastes and helps to stop microorganisms growing in water storage tanks.

One solution to the problem of water pollution is to use filtration. This device for filtering drinking water is made locally in Cambodia. Where necessary, filtered water can also be treated with chlorine bleach before being used to wash and prepare food.

Risks of water treatment

Chlorination has helped to prevent the diseases associated with drinking water. But some scientists are concerned that there may be side-effects of chlorination. They are worried that chlorination can lead to the formation of a group of chemicals known as trihalomethanes (THMs).

THMs can form when chlorine reacts with naturally found **organic matter** in water, such as fragments of leaves. Some organic matter is found in surface water, such as the lakes and rivers used for drinking water. During the cleaning process, very small amounts of THMs may be formed. When people drink the water, these THMs may be absorbed into their bodies.

There is a suspicion that THMs could lead to some forms of cancer. However, research studies have not found any firm evidence to support this idea. The International Agency for Research on Cancer and the World Health Organisation both say there is not enough evidence to prove any strong link between cancer risks and THMs.

Where organic matter is detected in the raw water to be used for drinking, water companies have devised methods to limit the levels of THMs in the water supply. Ozone gas is used to break down the organic material, then carbon filters can remove it before disinfection by chlorine.

Questions

1 Use the graph to estimate the number of deaths from typhoid in the USA:
  a in 1900, when the population was 76 million
  b in 1940, when the population was 132 million.

2 Water for homes can come from underground aquifers and springs, or it can come from surface sources such as rivers and reservoirs.
  a After chlorination, which type of water is more likely to contain THMs?
  b Suggest reasons why the risk of THMs forming in tap water varies with the time of year.

Key words

- microorganisms
- chlorination
- organic matter

Dr Harriette Chick (1875–1977) carried out her research at the Lister Institute, a centre for the study of diseases that was then based in London. In 1908, she published the results of her study into the factors affecting how quickly chlorine kills bacteria and viruses in water. Her 'laws of disinfection' helped water companies understand how to use chlorine effectively.

Chemicals from salt – a better way

Find out about

- the use of electricity to make new chemicals
- the chemicals made by the electrolysis of brine
- the environmental impact of the chemical industry based on salt

Chemicals from salt

Chlorine is now made on a very large scale. Over 10 million tonnes of chlorine are produced from salt in Europe each year. Today electricity is used to make chlorine from salt. This process is very much cleaner than the old Leblanc process.

The Ineos Chlor chemical plant in Runcorn. Chlorine from the electrolysis of brine is used to make PVC and other chemicals.

Brine is a solution of sodium chloride (NaCl) in water (H_2O). So there are just four elements in the brine, which can be rearranged to make chlorine (Cl_2), sodium hydroxide (NaOH), and hydrogen (H_2).

The chemical changes happen when an electric current flows through the solution. The process is called **electrolysis**.

The chemical changes take place at the surface of the metals that conduct the electric current into, and out of, the solution. The equipment for the electrolysis of brine has to be carefully designed to keep the two main products, chlorine and sodium hydroxide, separate. These two chemicals have to be kept apart because they react with each other when they mix.

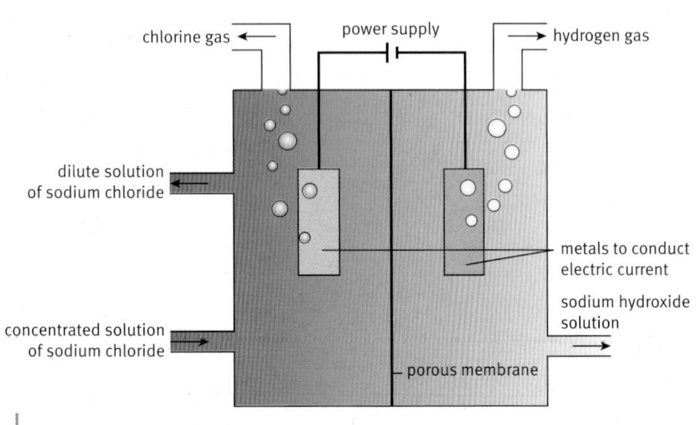

The electrolysis of brine. The porous membrane keeps the chlorine and the sodium hydroxide apart, but doesn't stop the electric current.

Uses of chemicals from salt

The chemicals produced from salt have many uses.

Chlorine	Sodium hydroxide	Hydrogen
• to treat drinking water and waste water • to make bleach. • to make hydrochloric acid • to make plastics including PVC • to make solvents	• to make bleach • to make soap and paper • to process food products • to remove pollutants from water • for chemical processing and products • to make fibres	• to make hydrochloric acid • as a fuel to produce steam

Key words

- electrolysis
- toxic

Sodium hydroxide and chlorine react to make bleach.

Chlorine and hydrogen react to make hydrogen chloride gas. The gas dissolves in water to make hydrochloric acid.

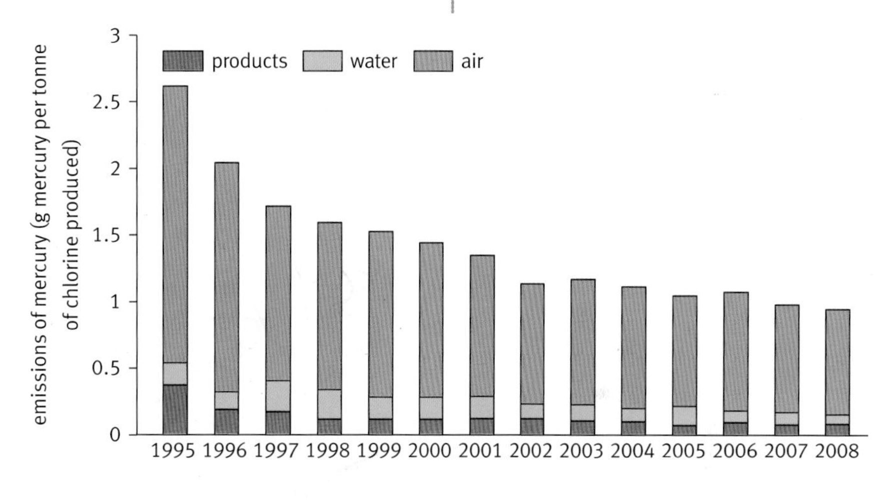

Changes in the emissions of mercury from European chemical plants for the electrolysis of brine from 1995 to 2008 (published in *Chlorine Industry Review 2008–09*). Some of the mercury ends up in the products, some in waste water, and some in the air.

Environmental impacts

Manufacturing chemicals from salt by electrolysis needs a lot of energy.

The chemical plant at Runcorn uses as much electricity as a city the size of Liverpool. At the moment, most of the electricity for the electrolysis of brine is generated using fossil fuels. However, the industry is moving towards producing much more of the electricity it needs from renewable sources. It is building a power plant to generate energy by burning household and industrial wastes that can't be recycled.

Until quite recently, the most common system in Europe for the electrolysis of brine used mercury as one of the metals in contact with the solution. Clever design means that chlorine can form in one part of the apparatus, while the flow of mercury circulating in the apparatus allows sodium hydroxide to form in another part. Unfortunately, this method produces products that are contaminated with very tiny amounts of mercury. Also, some of the **toxic** mercury escapes into the environment. As a result, the use of mercury is being phased out.

Increasingly, the industry uses equipment with a sophisticated polymer membrane to keep chlorine and sodium hydroxide apart. This uses less energy for electrolysis, but the sodium hydroxide formed is more dilute so it has to be concentrated by evaporation. Even so, it is the more efficient process.

Questions

1. Name the four chemical elements in brine.
2. How does a change in the method used to electrolyse brine account for the fall in the amount of mercury lost to the environment per tonne of chlorine in recent years?
3. What are the reasons for cutting down on the use of fossil fuels to generate electricity for the electrolysis of brine?

Find out about

- why there is a need to check on the safety of a very large number of chemicals
- the European Union's programme for testing
- the problem of persistent and harmful chemicals

Greenpeace activists hold banners reading 'Everyday Chemicals Harm My Sperm!' as they demonstrate in front of the Chancellery in Berlin in 2005.

REACH stands for the Registration, Evaluation and Authorisation of Chemicals.

Untested chemicals

Campaigns by environmental pressure groups, such has the World Wide Fund for Nature (WWF) and Greenpeace, have made many people fearful of **synthetic** (man-made) chemicals. The campaigns have highlighted evidence suggesting that chemicals, such as those found in plastics and pesticides, may cause cancer, or lead to defects in new-born babies.

Most scientists who study toxic chemicals agree that some commonly used synthetic chemicals can be harmful in large doses, but not at the concentrations usually found in people's bodies. Very sensitive chemical tests have typically found only very tiny amounts of these chemicals, less than one part per billion, in human blood. Scientists argue that campaigners are confusing risks with hazards. The chemicals may not be completely safe, but there is no evidence that such tiny traces of them are unsafe.

REACH

Up until 2007, many environmentalists campaigned to try to make the new European Union (EU) laws about chemicals as strict as possible.

European industry produces or uses 30 000 different chemicals a year – a tonne or more of each one. But information about their environmental and health effects is available for only a very small proportion of these compounds. European countries and the USA have been safety-testing all new chemicals since 1981, but this only accounts for about 3% of those in use.

In 2007, the EU introduced the REACH system to collect information about the hazards of chemicals and to assess the risks. REACH switches most of the responsibility for control and safety of chemicals from the authorities to the companies that make them, or use them. Now industry has to manage the risks of chemicals for human health and the environment.

POPs and pollution

There are some synthetic chemicals that everyone agrees are harmful even in very small amounts. These are chemicals that do not break down in the environment for a very long time. This means they can spread widely around the world in air and water. They build up in the fatty tissues of animals, including humans. So they can harm people and wildlife.

This set of chemicals is sometimes called the 'dirty dozen'. They are:

- eight pesticides (two of them are DDT and DDE)
- two types of compounds used in industry (including PCBs)
- two by-products of industrial activity (including dioxins).

All the chemicals in the 'dirty dozen' list are classified as **persistent organic pollutants** (POPs). Many of them are compounds containing chlorine. They are a particular problem for people living in the Arctic, where traditional diets are often high in fat. POPs tend to **accumulate** in fatty tissue of animals, which people then eat.

Experts at a conference in Stockholm in 2001 agreed a Convention to deal with POPs. It became effective in 2004, and about 150 countries have agreed to outlaw the 'dirty dozen' chemicals.

The 10 pesticides and industrial chemicals listed in the Convention have all been banned in Britain for several years. The two other POPs have never been produced intentionally, but may be formed as by-products during combustion of wastes, or in some industrial processes.

Key words

- accumulate
- synthetic
- persistent organic pollutants

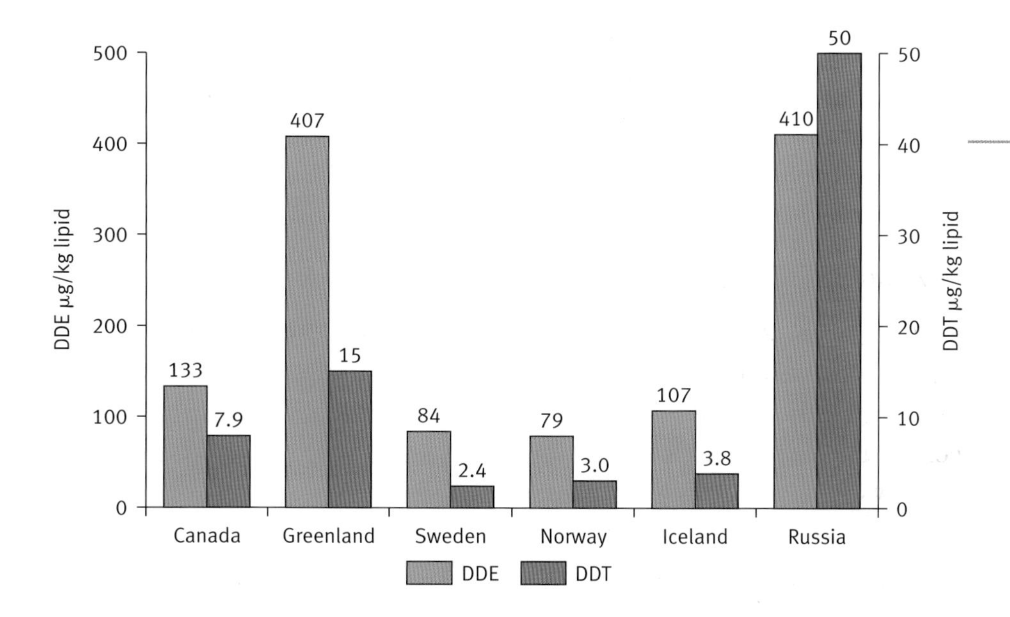

DDT and DDE levels in the blood plasma of mothers in different Arctic regions (published by the Arctic Council Secretariat). The units are micrograms per kilogram of fatty chemicals (lipids). A microgram is a millionth of a gram.

Questions

1. Give a chemical example to explain the difference between a hazard and a risk.
2. Why can't scientists say for sure that small traces of permitted chemicals are completely safe?
3. Manufacturers and importers will pay for most of the cost of REACH. The cost for the whole EU will be about €5 billion over the first 11 years of the testing programme. The EU has a population of about 500 million people.
   a. Do you think that such a large cost is justified? Explain your answer.
   b. Is it right that industry should have to organise and pay for the testing? Explain your answer.
4. Suggest reasons why the Stockholm convention allows the use of the insecticide DDT to control mosquitoes in parts of the world where malaria is a serious problem.

Stages in the life of PVC

Find out about

- the stages in the production, use, and disposal of PVC products
- risks involved in making and disposing of PVC

PVC is a synthetic polymer. It is strong, easy to mould and quite cheap. It is also hard-wearing, durable, and can be used to make a wide range of products. The stages in the life of PVC products include production, use, and disposal.

Chemicals from raw materials

PVC is made from two chemicals: ethene and chlorine. In the first stage of the process, chlorine and ethene are combined to make a chemical that is commonly called vinyl chloride. This is a hazardous compound because it can cause cancer. PVC manufacturers take great care to make sure that workers are not exposed to this chemical.

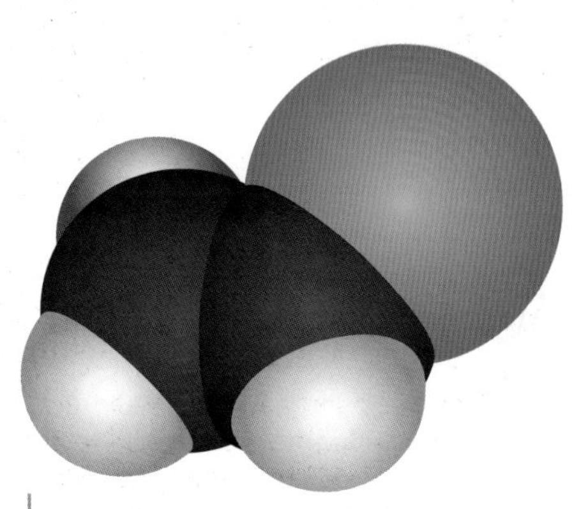

A molecule of vinyl chloride, C_2H_3Cl. Carbon atoms are shown as black, hydrogen atoms white, and the chlorine atom green.

Making PVC from chemicals

Vinyl chloride is a liquid made up of small molecules. These small monomer molecules are joined together to make long chains of poly(vinyl chloride) (PVC) by polymerisation. PVC molecules are made up of three different elements: carbon, hydrogen, and chlorine.

The polymer can be supplied as small plastic beads ready for moulding into products.

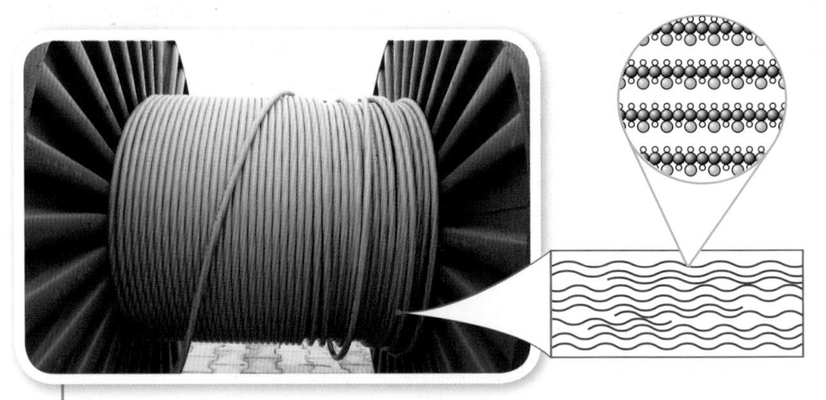

PVC is used to insulate electric cables. PVC is a polymer. The red lines represent the molecules, which are very long chains. These long chains are made up of carbon, hydrogen, and chlorine.

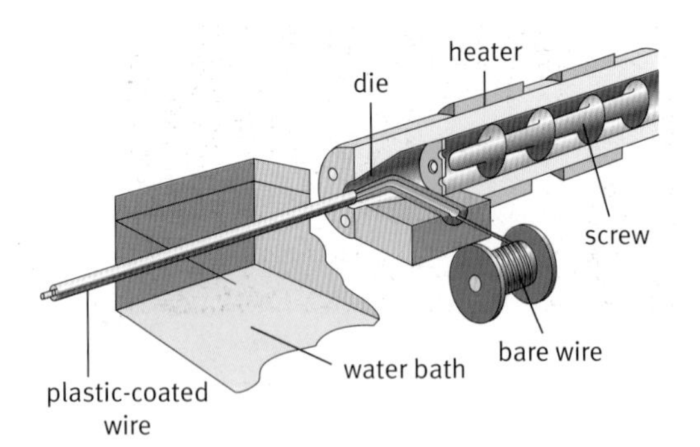

Extrusion is a way of coating copper wire with plastic. The machine heats the polymer to melt it. Then the screw forces the softened plastic though a die so that it coats the wire being pulled through the die at the same time. This process needs energy to heat the plastic and run the machinery. Water is used to cool the mould.

Making products from PVC

PVC granules are sent to factories to be moulded under heat and pressure. For example, the hot plastic can be **extruded** into pipes or blown to make bottles.

Using PVC products

About half of the PVC made each year is used to make pipes. Many of these pipes are underground, carrying drinking water, sewage, and gas. Much of the rest of the PVC produced is used in building, for gutters and window frames. Many commercial signs are made from sheets of PVC.

A softer type of PVC is used to make clothing, garden hoses and insulation for electric wires. PVC film is used for packaging, blood bags, and the bags for intravenous drips.

Disposing of PVC products

Recycling

The best way of getting rid of old PVC products is recycling. It is sometimes possible to grind the waste into pellets. The pellets can be reheated and moulded into new products.

Recycling cuts down the amount of raw materials used to make new PVC. It also reduces the amount of waste from PVC products that have reached the end of their life.

A major problem with recycling PVC is that it is often mixed with other materials. This can make separating, sorting, and recycling difficult and expensive. New methods of recycling are being developed for these mixed sources of PVC.

A woman sorting plastic waste in Mumbai, India. She is separating out pieces of PVC and putting them in baskets. She picks the pieces out of a barrel of water. PVC is denser than water so it sinks. Other plastics are less dense than water so they float.

Energy recovery

Some polymer waste can be burnt. The energy released can be used to generate electricity. This is done in special **incinerators**.

Plastics have to be burnt at a very high temperature to avoid releasing hazardous chemicals. This is a particular problem with PVC, because it produces hydrogen chloride gas when it burns. Acid gases can be removed from the fumes produced by burning before they are released into the air. Burning PVC may also produce toxic dioxins if the conditions in the incinerator are not controlled correctly.

Landfill

Unfortunately, much polymer waste still ends up being tipped into holes in the ground. We call this **landfill**. This really is a waste.

Key words

- extruded
- incinerator
- landfill

Questions

1 What is the raw material needed to make:
   a ethene?
   b chlorine?

2 Which three chemical elements are present in PVC?

3 It might seem better to re-use articles made of PVC rather than recycling them or throwing them away. Why might this be impossible or undesirable?

4 Why is it important that waste incinerators do not release hydrogen chloride into the air?

5 People often campaign when there are plans to build a waste incinerator near to where they live.
   a Suggest risks that the campaigners might be worried about.
   b What is it about the possible risks from burning waste that make people so worried?
   c Suggest arguments that might be used to defend the setting up of a waste incinerator.

K Benefits and risks of plasticisers

Find out about

- the chemical used to plasticise PVC
- why plasticisers may be harmful
- what the regulators are doing about the risks

Worries about plasticisers

Toymakers like to use PVC because it is very versatile:

- it can be either flexible or rigid
- it can be mixed with pigments to give bright colours
- it stands up to rough play
- it is easy to keep clean.

Plasticisers are chemicals that make PVC soft and flexible. The most common plasticisers for PVC are **phthalates**.

Plasticisers are made up of quite small molecules, which can escape from the plastic and dissolve in liquids that are in contact with it. For example, plasticisers can escape from a PVC toy into the saliva of a baby that chews it. They can also **leach** out of the plastic used to make blood bags, or the bags for intravenous drips, and so enter patients' blood.

This child is sliding down a rigid plastic slide into a flexible plastic paddling pool.

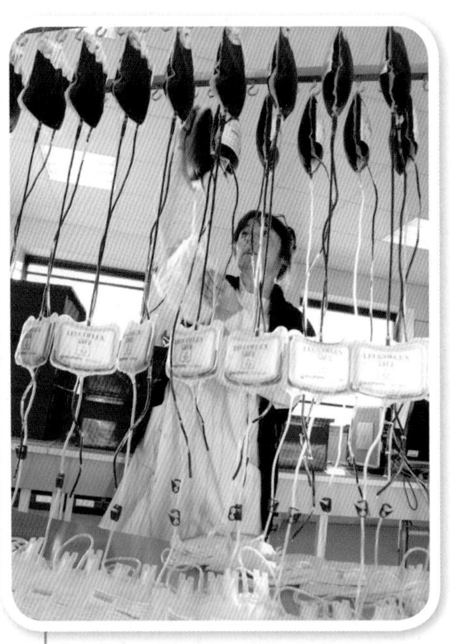

Blood from donors is stored in plasticised PVC bags. The tubing carrying the blood to patients is also made of PVC, which is usually plasticised with DEHP.

'Ducki' was displayed for the first time at a New York toy fair in 2010. The toy is advertised as being free of PVC and phthalates. It was designed by a pop-art sculptor who began making toys for his daughter because he was worried about the safety of plastic toys.

Key words

- plasticiser
- leach
- phthalates

Some campaigners argue that phthalates should be banned. They think that a ban is justified because of evidence linking plasticisers with health problems such as cancer, liver problems, and infertility.

Makers of PVC and PVC products point out that phthalates have been in use for over 50 years. They say that, in all that time, there has not been a single known case of anyone ever having been harmed as a result of the use of phthalates.

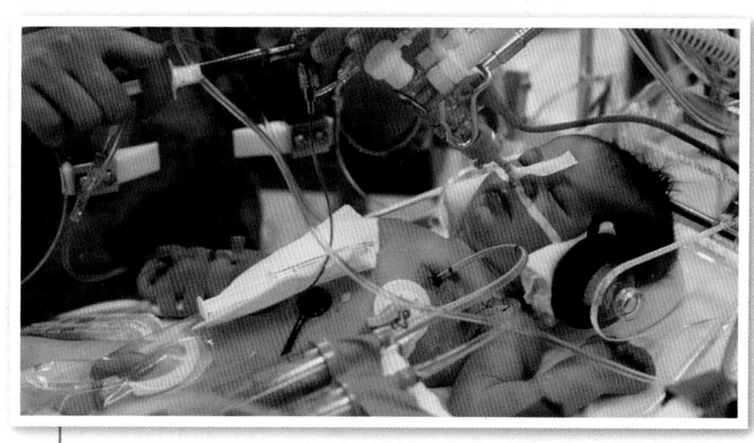

Flexible plastic tubing helps to keep babies alive in intensive care.

What the regulators say

Regulators in Europe and the USA are concerned about the possible effects of some plasticisers on young children and new-born babies. Since 2007 the European Union has restricted two common phthalate plasticisers to toys that cannot be placed in the mouth. A third plasticiser (DEHP) has been banned completely from toys.

Regulators are particularly worried about DEHP because it has been shown to affect the development of the reproductive system and sperm in young male animals. These effects have not been found in human babies but it has not been possible to show that there is no risk. As a precaution, regulators have issued warnings about the use of DEHP in PVC products for medicine.

Plasticisers in medical devices

PVC plasticised with DEHP is used in many medical devices because it has exactly the right combination of properties. It is flexible, strong, and transparent, and keeps its properties at the high temperatures needed for sterilisation (to kill microorganisms) and the low temperatures used for cold storage. It is one of the few materials that has all the right properties and is also affordable.

However, DEHP can leach out of PVC into liquids used to treat patients. Seriously ill people often need treatment for a long time. This can increase their exposure to DEHP. These patients include people who have regular dialysis to treat kidney failure. Others exposed to risk are new-born babies or young children needing repeated blood transfusions.

There are alternatives to DEHP but these are expensive and not always available. This means that if DEHP was banned from medical devices some patients might not receive the treatment they need. Regulators in the USA have told doctors that they should not avoid carrying out medical treatments using plasticised PVC. The regulators say that the risk of not treating a sick patient is far greater than the very small risk from exposure to the plasticiser. However, as a precaution the regulators have recommended that alternatives are used when treating male, new-born babies and women who are known to be pregnant with male fetuses.

Questions

1. Why does it makes sense for regulators to ban the use of the plasticiser DEHP in toys, but only to issue warnings and advice about the use of medical equipment made with PVC softened with the same plasticiser?
2. Why is it so hard to prove that there is no risk when people have fears about possible dangers from a chemical?
3. Some people in medical care are more at risk than others from the possible harm from the plasticiser DEHP.
   - a Identify two groups of people who are more at risk and for each group explain why.
   - b Give one benefit for patients of using DEHP in medical equipment.
   - c If alternatives to DEHP became more affordable and available, how might this affect decisions by hospitals about which material to use?

Find out about

- the life of products from cradle to grave
- the impacts of the products you use

A 1970s TV set. It contains glass, metals, plastics and wood.

The products we buy and use affect the environment. Environmental scientists add up all the effects of a product from cradle to grave. A life cycle assessment can show whether it is better to use a shopping bag made of a natural fibre or a bag made of plastic.

Lives or life cycles

At home, you are surrounded by many different manufactured products – furniture, clothes, carpets, china and glass, televisions, and mobile phones.

The life of each of these products has four distinct phases:

1. The materials are made from natural raw materials.
2. Manufacturers make products from the materials.
3. People use them.
4. People throw them away.

Imagine an old television that was bought in 1970 and thrown away some years ago. It contains glass, metals, plastics, and wood. It is now buried under many tonnes of rubble in a landfill.

The wood will eventually rot because it is **biodegradable**. But the rest of the materials are there forever. This is not sustainable because the materials cannot be used again. The materials had a life, but not a life cycle.

Once the life of a product is over, its materials should go back into another product. This is recycling.

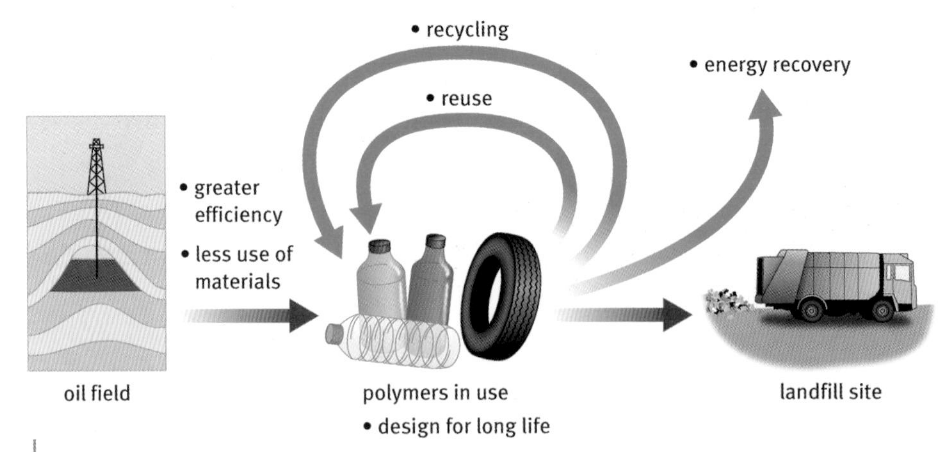

Oil and products from oil, such as polymers, are very valuable. They lose value as they are used and end up as waste. The aim is to slow down the journey of materials from natural resources (cradle) to landfill sites or incinerators (grave).

Life cycle assessment

Manufacturers can assess what happens to the materials in their products. This **life cycle assessment** (LCA) is part of legislation to protect the environment. The aim is to slow the rate at which we use up natural resources that are not renewable. At each stage in the life of a product, raw materials, water, and energy may be used:

CRADLE

- **raw materials** obtained and processed to make useful materials
- **materials** used to make the product
- **energy and water** used in processing and manufacturing

USE

- **energy** needed to **use** the product (eg electricity for a computer)
- **energy** needed to **maintain** the product (eg cleaning, mending)
- **water** and **chemicals** needed to maintain it

GRAVE

- **energy** needed to **dispose** of the product
- **space** needed to dispose of it.

Key words

- life cycle assessment
- biodegradable

An LCA involves collecting data about each stage in the life of a product. The assessment includes the use of materials and water, energy inputs and outputs, and environmental impact. An assessment of this kind can show, for example, whether it is better to make window frames from the traditional material (wood) or the modern material (PVC).

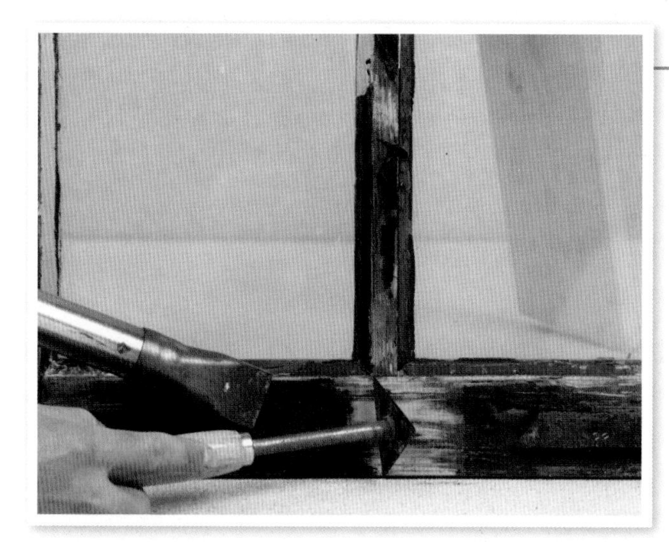

Restoring the paintwork on a wooden window. Wood can rot if it is not properly looked after. It needs regular painting. The advantage of painting is that you can change the colour easily. With double glazing, wood gives good thermal insulation.

PVC windows need little maintenance. Unlike wooden windows, they are not affected by moisture. Combined with double-glazing, they provide good thermal insulation.

Questions

1. Give two reasons why it is not a good idea to put products in landfill once we have used them.
2. Suggest examples of how to slow down the flow of materials from raw materials to waste. Include examples of:
   a. re-use
   b. recycling
   c. recovering energy.
3. Choose a product that has been designed to reduce its impact on the environment.
   a. Describe the product.
   b. Explain how its environmental impact has been reduced.

Science Explanations

Salt, limestone, coal, gas, and oil have been the basis of the chemical industry for many years. The use of manufactured chemicals has brought many benefits but they are not without risk.

You should know:

- how geologists explain the past history of the surface of the Earth from what they know about processes happening now
- that the parts of ancient continents that now make up Britain have moved over the surface of the Earth as a result of plate tectonics
- how magnetic clues in rocks help geologists to track the very slow movement of the continents
- how processes such as mountain building, erosion, sedimentation, dissolving, and evaporation have led to the formation of valuable minerals
- that clues to the conditions under which rocks were formed come from fossils, shapes of sand grains, and ripples made by flowing water
- why a chemical industry grew up in northwest England
- that salt (sodium chloride) is important for preserving and processing food, as a source of chemicals, and to treat roads in winter
- that salt comes from the sea or from underground salt deposits
- why the methods used to obtain salt may depend on how it is to be used
- why extracting salt may have an impact on the environment
- why alkalis were needed in pre-industrial times
- that the traditional sources of alkali included burnt wood or stale urine
- that examples of alkalis include soluble metal hydroxides and metal carbonates
- that alkalis neutralise acids to make salts
- how industrialisation led to a shortage of alkali in the 19th century
- why the first process to meet a growing demand for alkali was highly polluting
- how the pollution problems of the old process were reduced by producing useful chemicals such as chlorine
- why using chlorine to kill microorganisms in domestic water supplies has made a major contribution to public health
- that electrolysis is the process now used to make new chemicals, such as sodium hydroxide and chlorine and hydrogen, from salt solution
- that PVC is a polymer containing chlorine
- how the properties of PVC can be altered by plasticisers.

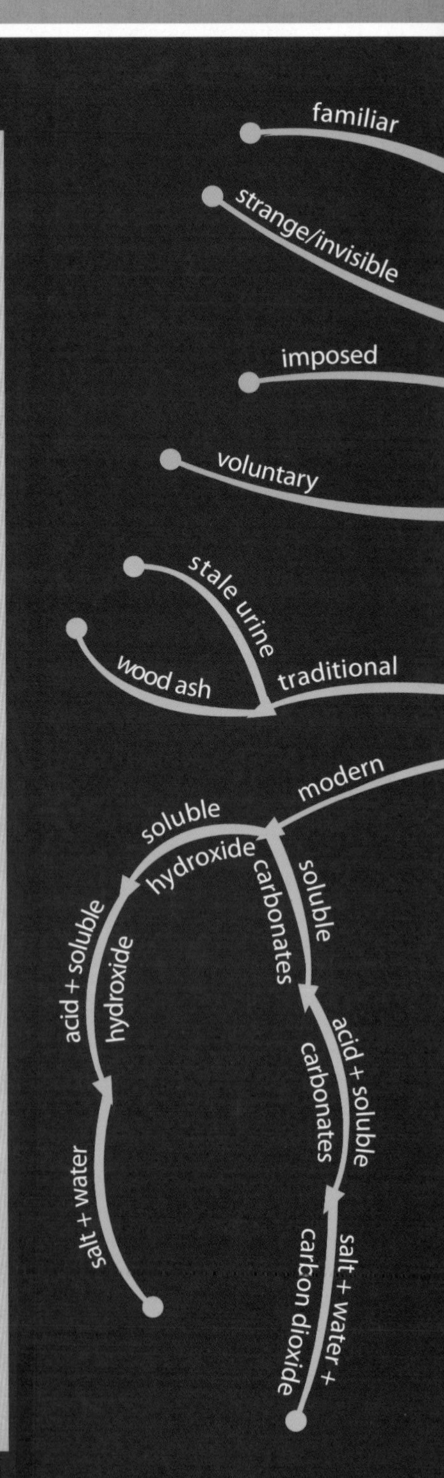

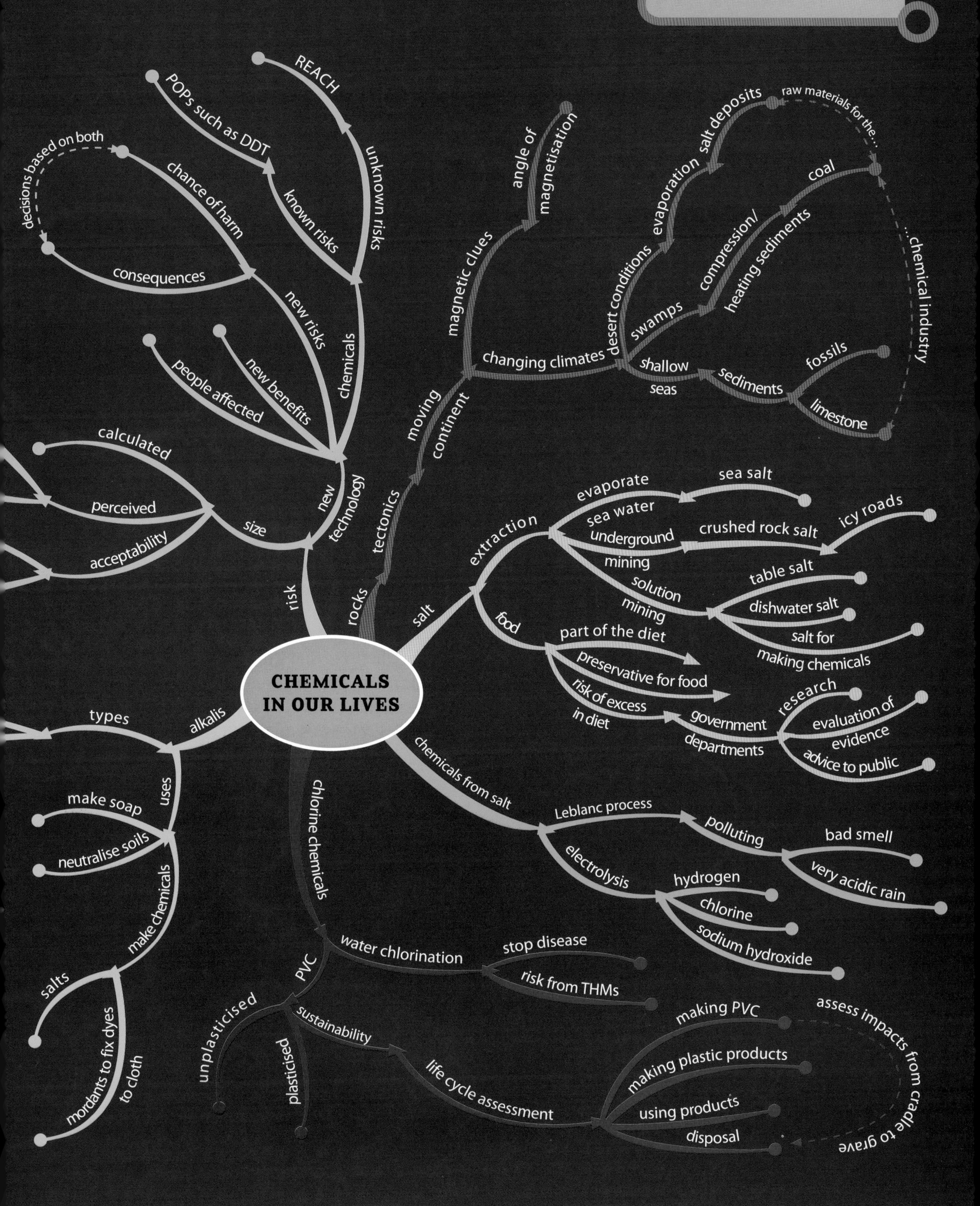
CHEMICALS IN OUR LIVES
risk
new technology
size
calculated
perceived
acceptability
chemicals
people affected
new benefits
new risks
chance of harm
consequences
decisions based on both
known risks
POPs such as DDT
unknown risks
REACH
rocks
tectonics
moving continent
changing climates
magnetic clues
angle of magnetisation
desert conditions
evaporation
salt deposits
raw materials for the...
...chemical industry
swamps
compression/ heating sediments
coal
shallow seas
sediments
fossils
limestone
salt
extraction
evaporate sea water
sea salt
underground mining
crushed rock salt
icy roads
solution mining
table salt
dishwater salt
salt for making chemicals
food
part of the diet
preservative for food
risk of excess in diet
government departments
research
evaluation of evidence
advice to public
chemicals from salt
Leblanc process
polluting
bad smell
very acidic rain
electrolysis
hydrogen
chlorine
sodium hydroxide
chlorine chemicals
water chlorination
stop disease
risk from THMs
PVC
unplasticised
plasticised
sustainability
life cycle assessment
making PVC
making plastic products
using products
disposal
assess impacts from cradle to grave
alkalis
types
uses
make soap
neutralise soils
make chemicals
salts
mordants to fix dyes to cloth

Ideas about Science

Scientists seek explanations to account for their findings, such as the data collected by studying rocks. You should be able to:

- explain how magnetic data and other clues in rocks support the theory that the continents have moved.

New technologies and processes based on scientific advances sometimes introduce new risks. Some people are worried about the effects arising from the use of chemicals. You should be able to:

- explain why nothing is completely safe
- identify examples of risks that arise from the use of chemicals
- interpret information on the size of risks, presented in different ways
- describe ways of reducing risks from hazardous chemicals
- discuss a given risk, taking into account both the chances of it happening and the consequences if it did
- identify risks and benefits, for the different individuals and groups involved, arising from uses of chemicals
- suggest why people accept (or reject) the risk of a certain activity, for example, eating a diet with more salt than is recommended
- recognise that people's perception of the size of a risk is often very different from the scientific assessment of the risk
- illustrate the idea that people tend to overestimate the risk of unfamiliar things and things that have an invisible effect.

Governments and public bodies assess what level of risk is acceptable. Treaties, regulations, and laws control scientific research and the applications of science. The decision to regulate may be controversial, especially if those most at risk are not those who benefit. You should understand that governments and regulators are responding to concerns that:

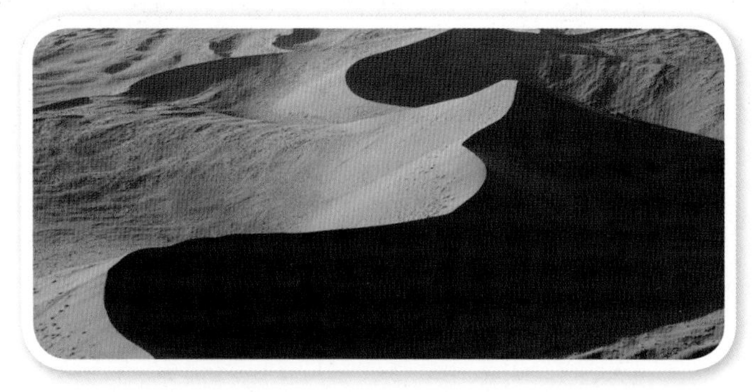

- many people are putting their health at risk by eating too much salt
- there are possible disadvantages of chlorinating drinking water, including possible health problems
- some toxic chemicals persist in the environment – they can be carried over large distances and may accumulate in food and human tissues
- the plasticisers added to PVC can leach out from the plastic into the surroundings where they may have harmful effects.

Science helps to find ways of using natural resources in a more sustainable way. You should understand that:

- a life cycle assessment (LCA) tests:
  - a material's fitness for purpose
  - the effects of using material products from production to final disposal.
- an LCA involves consideration of the use of resources including water, the energy input or output, and the environmental impact, of each of these stages:
  - making materials from natural raw materials
  - making useful products from materials
  - using the products
  - disposing of the products.

When given appropriate information from an LCA, you should be able to compare and evaluate the use of different materials for the same job.

Review Questions

1 **a** The table shows the mass of salt in different foods.

Food	Mass of salt in 100 g of the food (g)
White bread	1.2
Cornflakes	1.8
Ham	3.1
Crisps	2.5
Chocolate muffin	1.7

i Which food in the table contains the most salt in 100 g?

ii Emma eats a 50 g packet of crisps. Ben eats a 100 g chocolate muffin. Who has eaten more salt? Show how you work out your answer.

b Health experts recommend that adults should eat no more than 6 g of salt each day.

i Identify two risks linked to eating too much salt.

ii Suggest why some people eat more than 6 g of salt each day, even though they know about the risks of eating too much salt.

2 **a** In 1991, there was an outbreak of cholera in South America. The table shows data from the time of the outbreak.

Village	Number of people who caught cholera	Did the village add chlorine to its drinking water?
A	0	no
B	27	no
C	1	yes
D	31	no
E	42	no

Cholera is a disease that is caused by a microorganism. It spreads through contaminated water. Vincent suggests an explanation for the data. He says that adding chlorine to water kills the microorganism that causes cholera.

i Identify the data that is accounted for by this explanation.

ii Identify the data that conflicts with the explanation.

b Describe one possible disadvantage of adding chlorine to water.

c Give the name of the process by which chlorine is produced from salt (sodium chloride).

3 Sulfuric acid is an acid in acid rain. Sulfuric acid damages limestone buildings. Limestone is mainly calcium carbonate.

Write a word equation for the reaction of sulfuric acid with calcium carbonate.

4 Match each piece of evidence to one or more pieces of information that the evidence can provide.

Evidence	Information
shape of the grains in sandstone	latitude at which the rocks were originally formed
angle at which crystals are magnetised in the rock	whether the rock was formed under the sea
presence of shell fragments	age of the rock
types of fossilised living organisms in the rock	whether the rock was formed from materials from deserts or riverbeds.

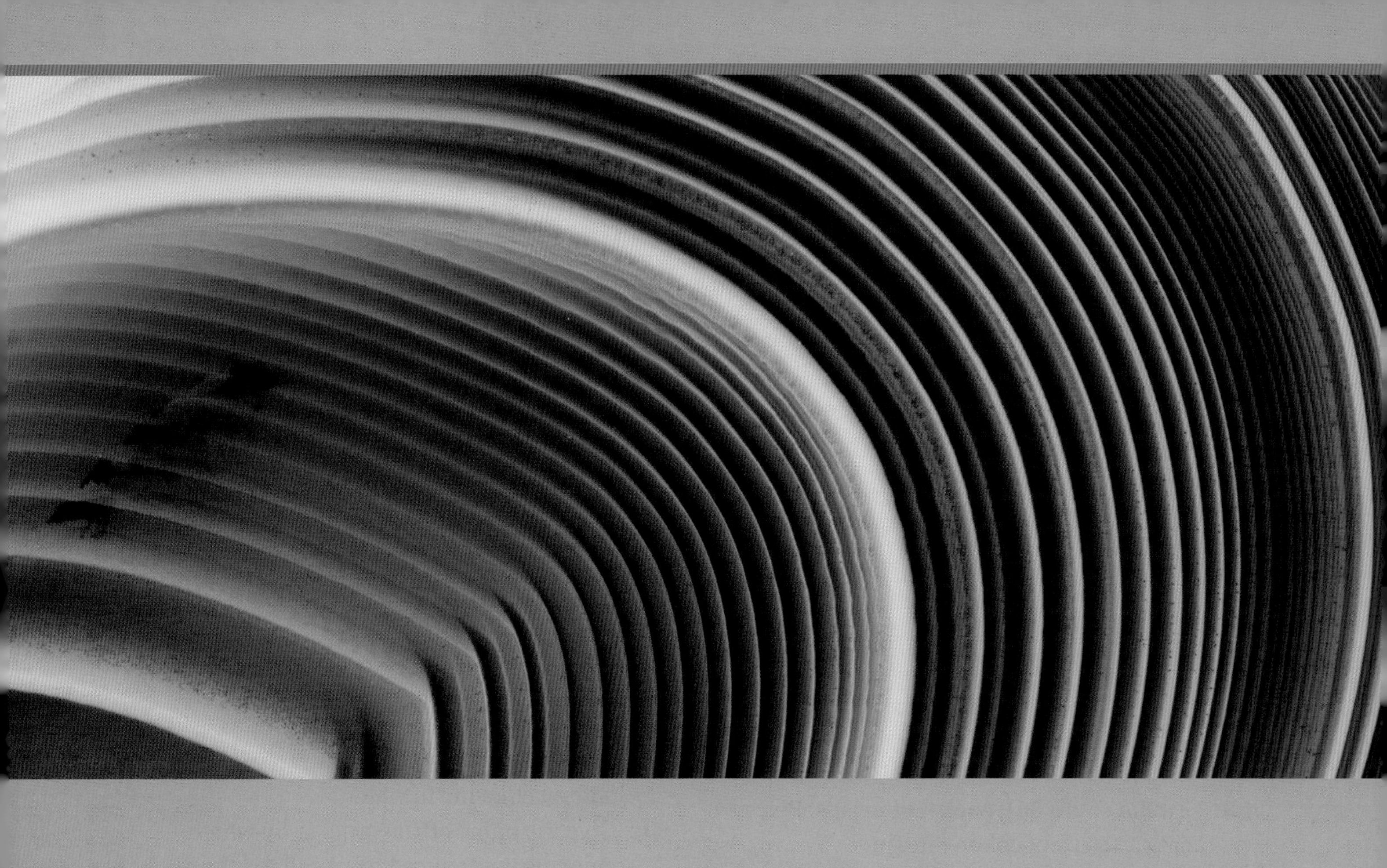

C4 Chemical patterns

Why study chemical patterns?

The periodic table is important in chemistry because it helps to make sense of the mass of information about all the elements and their compounds. The table offers a framework that can give meaning to all the known facts about properties and reactions.

What you already know

- Elements are made up of just one type of atom.
- A molecule is a group of atoms chemically joined together.
- During a chemical reaction atoms are conserved and mass is conserved.
- The properties of the reactants and products of chemical changes are different.
- There are patterns in the way chemicals react.
- Acids neutralise alkalis to form salts.
- When an acid reacts with a metal carbonate, the products are a salt, water, and carbon dioxide.
- Chlorine can be used to kill microorganisms.
- A chemical change caused by the flow of an electric current is called electrolysis.

Find out about

- the chemistry of some very reactive elements
- the patterns in the periodic table
- how scientists can learn about the insides of atoms
- the use of atomic theory to explain the properties of chemicals
- the ways in which atoms become charged and turn into ions.

The Science

Chemists use ideas about atomic structure to explain the periodic table and the properties of elements. Spectroscopy led to the discovery of new elements. Spectroscopy is now used to study chemicals and their reactions.

Ideas about Science

For a scientific explanation to be accepted and become a theory, it must be tested by other scientists to check it is supported by evidence and can be used to make predictions. Atomic theory explains patterns in the periodic table. Ionic theory explains electrolysis. Without ionic theory, a method of extracting aluminium metal would not have been developed.

Find out about

- relative masses of atoms
- the development of the periodic table
- groups and periods

A scientist in the 1800s explaining his latest research. Those who first tried to link the properties of elements to their relative atomic masses weren't taken seriously by other scientists. Their explanations didn't include all the known elements.

Germanium – one of Mendeleev's missing elements. He used his version of the periodic table to predict that the missing element would be a grey metal that would form a white oxide with a high melting point. An element was discovered in 1886 that fitted into the space left by Mendeleev. Its properties were very similar to those he predicted.

Looking for patterns

Around the beginning of the 1800s, only about 30 elements were known, but by the end of that century almost all of the stable elements found on Earth had been discovered.

Finding so many elements, with such a wide range of properties, encouraged chemists to look for patterns, for example, patterns between the properties of elements and the masses of their atoms.

Relative atomic masses

In those days scientists could not measure the actual masses of atoms so they compared the masses of atoms with the mass of the lightest one, hydrogen. This is called the **relative atomic mass**.

In the early 1800s, Johann Döbereiner, a German scientist, noticed that there were several groups where three elements have similar properties (for example, calcium, strontium, and barium). For each group, the relative atomic mass of the middle element was the mean of the relative atomic masses of the other two.

Almost 50 years later an English chemist, John Newlands, arranged the known elements in order of their relative atomic masses. He saw that every eighth element had similarities – like an octave in music. But this seemed only to work for the first 16 known elements. This made other scientists reluctant to accept Newlands' ideas – they did not account for all the data.

Elements in order

Dmitri Mendeleev, a Russian scientist, produced patterns with real meaning when he lined up elements in order of their relative atomic mass. Mendeleev's inspiration was to realise that not all of the elements had been discovered. He left gaps for missing elements to produce a sensible pattern, and predicted what the properties of these missing elements would be.

With the elements in order of relative atomic mass, Mendeleev spotted that at intervals along the line there were elements with similar properties. Using elements known today, for example, you can see that the third, eleventh, and nineteenth elements (lithium, sodium, and potassium) are very similar.

Periodicity

A repeating pattern of any kind is a **periodic** pattern. The table of the elements gets its name from the repeating, or periodic, patterns you see when the elements are in order.

The periodic table now

In the periodic table, the elements are arranged in rows, one above the other. Each row is a **period**. The most obvious repeating pattern is from metals on the left to non-metals on the right. Every period starts with a very reactive metal in Group 1 and ends with an unreactive gas in Group 8. Elements with similar properties fall into a column. Each column is a **group** of similar elements.

Key words

- relative atomic mass
- periodic
- period
- group

period number	1	2											3	4	5	6	7	0
1				1 H hydrogen 1														4 He helium 2
2	7 Li lithium 3	9 Be beryllium 4											11 B boron 5	12 C carbon 6	14 N nitrogen 7	16 O oxygen 8	19 F fluorine 9	20 Ne neon 10
3	23 Na sodium 11	24 Mg magnesium 12											27 Al aluminium 13	28 Si silicon 14	31 P phosphorus 15	32 S sulfur 16	35.5 Cl chlorine 17	40 Ar argon 18
4	39 K potassium 19	40 Ca calcium 20	45 Sc scandium 21	48 Ti titanium 22	51 V vanadium 23	52 Cr chromium 24	55 Mn manganese 25	56 Fe iron 26	59 Co cobalt 27	59 Ni nickel 28	63.5 Cu copper 29	65 Zn zinc 30	70 Ga gallium 31	73 Ge germanium 32	75 As arsenic 33	79 Se selenium 34	80 Br bromine 35	84 Kr krypton 36
5	86 Rb rubidium 37	88 Sr strontium 38	89 Y yttrium 39	91 Zr zirconium 40	93 Nb niobium 41	96 Mo molybdenum 42	98 Tc technetium 43	101 Ru ruthenium 44	103 Rh rhodium 45	106 Pd palladium 46	108 Ag silver 47	112 Cd cadmium 48	115 In indium 49	119 Sn tin 50	122 Sb antimony 51	128 Te tellurium 52	127 I iodine 53	131 Xe xenon 54
6	133 Cs caesium 55	137 Ba barium 56	57–71	178 Hf hafnium 72	181 Ta tantalum 73	184 W tungsten 74	186 Re rhenium 75	190 Os osmium 76	192 Ir iridium 77	195 Pt platinum 78	197 Au gold 79	201 Hg mercury 80	204 Tl thallium 81	207 Pb lead 82	209 Bi bismuth 83	209 Po polonium 84	210 At astatine 85	222 Rn radon 86
7	223 Fr francium 87	226 Ra radium 88	89–103	261 Rf rutherfordium 104	262 Db dubnium 105	266 Sg seaborgium 106	264 Bh bohrium 107	277 Hs hassium 108	268 Mt meitnerium 109	271 Ds darmstadtium 110	272 Rg roentgenium 111							

lanthanoid metals	139 La lanthanum 57	140 Ce cerium 58	141 Pr praseodymium 59	144 Nd neodymium 60	145 Pm promethium 61	150 Sm samarium 62	152 Eu europium 63	157 Gd gadolinium 64	159 Tb terbium 65	162 Dy dysprosium 66	165 Ho holmium 67	167 Er erbium 68	169 Tm thulium 69	173 Yb ytterbium 70	175 Lu lutetium 71
actinoid metals	227 Ac actinium 89	232 Th thorium 90	231 Pa protactinium 91	238 U uranium 92	237 Np neptunium 93	242 Pu plutonium 94	243 Am americium 95	247 Cm curium 96	247 Bk berkelium 97	249 Cf californium 98	254 Es einsteinium 99	253 Fm fermium 100	256 Md mendelevium 101	254 No nobelium 102	257 Lr lawrencium 103

other non metals | noble gases | halogens | transition metals | alkaline earth metals | alkali metals

Key

relative atomic mass

1 H hydrogen 1

name — symbol: **black** = solid, **blue** = liquid, **red** = gas, **white** = synthetically prepared

proton number

The periodic table. Over three-quarters of the elements are metals. They lie to the left of the table.

Questions

1. In the periodic table, identify and name:
   - a a liquid in Group 7
   - b a Group 1 metal that does not occur naturally
   - c a gaseous element with properties similar to sulfur
   - d a solid element similar to chlorine
   - e a liquid metal with properties similar to zinc.
2. How many times heavier is:
   - a a magnesium atom than a carbon atom?
   - b a sulfur atom than a helium atom?
3. Explain how Mendeleev could predict the properties of the unknown element germanium from what was known about other elements such as silicon and tin.
4. Give two reasons why scientists accepted Mendeleev's method of arranging the elements.

B The alkali metals

Find out about

- group 1 metals
- reactions with water and chlorine
- similarities and differences between Group 1 elements

Cutting a lump of sodium to show a fresh, shiny surface of the metal.

Pellets of sodium hydroxide. The traditional name is caustic soda. Caustic substances are corrosive. They attack skin. Alkalis such as NaOH are more damaging to skin and eyes than many acids.

Key words

- alkali metal
- tarnish
- crystalline

The metals in Group 1 of the periodic table are very reactive. They are so reactive that they have to be kept under oil to stop them reacting with oxygen or moisture in the air.

Chemists call these elements the **alkali metals** because they react with water to form alkaline solutions. It is the compounds of these metals that are alkalis – not the metals themselves.

There are six elements in Group 1. Two of them, rubidium and caesium, are so reactive and rare that you are unlikely to see anything of them except on video. A third, francium, is highly radioactive. Its atoms are so unstable that it does not occur naturally. As a result, the study of Group 1 usually concentrates on lithium (Li), sodium (Na), and potassium (K).

The alkali metals are corrosive and highly flammable. For this reason forceps should be used when handling these metals and goggles must be worn.

Strange metals

Most metals are hard and strong. The alkali metals are unusual because you can cut them with a knife. Cutting them helps to show up one of their most obvious metallic properties: they are very shiny. However, they **tarnish** quickly in the air – the shiny surface becomes dull with the formation of a layer of oxide. Group 1 elements, like other metals, are good conductors of electricity.

Most metals are dense and have a high melting point. Again the alkali metals are odd: they float on water and melt on very gentle heating.

Reactions with water

The reactions with water should be carried out behind a safety screen, because the reactions can be violent and particles can be thrown out from the surface of the water. Drop a small piece of grey lithium into water and it floats, fizzes gently, and disappears as it forms lithium hydroxide (LiOH). It dissolves, making the solution alkaline. It is possible to collect the gas and use a burning splint to show that it is hydrogen.

$$\text{lithium} + \text{water} \longrightarrow \text{lithium hydroxide} + \text{hydrogen}$$

The reaction of sodium with water is more exciting. The reaction gives out enough energy to melt the sodium, which skates around on the surface of the water. It fizzes as hydrogen is formed. Like lithium, the sodium forms its hydroxide (sodium hydroxide, NaOH), which dissolves to give an alkaline solution.

The reaction of potassium with water is very violent. The hydrogen given off catches fire at once, and the metal moves around the surface of the water quickly. The result is an alkaline solution of potassium hydroxide (KOH). All the alkalic metals form hydroxides with the formula MOH, where M is the symbol for the metal.

Reactions with chlorine

Hot sodium burns in chlorine gas with a bright yellow flame. It produces clouds of sodium chloride crystals (NaCl). This is everyday table salt, used to flavour food.

The other alkali metals react with chlorine in a similar way. Lithium produces lithium chloride (LiCl). Potassium produces potassium chloride (KCl). Like everyday salt, these compounds are also colourless, **crystalline** solids that dissolve in water.

Chemists use the term **salt** to cover all the compounds of metals with non-metals. So the chlorides of lithium, sodium, and potassium are all salts.

Trends

The alkali metals are all very similar, but they are not identical. Their reactions with water show that there is a **trend** in their **chemical properties**. The alkali metals increase in reactivity down the group from lithium to sodium to potassium. There are also trends in their **physical properties**.

	Melting point (°C)	Boiling point (°C)	Density (g/cm^3)
lithium, Li	181	1342	0.53
sodium, Na	98	883	0.97
potassium, K	63	760	0.86

Physical properties of lithium, sodium, and potassium.

Lithium, sodium, and potassium all have a low density. The melting points and boiling points decrease down the group.

Compounds of the alkali metals

The compounds of the alkali metals are very different to the elements. The elements are dangerously reactive. But chlorides of sodium and potassium, for example, have a vital role to play in our blood and in the way in which our nerves work.

Many compounds of alkali metals are soluble in water. Soluble sodium compounds, in particular, make up a number of common everyday chemicals including sodium hydroxide (in oven cleaners), sodium hypochlorite (in bleach), and sodium hydrogencarbonate (as the bicarbonate of soda in antacid remedies and baking powder).

Sodium burning in chlorine gas.

Key words

- salt
- trends
- physical properties
- chemical properties

Questions

1. Predict these properties of rubidium (Rb).
   a. Its melting point
   b. What happens to a freshly cut surface of the metal in the air?
   c. What happens if you drop a small piece of rubidium onto water?
2. For the hydroxide of rubidium predict:
   a. its formula
   b. whether or not it is soluble in water.
3. For the chloride of caesium (Cs) predict:
   a. its colour
   b. its formula
   c. whether or not it is soluble in water.
4. Suggest and explain the precautions necessary when potassium reacts with water.

Chemical equations

Find out about

- chemical symbols
- formulae
- balanced equations

Equations are important because they do for chemists what recipes do for cooks. They allow chemists to work out how much of the starting materials to mix together and how much of the products they will get.

Chemical models

In a **chemical change** there is no change in mass because the number of each type of atom stays the same. The atoms rearrange, but no new ones appear and no atoms are destroyed during a chemical reaction.

Hydrogen burns in oxygen to form **molecules** of water. The models in the diagram below show what happens. In each water molecule there is only one oxygen atom. So one oxygen molecule (= two red atoms in the model) reacts with two molecules of hydrogen (each molecule = two white atoms in the model) to make two water molecules. There are equal numbers of hydrogen atoms and oxygen atoms on each side of the arrow. This is a model of a **chemical equation**.

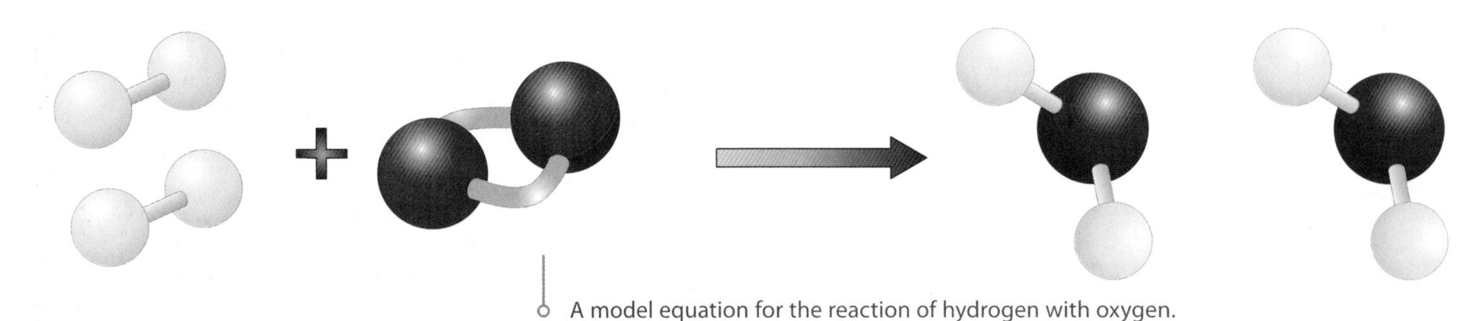

A model equation for the reaction of hydrogen with oxygen.

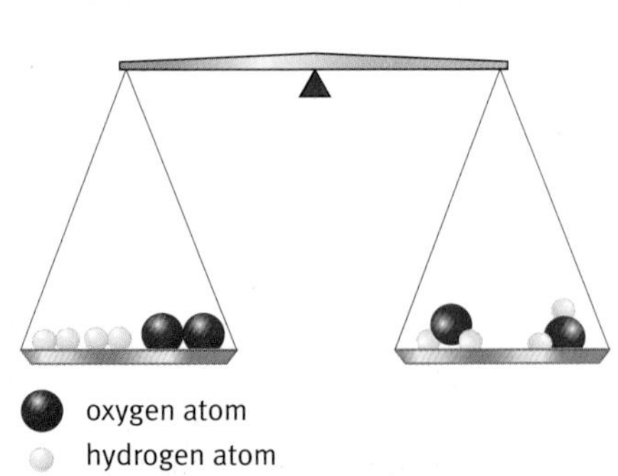

The reactants and the products of a reaction 'balance'.

Chemical symbols

Using models to describe every reaction would be very tiresome. Instead, chemists write symbol equations to show the numbers and arrangements of the atoms in the reactants and the products.

When written in symbols, the equation in the figure above becomes:

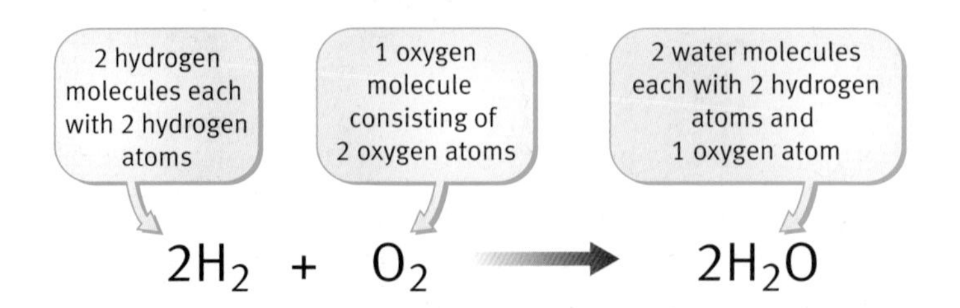

$$2H_2 + O_2 \rightarrow 2H_2O$$

The equation is balanced because it has the same number of atoms of each type on the left and the right of the arrow. Because no atoms have been gained or lost the reactants have the same mass as the products.

Formulae

You can't write an equation unless you first know:

- all the starting chemicals (the reactants)
- everything that is formed during the change (the products).

When writing an equation, you have to write down the correct chemical **formulae** for the reactants and products. Chemists have worked these out by experiment, and you can look them up in data tables.

If the element or compound is molecular, you write the formula for the molecule in the equation. This applies to most non-metals (O_2, H_2, Cl_2) and most compounds of non-metals with non-metals (H_2O, HCl, NH_3).

Not all elements and compounds consist of molecules. For all metals, and for the few non-metals that are not molecular (C, Si), you just write the symbol for a single atom.

The compounds of metals with non-metals are also not molecular. For these compounds you write the simplest formula for the compound, such as $LiOH$, $NaCl$, or K_2CO_3.

Writing balanced equations

Follow the four steps shown in the margin to write **balanced equations**.

Worked example

Write a balanced equation to show the reaction of natural gas (methane, CH_4) with oxygen.

Step 1 Describe the reaction in words:

methane + oxygen ⟶ carbon dioxide + water

Step 2 Write down the formulae for the reactants and products:

$$CH_4 + O_2 \longrightarrow CO_2 + H_2O$$

Step 3 Balance the equation.

You must not change any of the formulae. You balance the equation by writing numbers in front of the formulae. These numbers then refer to the whole formula.

$$CH_4 + 2O_2 \longrightarrow CO_2 + 2H_2O$$

Step 4 Add state symbols.

State symbols usually show the states of the elements and compounds at room temperature and pressure. The chemicals in an equation may be solid (s), liquid (l), gaseous (g), or dissolved in water (aq, for aqueous).

$$CH_4(g) + 2O_2(g) \longrightarrow CO_2(g) + 2H_2O(l)$$

Key words

- ✔ chemical change
- ✔ molecules
- ✔ chemical equation
- ✔ formulae
- ✔ balanced equation

RULES FOR WRITING BALANCED EQUATIONS

STEP 1 Write down a word equation.

STEP 2 Underneath, write down the correct formula for each reactant and product.

STEP 3 Balance the equation, if necessary, by putting numbers in front of the formulae.

STEP 4 Add state symbols.

NEVER change the formula of a compound or element to balance the equation.

Question

1 Write balanced symbol equations for these reactions of the alkali metals:

a sodium with water

b potassium with water

c sodium with chlorine

d lithium with chlorine.

D The halogens

Find out about

- Group 7 elements
- halogen molecules
- similarities and differences between Group 7 elements

Crystals of the mineral fluorite (calcium fluoride).

Salt formers

The Group 7 elements are all very reactive non-metals. They include fluorine, chlorine, bromine, and iodine. Group 7 elements are also known as the **halogens**. The name 'halo-gen' means 'salt-former'. These elements form salts when they combine with metals. Examples include everyday 'salt' itself, which occurs as the mineral halite (NaCl), and fluorite (CaF_2). A type of fluorite called Blue John is found in Derbyshire and is used in jewellery.

The halogens are interesting because of their vigorous chemistry. As elements they are hazardous because they are so reactive. This is also why they are not found existing freely in nature, but occur as compounds with metals.

You must wear eye protection when working with any of the halogens. You should use chlorine and bromine in a fume cupboard because they are **toxic** gases. You must wear chemical-resistant gloves when using liquid bromine, which is **corrosive**. It is not normally possible to study fluorine because it is so dangerously reactive.

Non-metal patterns

Like most non-metals, the halogens are molecular. They each consist of **diatomic** molecules with the atoms joined in pairs: Cl_2. Br_2 and I_2. They have low melting and boiling points because the forces between the molecules are weak, and so it is easy to separate them and turn the halogens into gases. The melting points and boiling points of the halogens increase down the group.

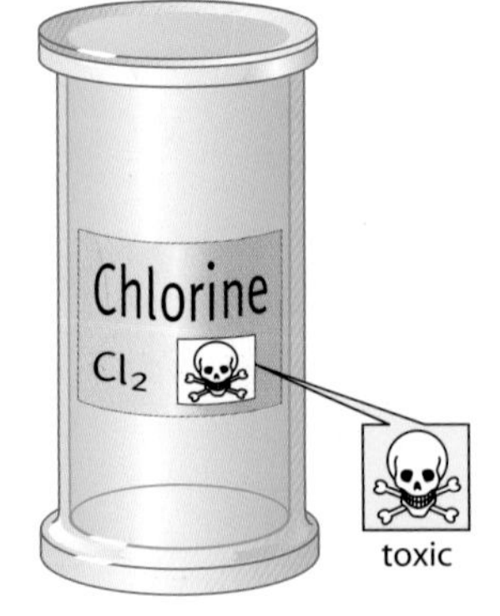

- dense, pale-green gas
- smelly and poisonous
- occurs as chlorides, especially sodium chloride in the sea
- melting point −101 °C
- boiling point −35 °C

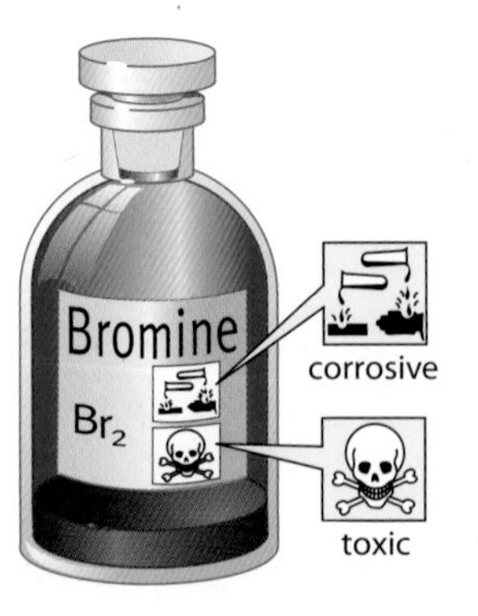

- deep red liquid with red–brown vapour
- smelly and poisonous
- occurs as bromides, especially magnesium bromide in the sea
- melting point −7 °C
- boiling point 59 °C

- grey solid with purple vapour
- smelly and harmful
- occurs as iodides and iodates in some rocks and in seaweed
- melting point 114 °C
- boiling point 184 °C

Key words

- halogens
- diatomic
- toxic
- corrosive
- bleach
- displacement reactions
- harmful

Halogen patterns

All the halogens can harm living things. They can all kill bacteria. Domestic **bleach** is a solution of chlorine in sodium hydroxide, sold to disinfect worktops and toilets. Iodine solution can be used to prevent infection of wounds. These uses illustrate the decreasing reactivity of halogens down the group. Chlorine is much too reactive for its solution to be used to treat wounds.

The reactions of the halogens with an alkali metal such as sodium show the same decrease in reactivity down the group. The reactions with another metal, iron, show the same trend. Hot iron glows brightly in chlorine gas. The product is iron chloride ($FeCl_3$), a rust-brown solid. Iron also glows when heated in bromine vapour, but less brightly. Iron does not even glow when heated in iodine vapour.

The trend in reactivity of the halogens is also shown by **displacement reactions**. For example, if a pale-green solution of chlorine is added to a colourless solution of sodium bromide, the solution immediately turns red because bromine has been formed. Chlorine is more reactive than bromine, so it displaces bromine from its salt, sodium bromide:

$$Cl_2(aq) + 2NaBr(aq) \longrightarrow 2NaCl(aq) + Br_2(aq)$$

In the same way, a solution of bromine will displace iodine from sodium iodide.

Practical importance

Compounds of the halogens are very important in everyday life.

The chemical industry turns everyday salt (NaCl) into chlorine (and sodium or sodium hydroxide). The chlorine is used to make plastics such as PVC. It is also used in water treatment to stop the spread of diseases. But some chlorine compounds, such as chlorofluorocarbons (CFCs), which were used as refrigerants in the past, have damaged the ozone layer in the upper atmosphere.

Most of the bromine we use comes from the sea. Liquid bromine is extremely corrosive. However, the chemical industry makes important bromine compounds including medical drugs, and pesticides to protect crops.

Traces of iodine compounds are essential in our diet. Where there is little or no natural iodine, it is usual to add potassium iodide to everyday table salt or to drinking water. This prevents disease of the thyroid, a gland in the neck. Iodine and its compounds are starting materials for the manufacture of medicines and dyes.

Hot iron in a jar of chlorine.

Questions

1 Write balanced equations for the reactions of:
   a iron with chlorine to form $FeCl_3$
   b potassium with bromine
   c lithium with iodine.

2 Fluorine (F_2) is the first element in Group 7.
   a Predict the effect of passing a stream of fluorine over iron. Write an equation for the reaction.
   b What would you expect to see if fluorine gas is passed into a solution of sodium bromide? Write an equation for the reaction.
   c Astatine is a halogen and is below iodine in the periodic table. Would you expects its melting point to be higher or lower than the melting point of iodine?

The discovery of helium

Find out about

- flame colours
- line spectra
- discovery of the elements

Robert Bunsen (1811–1899), who discovered the flame colours of elements with the help of his new burner.

The bright red flame produced by lithium compounds. The compounds of other elements also produce colours in a flame:

- sodium – bright yellow
- potassium – lilac
- calcium – orange–red
- barium – green

A new burner for chemistry

Robert Bunsen moved to the University of Heidelberg in Germany in 1852. Before taking up the job as professor of chemistry, he insisted on having new laboratories. He also demanded gas piping to bring fuel from the gas works, which had just opened to light the city streets.

Existing burners produced smoky and yellow flames. Bunsen wanted something better. In 1855 he invented the type of burner that is still used today in laboratories all over the world.

The great advantage of Bunsen's burner is that it can be adjusted to give an almost invisible flame. Bunsen used his burner to blow glass. He noticed that whenever he held a glass tube in a colourless flame, the flame turned yellow.

Flame colours

Soon Bunsen was experimenting with different chemicals, which he held in the flame at the end of a platinum wire. He found that different chemicals produced characteristic **flame colours**.

Bunsen thought that this might lead to a new method of chemical analysis, but he soon realised that it seemed only to work for pure compounds. It was hard to make any sense of flames from mixtures. So he mentioned his problem to Gustav Kirchhoff, who was the professor of physics.

Flame spectra

'My advice as a physicist,' said Kirchhoff, 'is to look not at the colour of the flames, but at their spectra.'

Kirchhoff built a spectroscope, by putting a glass prism into a wooden box and inserting two telescopes at an angle. Light from a flame entered through one telescope. It was split into a spectrum by the prism and then viewed with the second telescope.

Bunsen and Kirchhoff soon found that each element has its own characteristic spectrum when its light passes through a prism. Each spectrum consists of a set of lines. With their spectroscope, they were able to record the **line spectra** of many elements.

Seen using spectroscopy, cadmium's main lines are red, green, and blue (many fainter lines aren't visible in the photograph). The lines give cadmium its unique fingerprint.

Using **spectroscopy**, Bunsen discovered two new elements in the waters of Durkheim Spa. He based their names on the colours of their spectra. He called them caesium and rubidium, from the Latin for 'sky blue' and 'dark red'.

A Sun element

In 1868 there was a total eclipse of the Sun. Normally, the blinding light from the centre of the Sun makes it impossible to see the much fainter light from the hot gases around the edges of the star. During an eclipse, the Moon hides the whole bright disc of the Sun but not the much fainter light from the hot gases around the edges. This makes it possible to study the light from these gases.

Pierre Janssen, a French astronomer, took very careful observations of the Sun's spectrum during the 1868 eclipse. In the spectrum of the light, he saw a yellow line where no yellow line was expected to be.

Excited by these observations, both Janssen and an English astronomer, Joseph Lockyer, developed new methods to study the light from the Sun's gases. They worked independently, but both came to the same conclusion. There must be an unknown element in the Sun, producing the unexpected yellow line in the spectrum. Janssen and Lockyer published their findings at almost the same time – a coincidence that led to their becoming good friends.

The new element was called 'helium' from the Greek word *helios*, meaning Sun. Both astronomers were still alive in 1895 when William Ramsay, a British chemist, used spectroscopy to discover helium on Earth. The gas came from boiling up a rock containing uranium with acid.

A solar eclipse in 1868 helped scientists to discover helium. During an eclipse it is possible to study the spectra of the light from the hot gases around the edges of the Sun.

WARNING!

Never look directly at the Sun, even during an eclipse. You can damage your eyes or even be blinded!

Key words

- flame colour
- line spectra
- spectroscopy

Questions

1. Why was it important for Bunsen to have a burner with a colourless flame?
2. Why is it not possible to analyse chemical mixtures simply by looking at their flame colours?
3. Use your knowledge of Group 1 chemistry to suggest an explanation for the fact that rubidium and caesium were not discovered until the technique of spectroscopy was developed.
4. In the story of Janssen and Lockyer, identify a statement that is data and a statement that is an explanation of the data.
5. Janssen and Lockyer discovered helium at about the same time. Explain why this meant that other scientists were more likely to accept their findings.

Find out about

- atomic theory
- the nuclear model of the atom
- protons, neutrons, and electrons

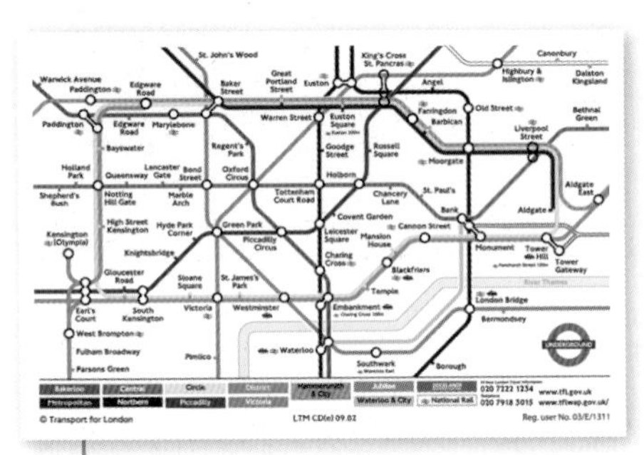

Part of the map of the London Underground.

Atomic models

A picture, or model, of an atom can be used to understand how atoms join together to form compounds and how atoms rearrange during chemical reactions.

Scientists use different models to solve different problems. There is not one model that is 'true'. Each model can represent only a part of what we know about atoms.

It is like using maps to travel through London. The usual map of the Underground is a very useful guide for getting from one tube station to another. It is 'true' in that it shows how the lines and stations connect, but it can't solve all of a traveller's problems. The map doesn't show how the tube stations relate to roads and buildings on the surface. For that you need a street map.

Atomic models from 1800 to the present. The diameter of an atom is about ten million times smaller than a millimetre. These diagrams are distorted. For atoms at this scale, the nuclei would be invisibly small.

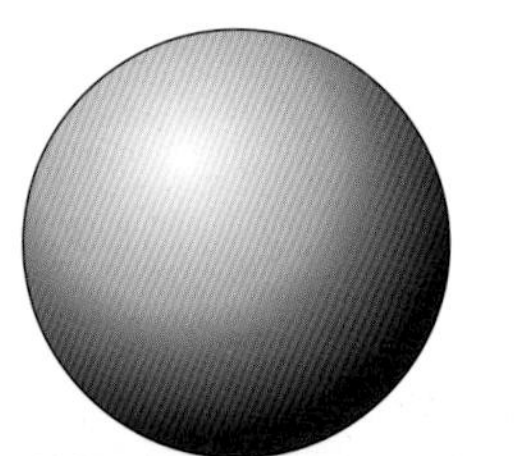

1804
Dalton's solid atom.

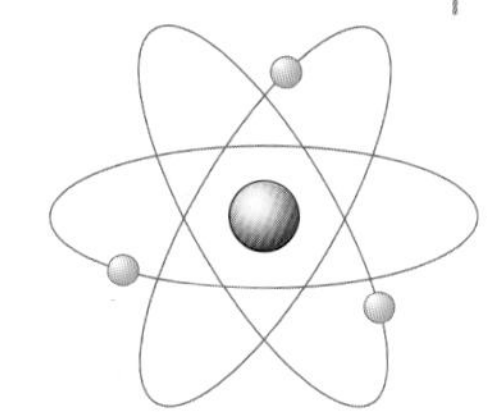

1913
The Bohr–Rutherford 'Solar System' atom, in which electrons orbit round a very small nucleus.

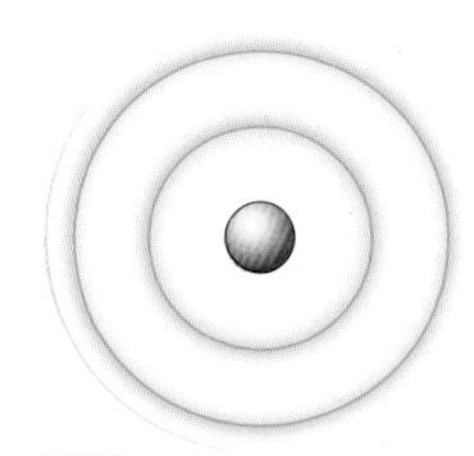

1924
A model of the atom in which the electrons are no longer treated as particles but pictured as occupying energy levels, which give rise to regions of negative charge around the nucleus (charge clouds).

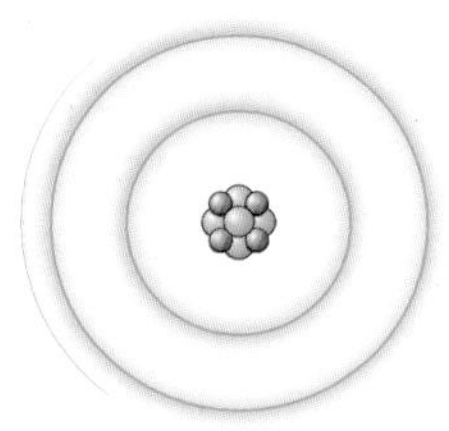

1932
The atom in which the nucleus is built up from neutrons as well as protons.

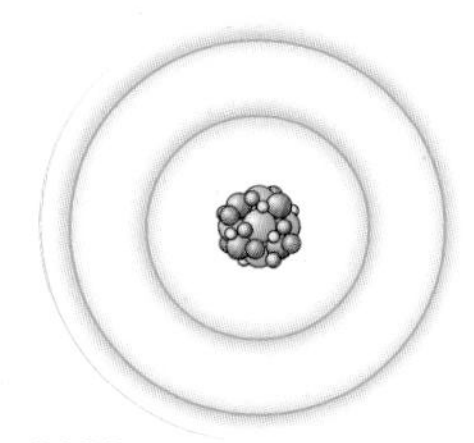

2000+
The present-day atom in which the nucleus is built up from many kinds of particles.

Key words

- nucleus
- protons
- neutrons
- electrons
- proton number
- sub-atomic particles

Dalton's atomic theory

The story of our modern thinking about atomic structure began with John Dalton. In 1803 John Dalton was studying mixtures of gases and their properties. He did not just summarise the data he collected, but used his creativity to think up an explanation that would account for the data. In Dalton's theory, everything is made of atoms that cannot be broken down. The very word 'atom' means 'indivisible'.

The main ideas in Dalton's theory still apply to chemistry today. So far as chemistry is concerned, each element does have its own kind of atom, and the atoms of different elements differ in mass. The idea that equations must balance is based on Dalton's view that atoms are not created or destroyed during chemical changes.

Even so, Dalton's theory is limited. It cannot explain the pattern of elements in the periodic table. Nor can it explain how atoms join together in elements and compounds. A better explanation was needed that would account for most, if not all, the data available.

The Large Hadron Collider at CERN in Switzerland is designed to collide beams of protons at very high energy. It was first switched on in 2008. After fixing some initial problems, CERN engineers have gradually increased the proton energy. When it is fully working, it will probe more deeply into matter than ever before.

Inside the atom

In time it became clear that atoms are not solid, indivisible spheres. From the middle of the nineteenth century, scientists began to find ways of exploring the insides of atoms. Still today, scientists are spending vast sums of money to build particle accelerators (atom-smashers), which work at higher and higher energies. They hope to discover more about the fine structure of atoms.

It is possible to explain much more about the chemistry of elements and compounds with the help of a model of atomic structure that includes **sub-atomic particles**, the particles that make up atoms.

A model for chemistry

In your study of chemistry you will be using an atomic model that dates back to 1932, when James Chadwick discovered the neutron. In this model the mass of the atom is concentrated in a tiny, central **nucleus**. The nucleus consists of **protons** and **neutrons**. The protons have a positive electric charge. Neutrons are uncharged.

Around the nucleus are the **electrons**. The electrons are negatively charged. The mass of an electron is so small that it can often be ignored. In an atom, the number of electrons equals the number of protons in the nucleus (the **proton number**). This means that the total negative charge equals the total positive charge, and overall an atom is uncharged.

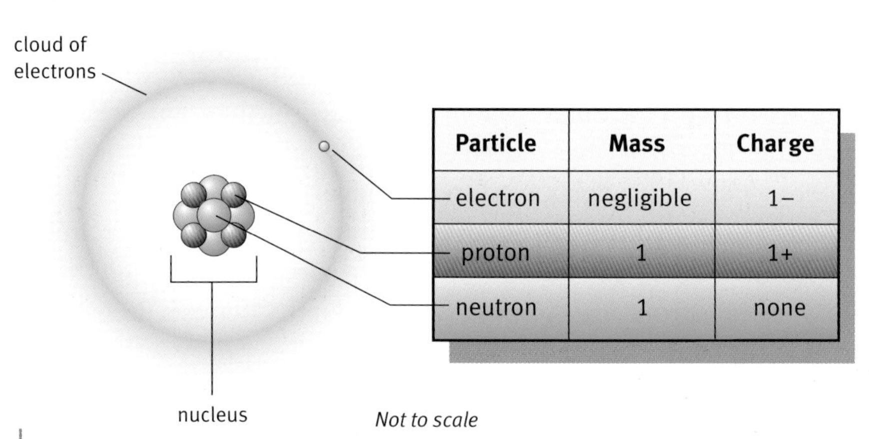

Particle	Mass	Charge
electron	negligible	1−
proton	1	1+
neutron	1	none

All atoms consist of these three basic particles. The nucleus of an atom is very, very small. The diameter of an atom is about ten million million times greater than the diameter of its nucleus.

Question

1 With the help of the periodic table on page 107 work out:
- a the element with one more proton in its nucleus than a chlorine atom
- b the element with one proton fewer in its nucleus than a neon atom
- c the number of protons in a sodium atom
- d the number of electrons in a bromine atom
- e the size of the positive charge on the nucleus of a fluorine atom
- f the total negative charge on the electrons in a potassium atom.

Find out about

- evidence for energy levels
- electrons in shells
- electron arrangements

Electrons in orbits

In 1913, the Danish scientist Niels Bohr came up with an explanation for the line spectra from atoms. He made a close study of the spectrum from hydrogen.

In the Bohr model for atoms, the electrons orbit the nucleus as the planets orbit the Sun. Bohr's idea was that heating atoms gives them energy. This forces the electrons to move to higher-energy orbits further from the nucleus. These electrons then drop back from outer orbits to inner orbits. They give out light energy as they do so. Each energy jump corresponds to a particular colour in the spectrum. The bigger the jump, the nearer the line to the blue end of the spectrum. Only certain energy jumps are possible, so the spectrum consists of a series of lines.

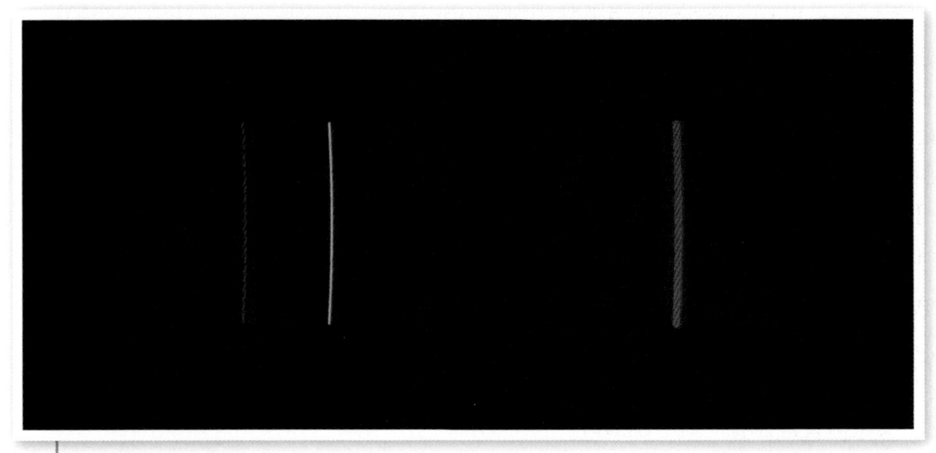

The line spectrum of hydrogen. Atomic theory can explain why this spectrum is a series of lines.

Bohr was able to use his theory to calculate sizes of the energy jumps. He could then deduce the energy levels of electrons in the various orbits.

Electrons in shells

The comparison with the Solar System and the use of the term 'orbit' can be misleading. The theory has moved on since Bohr's time. Scientists still picture the electrons at a particular **energy level**. However, in the modern theory the electrons do not orbit the nucleus like planets round the Sun. All that theory can tell us is that there are regions around the nucleus where electrons are most likely to be found. Chemists describe these regions as 'clouds' of negative charge.

Think of each electron cloud as a **shell** around the nucleus. Each shell is one of the regions in space where there can be electrons. The shells only exist if there are electrons in them. Electrons in the same shell have the same energy.

Key words

- energy level
- shell
- electron arrangement

Electron arrangements

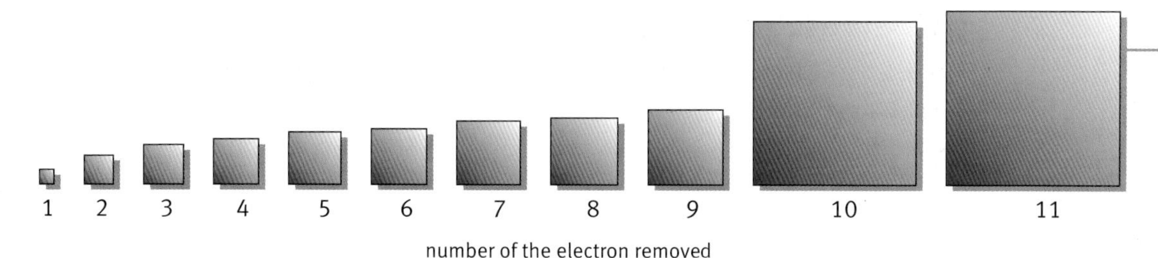

The areas of the squares are in proportion to the amount of energy needed to remove the electrons one by one from a sodium atom.

Each electron shell can contain only a limited number of electrons. The innermost shell with the lowest energy fills first. When it is full, the electrons go into the next shell. Evidence for this theory comes not only from spectra, but also from measurements of the energy needed to remove electrons from atoms.

There are 11 electrons in a sodium atom. Scientists have measured the quantities of energy needed to remove these electrons one by one. The values are represented by the areas of the squares in the picture above. It is quite easy to remove the first electron. The next eight are more difficult to remove. Finally it becomes really hard to remove the last two electrons, which are held very powerfully because they are in the shell closest to the nucleus.

This supports the idea that the electrons in a sodium atom are arranged in three shells, as shown in the diagram on the right. The diagram shows common representations of the **electron arrangement** of the element.

The first shell that is closest to the nucleus can hold up to two electrons. The second shell can hold eight. Once the second shell holds eight electrons, the third shell starts to fill.

If there are more electrons, they occupy further shells. After the first 20 elements, the arrangements become increasingly complex as the shells hold more electrons and the energy differences between shells get smaller.

Na
electron (e^-)
nucleus
electron shells

This diagram can be abbreviated to:

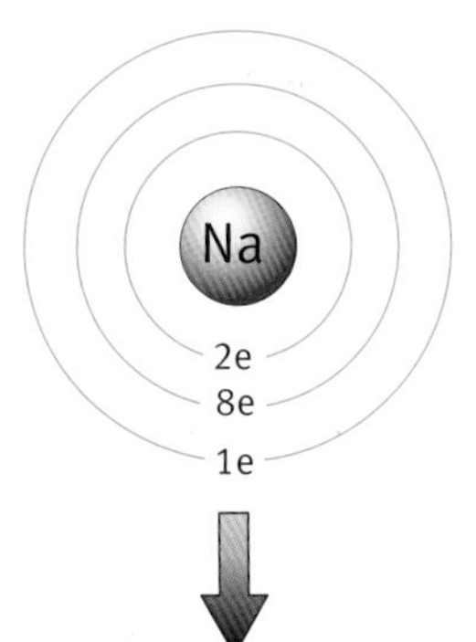

or even more simply to:
Na: 2e.8e.1e or 2.8.1

Two-dimensional representations of the electrons in shells in a sodium atom.

Questions

1 Draw diagrams to show the electrons in shells for these atoms:
a beryllium b oxygen c magnesium.
Refer to the periodic table on page 107 for the proton numbers, and therefore the number of electrons, in each atom.

2 How does the diagram at the top of this page support the electron arrangements of a sodium atom shown in the diagram on the right?

Electrons and the periodic table

Find out about

- atomic structure and periods
- electron arrangements and groups
- explaining similarities and differences between the elements

The periodic table then and now

Scientists discovered electrons in 1897, nearly 30 years after Mendeleev published his first periodic table. Mendeleev knew nothing about atomic structure, and he used the relative masses of atoms to put the elements in order.

A modern periodic table shows the elements in order of proton number, which is also the number of electrons in an atom. One convincing piece of evidence for the 'shell model' of atomic structure is that it can help to explain the patterns in the periodic table.

Periods

The diagram below shows the connection between the horizontal rows of the periodic table and the structure of atoms. From one atom to the next, the proton number increases by one and the number of electrons increases by one. So the electron shells fill up progressively from one atom to the next.

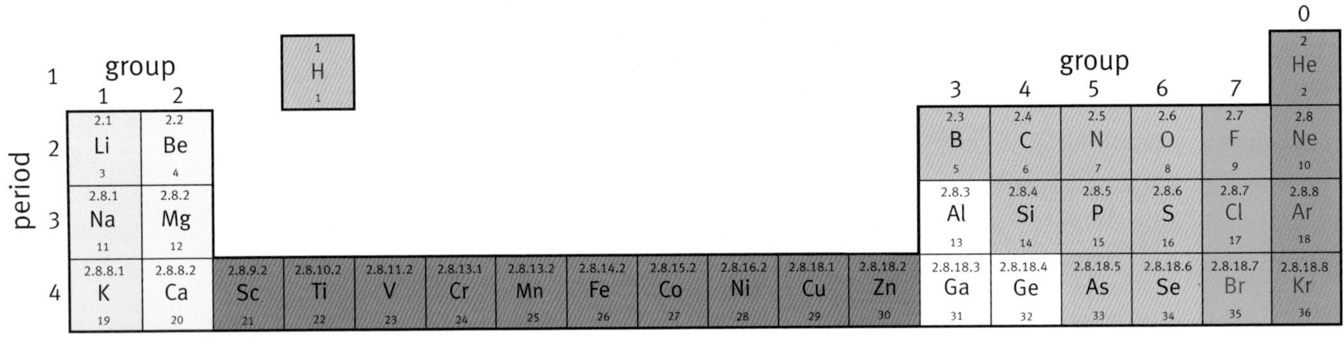

Key

2.4 — number of electrons in each shell
C — symbol
6 — proton number

Electron arrangements for the first 20 elements in the periodic table.

The first period, from hydrogen to helium, corresponds to filling the first shell. The second shell fills across the second period, from lithium (2.1) to neon (2.8). Eight electrons go into the third shell, from sodium (2.8.1) to argon (2.8.8), and then the fourth shell starts to fill from potassium to calcium.

In fact the third shell can hold up to 18 electrons. This shell is completed from scandium to zinc, before the fourth shell continues to fill from gallium to krypton. This accounts for the appearance of the block of transition metals in the middle of the table. Why this happens cannot be explained by the simple theory described here. You will find out the explanation if you go on to a more advanced chemistry course.

Groups

When atoms react, it is the electrons in their outer shells that get involved as chemical bonds break and new chemicals form. It turns out that elements have similar properties if they have the same number and arrangement of electrons in the outer shells of their atoms.

Three of the alkali metals appear in the diagram on the right. You can see that they each have one electron in the outer shell of their atoms. This is the case for the other alkali metals too. This helps to account for the similarities in their chemical properties.

The alkali metals are not all the same because their atoms differ in the number of inner full shells. A sodium atom has two inner filled shells, so it is larger than a lithium atom, and its outer electron is further away from the nucleus. As a result, the two metals have similar, but not identical, physical and chemical properties.

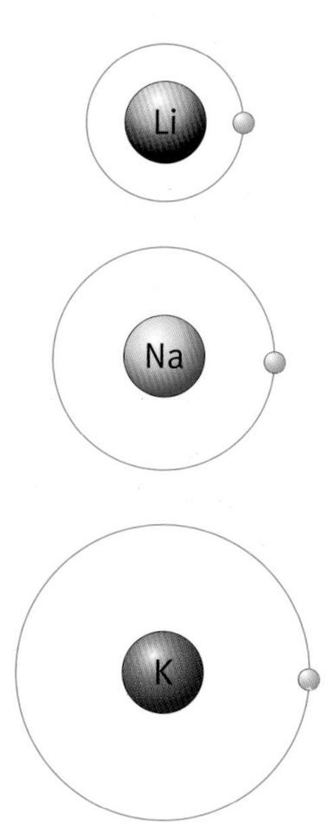

The trend in the size of the atoms of Group 1 elements reflects the increasing number of full, inner electron shells down the group. Only the outer shells are shown here.

Metals and non-metals

Elements with only one or two electrons in the outer shell are metals, with the exception of hydrogen and helium. Elements with more electrons in the outer shell are generally non-metals, though there are exceptions to this, such as aluminium, tin, and lead. The halogens are non-metal elements with seven electrons in the outer shell.

At the end of each period there is a noble gas. This is a group of very unreactive elements. The first member of the group is helium.

The term 'noble' has been used by alchemists and chemists for hundreds of years to describe elements that will not react easily. The chemical nobility stand apart from the hurly-burly of everyday reactions.

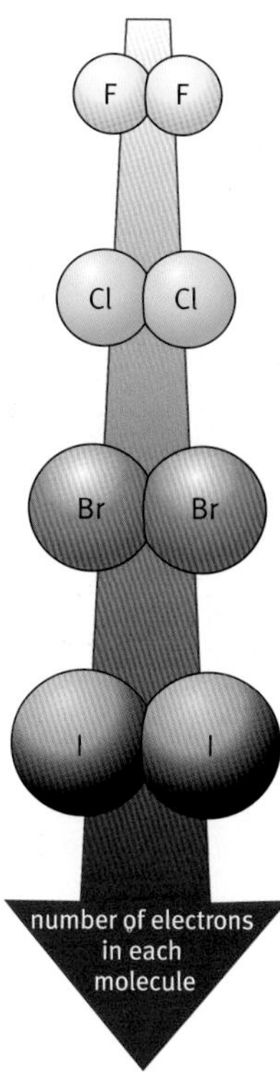

The trend in the size of the molecules of Group 7 elements reflects the increasing number of full, inner electron shells in the atoms down the group.

Questions

1 Explain the meaning of this statement: the electron arrangement of chlorine is (2.8.7).

2 a What are the electron arrangements of the elements beryllium, magnesium, and calcium?
 b In which group of the periodic table do these three elements appear?
 c Are these elements metals or non-metals?

Find out about

- salts
- properties of salts
- electricity and salts

Sodium chloride crystals. Sodium chloride is soluble in water. The chemical industry uses an electric current to convert sodium chloride solution into chlorine, hydrogen, and sodium hydroxide.

Crystals of the mineral galena, which is an ore of lead. Galena consists of insoluble lead sulfide.

Crystals of the mineral pyrite. Pyrite consists of insoluble iron sulfide.

Why are salts so different from their elements?

Compounds of metals with non-metals are salts. Chemists can explain the differences between a salt and its elements by studying what happens to the atoms and molecules as they react. A good example is the reaction between two very reactive elements to make the everyday table salt you can safely sprinkle on food.

A chemical reaction in pictures: sodium and chlorine react to make sodium chloride.

Salts

Salts such as sodium chloride are crystalline. The crystals of sodium chloride are shaped like cubes. So are the crystals of calcium fluoride shown in Section D.

Salts have much higher melting and boiling points than chemicals such as chlorine and bromine, which are made up of small molecules.

Chemical	Formula	Melting point (°C)	Boiling point (°C)
sodium	Na	98	890
chlorine	Cl_2	−101	−34
sodium chloride	$NaCl$	808	1465
potassium	K	63	766
bromine	Br_2	−7	58
potassium bromide	KBr	730	1435

Sodium chloride is an example of a salt that is soluble in water. There are many other examples of soluble salts, including most of the compounds of alkali metals with halogens.

Some salts are insoluble in water. Lithium fluoride is an example of a salt that is only very slightly soluble in water. Many minerals consist of insoluble salts. Fluorite (CaF_2) is one example. Others are galena (PbS) and the brassy-looking pyrite (FeS_2), sometimes called fool's gold.

Molten salts and electricity

The apparatus on the right is used to investigate whether or not chemicals conduct electricity. The crucible contains some white powdered solid. This is zinc chloride.

At first there is no reading on the meter, showing that the solid does not conduct electricity. This is true of all compounds of metals with non-metals; they do not conduct electricity when solid.

Heating the crucible melts the zinc chloride. As soon as the compound is **molten**, there is a reading on the meter. This shows that a current is flowing round the circuit. As a liquid, the compound is a conductor. That is not all. The electric current causes the compound to decompose chemically. The most obvious change is the bubbling around the positive electrode. The gas produced is chlorine.

After a while, it is possible to show that zinc has formed at the negative electrode. This is done by switching off the current, allowing the crucible to cool, and adding pure water to dissolve any remaining zinc chloride. Shiny pieces of zinc metal remain. So the electric current splits the compound into its elements: zinc and chlorine.

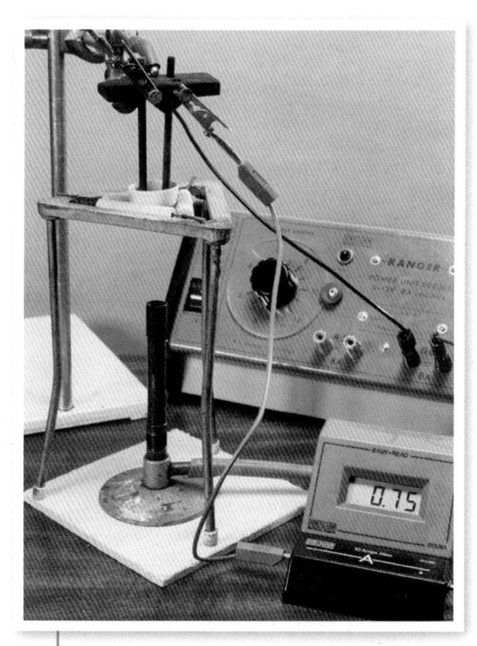

The crucible contains zinc chloride. The carbon rods dipping into the crucible are the electrodes. A current begins to flow in the circuit when the zinc chloride is hot enough to melt.

Salts in solution and electricity

Soluble salts also conduct electricity. This can be studied using the apparatus shown on the right. There are changes at the electrodes when an electric current flows.

The presence of water has an effect on the chemicals produced when a salt solution conducts electricity. The products are not always the same as the elements in the compound.

This apparatus is used to study the changes at the electrodes when a solution of a salt conducts electricity. In this example, the flow of an electric current is producing gases at the electrodes.

Questions

1. Draw up a table to compare the physical properties of sodium, chlorine, and sodium chloride.
2. Refer to the table of data on the left. Which of the chemicals is a liquid:
   a. at room temperature?
   b. at the boiling point of water?
   c. at 1000 °C?
3. Draw a two-dimensional line diagram and circuit diagram to represent the apparatus used to show that zinc chloride conducts electricity when hot enough to melt.

Key word

✓ **molten**

Ionic theory

Find out about

- ions
- ionic compounds
- explaining properties of salts

Michael Faraday lectured at the Royal Institution. He started the Christmas lectures, which continue today in the same lecture theatre.

Key words

- ions
- electrolysis

Questions

1 What big idea occurred to Faraday that enabled him to explain how solutions of salts and molten salts conduct electricity?

2 Why was it important for Faraday to communicate his ideas to other scientists?

Electrolysis

An electric current can split a salt into its elements, or other products, if it is molten or dissolved in water. This process is called **electrolysis**. The term 'electro-lysis' is based on two Greek words that mean 'electricity-splitting'.

The discovery of electrolysis was very important in the history of chemistry. Electrolysis made it possible to split up compounds, which previously no-one could decompose. In 1807 and 1808, an English chemist, Humphry Davy, used electrolysis to isolate for the first time the elements potassium, sodium, barium, strontium, calcium, and magnesium.

Faraday's theory

Michael Faraday worked with Humphry Davy. He began as an assistant, then established himself as a leading scientist in his own right. In 1833 he began to study the effects of electricity on chemicals. He had to think creatively to come up with an explanation that would account for his observations.

Faraday decided that compounds that can be decomposed by electrolysis must contain electrically charged particles. Since opposite electrical charges attract each other, he could imagine the negative electrode attracting positively charged particles and the positive electrode attracting negatively charged particles.

The charged particles move towards the electrodes. When they reach the electrodes, they turn back into atoms. This accounts for the chemical changes that decompose a compound during electrolysis.

Faraday named the moving, charged particles **ions**, from a Greek word meaning 'to go'.

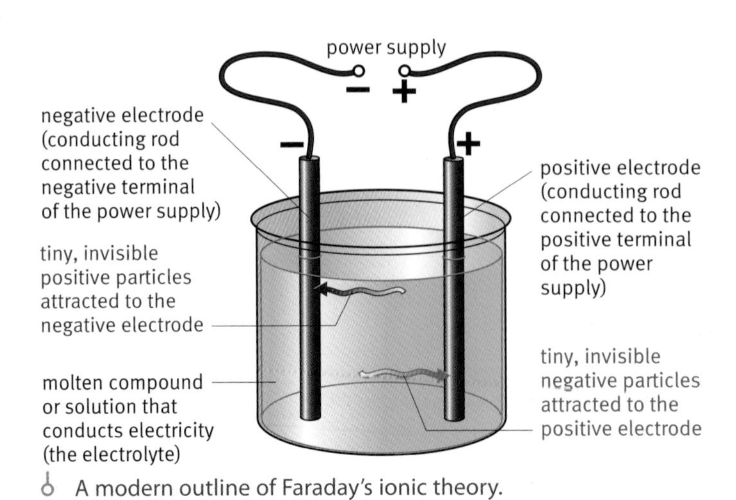

A modern outline of Faraday's ionic theory.

Explaining electrolysis

Chemists continue to use ionic theory to explain electrolysis. According to the theory, salts such as sodium chloride consist of ions.

Sodium chloride is made up of sodium ions and chloride ions. Sodium ions, Na^+, are positively charged. The chloride ions, Cl^-, carry a negative charge. These oppositely charged ions attract each other.

A crystal of sodium chloride consists of millions and millions of Na^+ and Cl^- ions closely packed together. In the solid, these ions cannot move towards the electrodes, and so the compound cannot conduct electricity. The ions can move when sodium chloride is hot enough to melt or when it is dissolved in water.

During electrolysis, the negative electrode attracts the **positive ions**. The positive electrode attracts the **negative ions**. When the ions reach the electrodes, they lose their charges and turn back into atoms.

Metals form positive ions, and non-metals generally form negative ions.

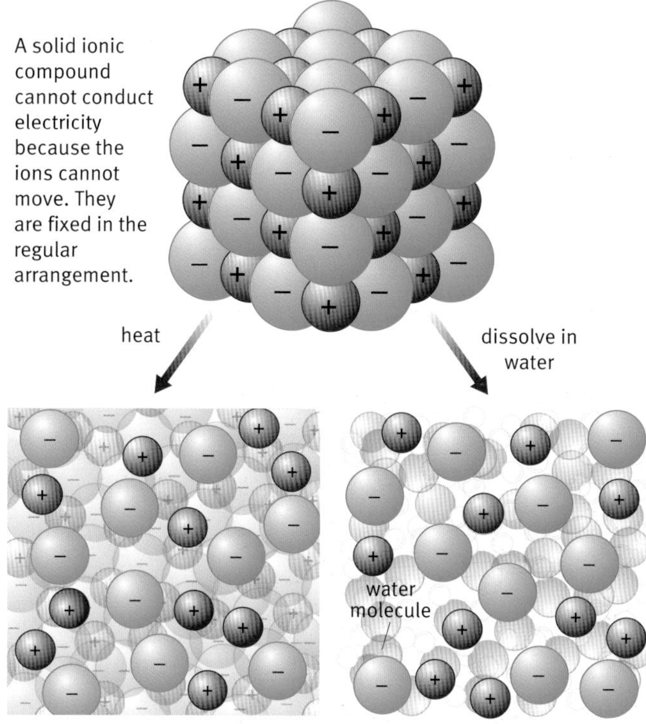

When an ionic compound is heated strongly, the ions move so much that they can no longer stay in the regular arrangement. The solid melts. Because the ions can now move around independently, the molten compound conducts electricity.

When an ionic compound has dissolved, it can conduct electricity because its ions can move independently among the water molecules.

Elements and compounds

Ionic theory can help to explain why compounds are so different from their elements. Sodium atoms and chlorine molecules are dangerously reactive. Sodium chloride is safe because its ions are much less reactive.

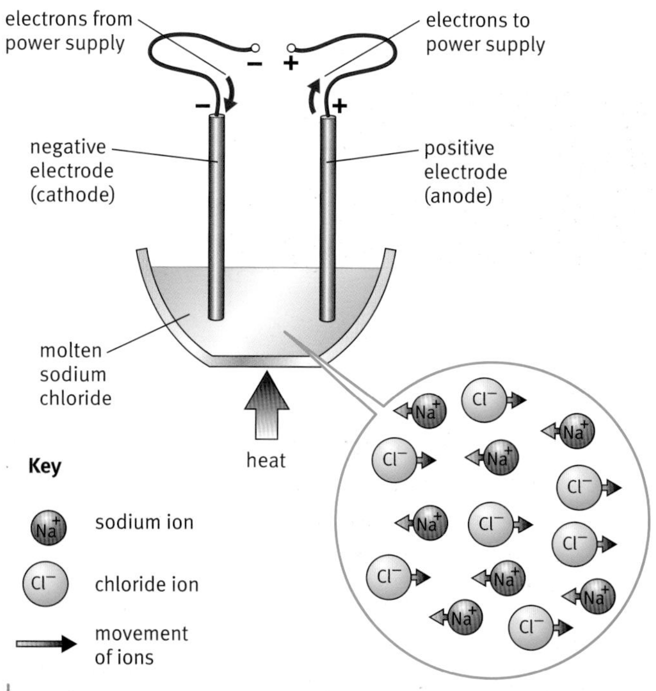

Sodium chloride conducts when molten because its ions can move towards the electrodes.

Questions

3 Why do solid compounds made of ions not conduct electricity?

4 Chemists sometimes call the negative electrode the cathode. Cations are the ions that move towards the cathode. What is the charge on a cation? Which type of element forms cations? Give an example of a cation.

5 Chemists sometimes call the positive electrode the anode. Anions are the ions that move towards the anode. What is the charge on an anion? Which type of element forms anions? Give an example of an anion.

Ionic theory and atomic structure

Find out about

- atoms and ions
- electron arrangements of ions
- the formulae of ionic compounds

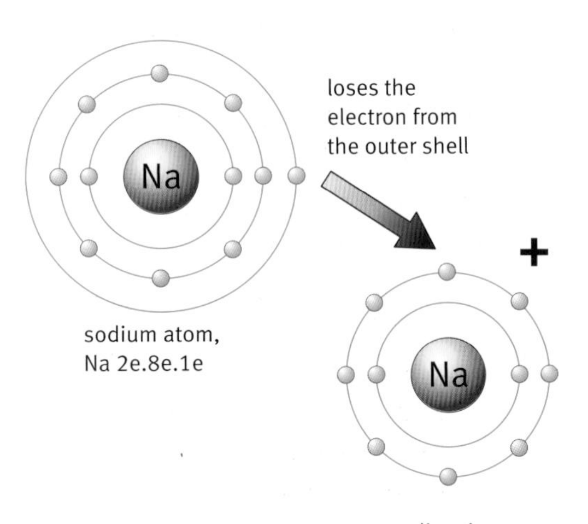

A sodium atom turns into a positive ion when it loses a negatively charged electron.

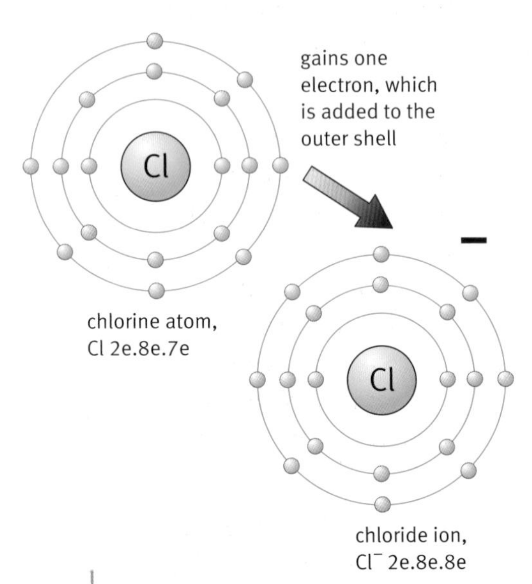

A chlorine atom turns into a negative ion by gaining an extra negatively charged electron.

Atoms into ions

Faraday could not explain how atoms turn into ions because he was working long before anyone knew anything about the details of atomic structure. Today, chemists can use the shell model for electrons in atoms to show how atoms become electrically charged.

The metals on the left-hand side of the periodic table form ions by losing the few electrons in their outer shell. This leaves more protons than electrons, and so the ions are positively charged.

All the metals in Group 1 have one electron in the outer shell. The diagram in Section G shows that removing the first electron from a sodium atom needs relatively little energy. The same is true for the other Group 1 metals, so they all form ions with a 1+ charge: Li^+, Na^+, and K^+, for example.

Chlorine gas consists of Cl_2 molecules. But it is easier to see what happens when chloride ions form by looking at one atom at a time, as shown in the diagram on the left. As each chlorine atom turns into an ion, it gains one electron and becomes negatively charged, Cl^-.

Electron arrangements of ions

Notice that when sodium and chlorine atoms turn into ions, they end up with the same electron arrangement as the nearest noble gas in the periodic table. This is generally true for simple ions of the first 20 or so elements in the periodic table. An explanation of why this is so involves an analysis of the energy changes that occur when metals react with non-metals. This is something you will study if you go on to a more advanced chemistry course.

Ions into atoms

Electrolysis turns ions back into atoms. Metal ions are positively charged, so they are attracted to the negative electrode. It is a flow of electrons from the battery into this electrode that makes it negative. Positive metal ions gain electrons from the negative electrode and turn back into atoms.

Non-metal ions are negatively charged, so they are attracted to the positive electrode. This electrode is positive because electrons flow out of it to the battery. Negative ions give up electrons to the positive electrode and turn back into atoms.

Formulae of ionic compounds

The formula of sodium chloride is NaCl because there is one sodium ion (Na^+) for every chloride ion (Cl^-). There are no molecules in everday table salt, only ions.

Not all ions have single positive or negative charges like sodium and chlorine. The formula of lead bromide is $PbBr_2$. In this compound there are two bromide ions for every lead ion. All compounds are overall electrically neutral, so the charge on a lead ion must be twice that on a bromide ion. A bromide ion, like a chloride ion, has a single negative charge (Br^-), so a lead ion must have a double positive charge (Pb^{2+}).

Compound	Ions present		Formula
	Positive	Negative	
magnesium oxide	Mg^{2+}	O^{2-}	MgO
calcium chloride	Ca^{2+}	Cl^- Cl^-	$CaCl_2$
aluminium oxide	Al^{3+} Al^{3+}	O^{2-} O^{2-} O^{2-}	Al_2O_3

Examples of formulae of ionic compounds.

Ions in the periodic table

The charges of simple ions show a periodic pattern. You can see this from the diagram below, showing the ionic symbols in the periodic table. Many of the transition metals in the middle block of the table can form more than one type of ion. Iron, for example, can form Fe^{2+} and Fe^{3+} ions, while copper can exist as Cu^+ and Cu^{2+} ions. Why this should be so is something you will study if you go on to a more advanced chemistry course.

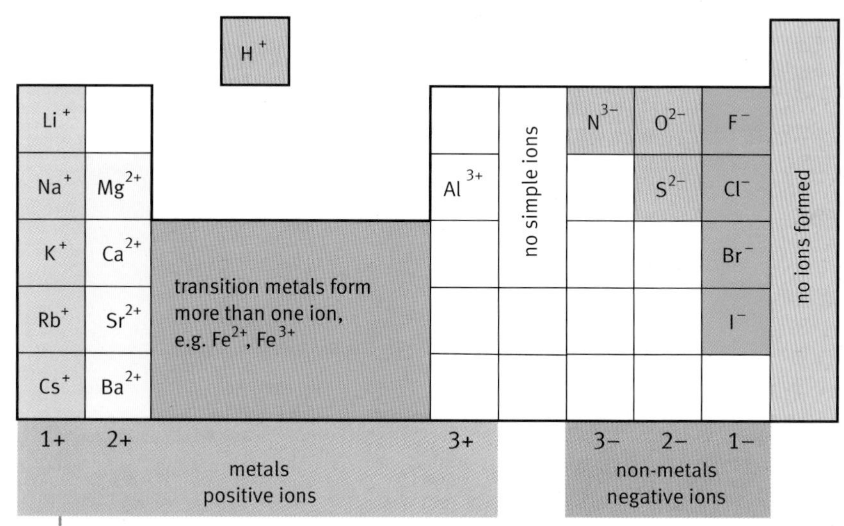

Simple ions in the periodic table.

Questions

1. Draw diagrams to show the number and arrangement of electrons in a lithium atom and in a lithium ion. What is the charge on a lithium ion?
2. Draw diagrams to show the number and arrangement of electrons in a fluorine atom and in a fluoride ion. What is the charge on a fluoride ion?
3. Write down the electron arrangements of:
   a. a fluoride ion (nucleus with 9 protons and 10 neutrons)
   b. a neon atom (nucleus with 10 protons and 10 neutrons)
   c. a sodium ion (nucleus with 11 protons and 12 neutrons).

   In what ways are a fluoride ion, a neon atom, and a sodium ion the same? How do they differ?
4. With the help of the table of ions, work out the formulae of these ionic compounds:
   a. potassium iodide
   b. calcium bromide
   c. aluminium chloride
   d. magnesium nitride
   e. aluminium sulfide.
5. Use the table of ions to work out the charge on the following ions:
   a. copper (Cu) in $CuCl_2$
   b. zinc (Zn) in ZnO
   c. iron (Fe) in Fe_2O_3.

Chemical species

Find out about

- atoms, molecules, and ions
- chemical species

In this module you have met the idea that the same element can take different chemical forms with distinct properties. Chemists describe these different forms as **chemical species**.

Species of chlorine

Chlorine has three simple species: atom, molecule, and ion. Each of these species of chlorine has distinct properties. Chlorine atoms (Cl) do not normally exist on their own. They rapidly pair up to form chlorine molecules (Cl_2). However, ultraviolet radiation can split chlorine (and chlorine compounds) into atoms. This is what happens to CFCs such as CCl_3F when they get into the upper atmosphere. In the full glare of the Sun's radiation, the molecules break up into atoms. The highly reactive free chlorine atoms rapidly destroy ozone. Lowering the concentration of ozone creates the so-called 'hole' in the ozone layer.

Chlorine gas at room temperature consists of chlorine molecules (Cl_2). These are very reactive, as illustrated by the chemistry of the alkali metals and halogens described in Section D. The chlorine molecules are reactive enough to do damage to human tissues, so the gas is given the label 'toxic'.

Chloride ions are quite different. They occur in compounds such as sodium chloride and magnesium chloride. Chloride ions in these salts are essential to life and occur in all living tissues. Chloride ions are chemically active in many ways, but they are not as reactive and harmful as the atoms or molecules of the element.

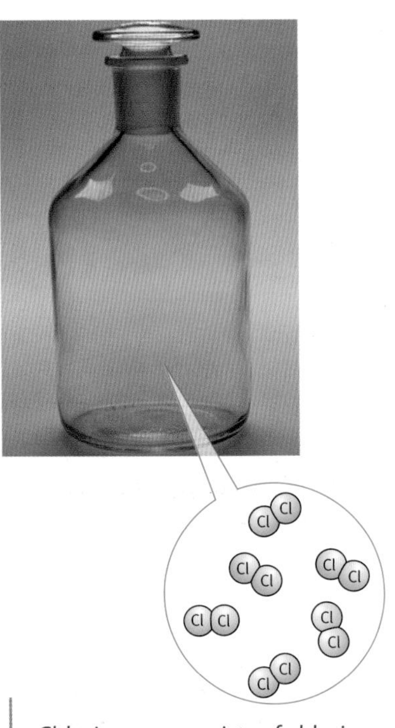

Chlorine gas consists of chlorine molecules. Chlorine molecules are chemically very reactive.

Polar stratospheric cloud formations over the Arctic (seen as thin orange and brown layers). Ice particles in these clouds provide a surface for the chemical reactions that release reactive chlorine atoms, which can then destroy ozone.

There are more complex species of chlorine, with the element joined to other atoms. This includes molecules that contain chlorine and other elements, such as tetrachloromethane (CCl_4).

Chlorine dioxide is oxidising, toxic, and corrosive. As a liquid it can be explosive. Its properties are different to those of chlorine, which is toxic, and oxygen, which is oxidising.

Species of sodium

There are only two species of sodium: atom and ion. The atoms in sodium metal are chemically very reactive. In the presence of other chemicals, the sodium atoms react to produce compounds containing sodium ions.

Sodium combines with chlorine to produce the ionic compound sodium chloride. This is made up of two chemical species: Na^+ and Cl^-. These two ions are quite unreactive. Sodium chloride is soluble in water, but its solution has a neutral pH. Water does not react with the ions.

When sodium reacts with water, it produces another ionic compound: sodium hydroxide, containing Na^+ and OH^-. The sodium hydroxide dissolves in the water to give a solution that is very alkaline. It is the hydroxide ions that make a solution of sodium hydroxide alkaline, not the sodium ions.

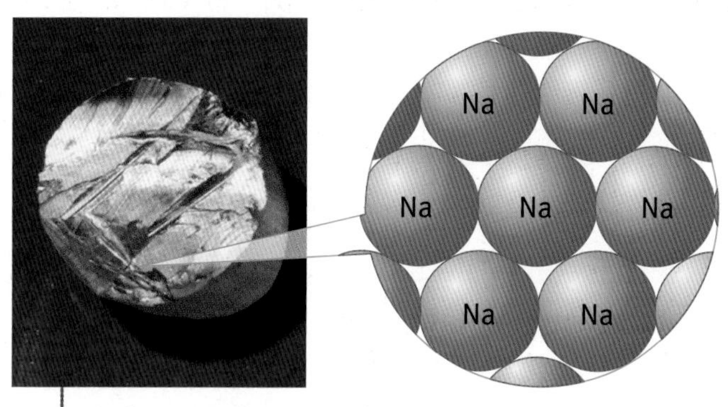

Sodium metal consists of sodium atoms. Sodium atoms are chemically very reactive.

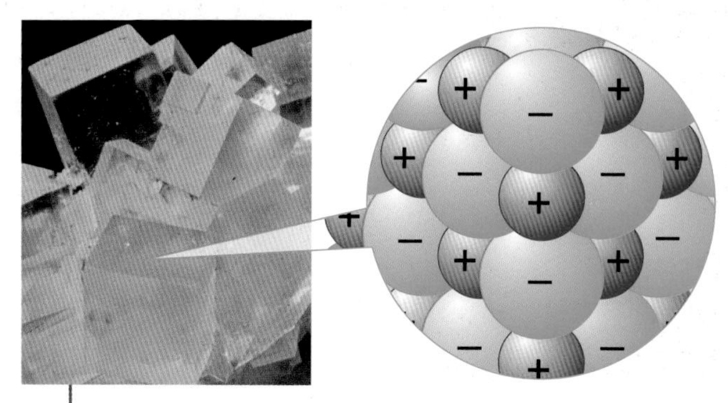

Sodium chloride consists of sodium ions and chloride ions. These ions are not very reactive.

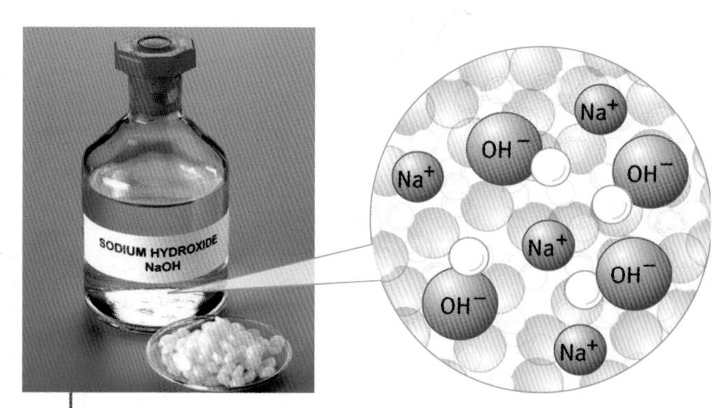

Sodium hydroxide is a strong alkali. It consists of sodium ions and hydroxide ions. In solution, the ions move around separately, mixed with water molecules.

Questions

1 Use the idea of chemical species to explain why the properties of sodium chloride are very different from the properties of its elements sodium and chlorine.

2 Give the name and formulae of all chemical species in:
  a potassium
  b bromine
  c potassium bromide.

3 a Identify four distinct chemical species in unpolluted air, giving their names and formulae.
  b Identify three more chemical species present in the polluted air of a busy city street.

Science Explanations

Chemists have identified patterns and come up with theories to make sense of the world, and to explain how roughly 100 elements can give rise to such a huge variety of chemical compounds.

You should know:

- that the chemists' model of the atom has a tiny central nucleus, containing protons and neutrons, which is surrounded by electrons
- that the number of electrons is equal to the number of protons and that the electrons in an atom have definite energies
- how the arrangement of electrons in shells is determined by the way that the electron shell with the lowest energy fills first until full, then the next shell starts to fill, and so on
- that the chemistry of an element is largely determined by the number and arrangement of the electrons in its atoms
- that elements in the periodic table are lined up in order of their proton numbers so that there are repeating patterns across each row (period) in the table
- that each column in the periodic table consists of a group of related elements that have similar chemistries because they have the same number of electrons in their outer shells
- that there are trends in the properties of the elements down a group because of the increasing number of inner full shells
- that Group 1 elements are the alkali metals, which react with moist air, water, and chlorine, becoming more reactive down the group
- that Group 7 elements are the halogens made up of diatomic molecules
- why there is a trend in the physical state of the halogens at room temperature
- that the halogens become less reactive down the group
- that chemists use word equations and symbol equations to describe reactions
- why safety precautions are important when working with hazardous chemicals such as Group 1 metals, alkalis, and the halogens
- how ionic compounds form when metals react with non-metals such that metal atoms lose electrons while the non-metal atoms gain electrons
- how a regular lattice of ions gives rise to the shape of the crystals of an ionic compound
- that the properties of an ionic compound are the properties of its ions, which behave in a different way to the atoms or molecules of its elements
- that ionic compounds conduct electricity when molten or when dissolved in water, because the ions are charged and they are able to move around in the liquid.

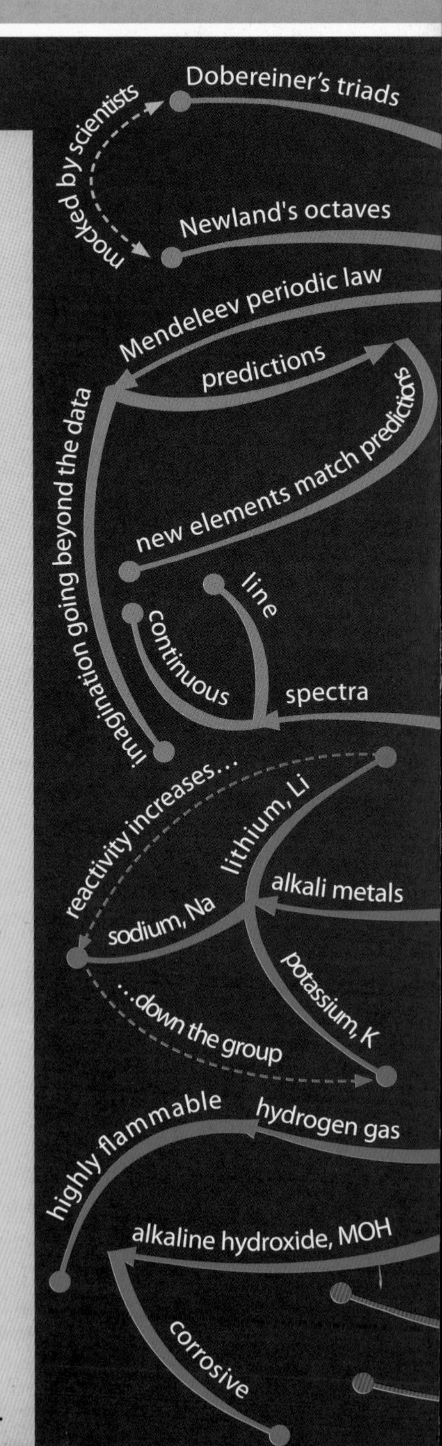

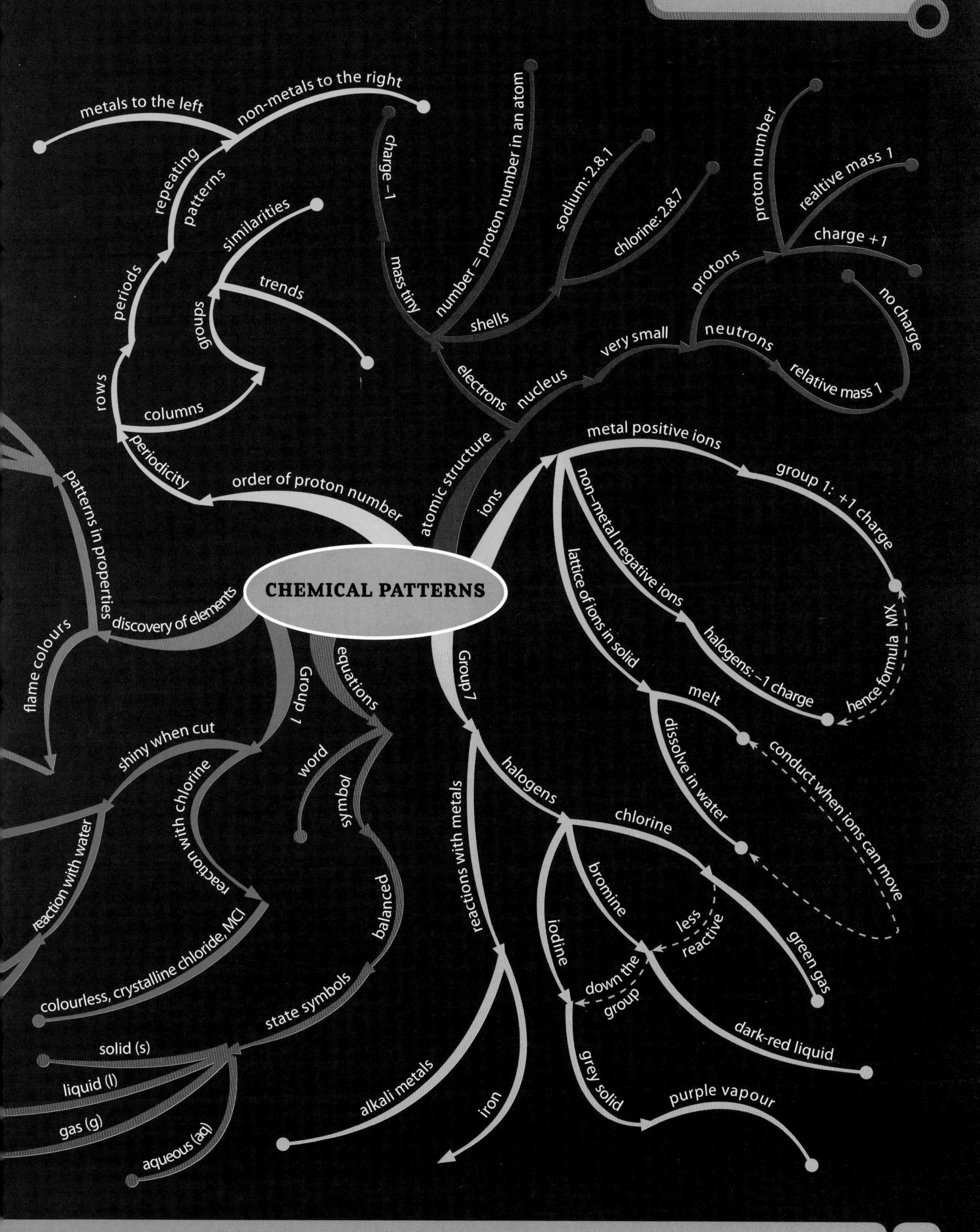

CHEMICAL PATTERNS
order of proton number
periodicity
rows
periods
repeating
patterns
metals to the left
non-metals to the right
columns
groups
similarities
trends
atomic structure
nucleus
very small
protons
proton number
realtive mass 1
charge +1
neutrons
no charge
relative mass 1
electrons
mass tiny
charge −1
number = proton number in an atom
shells
sodium: 2.8.1
chlorine: 2.8.7
ions
metal positive ions
group 1: +1 charge
hence formula MX
non-metal negative ions
halogens: −1 charge
lattice of ions in solid
melt
dissolve in water
conduct when ions can move
Group 7
halogens
chlorine
green gas
bromine
dark-red liquid
iodine
grey solid
purple vapour
less reactive
down the group
reactions with metals
alkali metals
iron
equations
word
symbol
balanced
state symbols
solid (s)
liquid (l)
gas (g)
aqueous (aq)
Group 1
shiny when cut
reaction with chlorine
colourless, crystalline chloride, MCl
reaction with water
discovery of elements
patterns in properties
flame colours

Ideas about Science

Scientific explanations are based on data but they go beyond the data and are distinct from them. An explanation has to be thought up creatively to account for the data. A new explanation may explain a range of phenomena not previously thought to be linked. The explanation should also allow predictions to be made about new situations or examples.

In the context of the discovery of new elements and the development of the periodic table you should be able to:

- give an account of scientific work and distinguish statements that report data from statements of explanatory ideas (hypotheses, explanations, and theories)
- recognise that an explanation may be incorrect even if the data is correct
- identify where creative thinking is involved in the development of an explanation, such as Mendeleev's insight that he had to leave gaps for undiscovered elements
- recognise data or observations that are accounted for by (or conflict with) an explanation, such as the data from spectra that could be explained by the existence of new elements
- give good reasons for accepting or rejecting a proposed scientific explanation, such as the way that the shell model of atomic structure accounts for the arrangement of elements in the periodic table
- identify the best of given scientific explanations for a phenomenon, such as Mendeleev's use of his periodic table to predict the existence of unknown elements
- understand that when a prediction agrees with an observation this increases confidence in the explanation on which the prediction is based, but does not prove it is correct. For example, Mendeleev correctly predicted the properties of missing elements in his periodic table

- understand that when a prediction disagrees with an observation this indicates that one or the other is wrong and decreases confidence in the explanation on which the prediction is based.

Scientists report their claims to other scientists through conferences and journals. Scientific claims are only accepted once they have been evaluated critically by other scientists. Scientists are usually sceptical about claims that cannot be repeated by anyone else and about unexpected findings until they have been replicated (by themselves) or reproduced (by someone else). You should be able to:

- broadly outline the peer review process, in which new scientific claims are evaluated by other scientists
- recognise that new scientific claims that have not yet been evaluated by the scientific community are less reliable than well-established ones; an example is the lack of acceptance of early attempts to find connections between the chemical properties of the elements and their relative atomic mass
- identify the fact that a finding has not been reproduced by another scientist as a reason for questioning a scientific claim.

Review Questions

1 A teacher reacts sodium with chlorine. The product is sodium chloride, NaCl.

a The sodium bottle has two hazard symbols.

i Give the meaning of each symbol.

ii List two safety precautions the teacher must take when using sodium. Link each to a hazard.

b Write a balanced symbol equation for the reaction of sodium with chlorine, including state symbols.

c Sodium chloride is made up of two types of ion.

i Give the formulae of the two ions.

ii State the number of protons and electrons in each ion.

iii Draw diagrams to show the electron arrangements of each ion.

2 The table shows part of the periodic table, with only a few symbols included.

period \ group	1	2	3	4	5	6	7	0
1								He
2						O	F	
3	Na	Mg	Al	Si		S	Cl	
4				Ge			Br	
5								
6	Cs							

a Using only the elements in the table, write down symbols for:

i a metal that floats on water

ii an element with similar properties to silicon (Si)

iii three elements that have molecules made up of two atoms

iv the most reactive metal.

b Predict the state of fluorine, F, at room temperature. Give a reason for your prediction.

3 The table shows the properties of three halogens.

Halogen	Boiling point (°C)	Formula of compound with iron
chlorine	–34.7	$FeCl_3$
bromine	58.8	$FeBr_3$
iodine	184	

a **i** Describe the trend in boiling points.

ii Predict the formula of iron iodide.

iii The formula of a bromide ion is Br^-. Work out the charge on the iron ion.

b Describe the pattern of reactivity of the halogens going down the group. Give examples to illustrate your answer. Use ideas about electron arrangement and the formation of ions to help you explain the pattern.

c Give an example to show how the model of atomic structure with electrons in shells can explain the similarities between the halogens.

C5 Chemicals of the natural environment

Why study chemicals of the natural environment?

Conditions on Earth are special. The temperature is just right for most water to be liquid. The atmosphere has enough oxygen for living things to breathe, but not so much that everything catches fire. Rocks are the source of many of the chemicals that meet our daily needs.

What you already know

- The atmosphere is a protective blanket of gases around the Earth, which supports life.
- Bonds between atoms in molecules are strong, while forces between molecules are very weak.
- An atom consists of a central nucleus surrounded by electrons.
- Ionic compounds have a lattice structure.
- Ionic compounds conduct electricity when molten or in solution, because the ions are free to move.
- During a chemical reaction, atoms and mass are both conserved.
- A chemical change caused by the flow of an electric current is called electrolysis.
- Making, using, and disposing of products can affect the environment.

Find out about

- the chemicals in spheres of the Earth
- theories of structure and bonding
- tests for analysing water quality
- methods used to extract metals from their ores
- how to calculate chemical quantities.

The Science

Theories of structure and bonding explain how atoms are arranged and held together in all the chemicals that make up our Earth. There are three types of strong bonding that give rise to useful materials. There are weaker forces of attraction that allow molecules to stick to each other enough to make liquids and solids.

Ideas about Science

The properties of metals make them very useful and important in our lives. But mining, mineral processing, and metal extraction can all have a serious impact on the environment. Scientists are applying their understanding of chemistry in the environment to work out how we can use natural resources in a sustainable way.

Chemicals in spheres

Find out about

- naturally occurring elements and compounds
- abundances of elements in different spheres
- cycling of elements and compounds between the spheres

People who study the Earth often think of it as being made up of spheres (see the diagram below). Starting from the middle, first comes the core, then the mantle, then the crust.

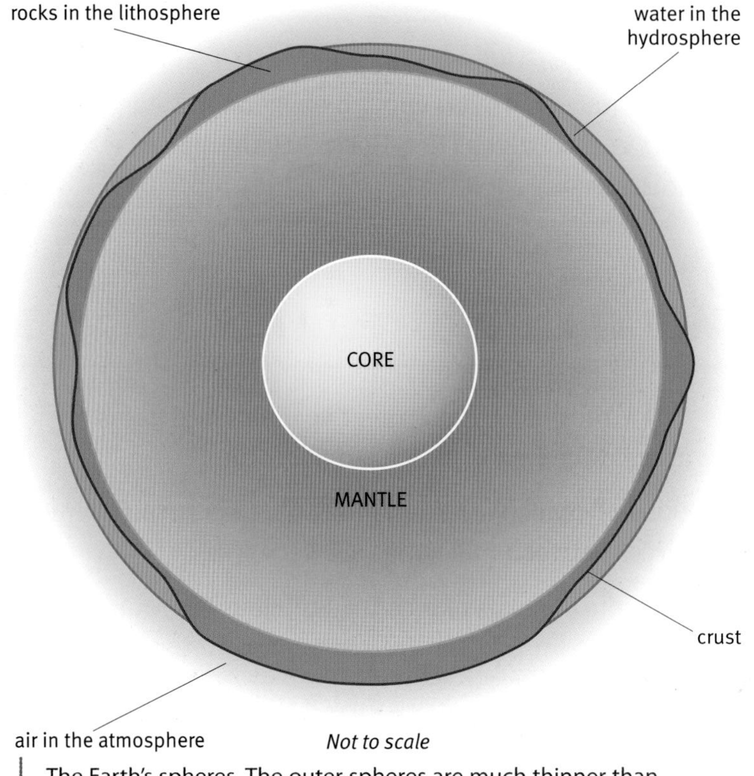

The Earth's spheres. The outer spheres are much thinner than this relative to the core and mantle.

This module looks at the spheres that make up the Earth's surface and provide us with natural resources: the lithosphere, hydrosphere, and atmosphere.

The **lithosphere** is broken into giant plates that fit around the globe like puzzle pieces. These are the tectonic plates, which move a little bit each year as they slide on top of the upper part of the mantle. The **crust** and upper **mantle** make up the lithosphere. The lithosphere is about 100 km thick.

The oceans and rivers make up the **hydrosphere**. This is not a complete sphere, but the oceans cover two-thirds of the globe, so it almost is.

Finally, wrapped like a big fluffy duvet around the Earth, keeping it warm, is the layer of air we call the **atmosphere**.

There are living things in parts of all three outer spheres.

Key words

- lithosphere
- crust
- mantle
- hydrosphere
- atmosphere

Elements in the spheres

The spheres vary greatly in their chemical composition. The atmosphere consists mainly of two uncombined elements: nitrogen and oxygen. The hydrosphere is almost entirely the compound water.

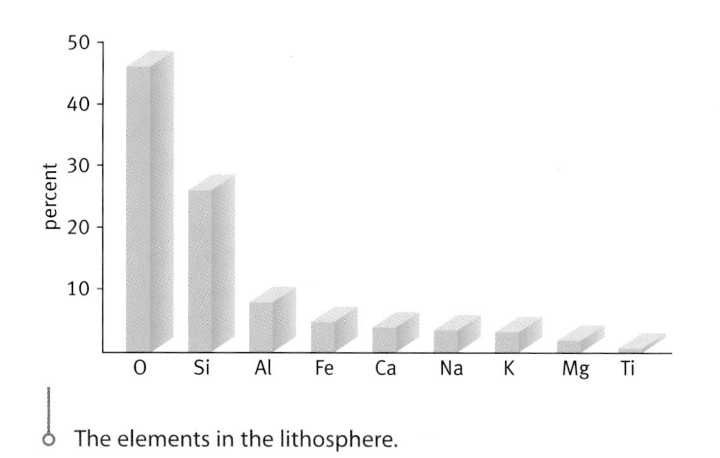

The elements in the lithosphere.

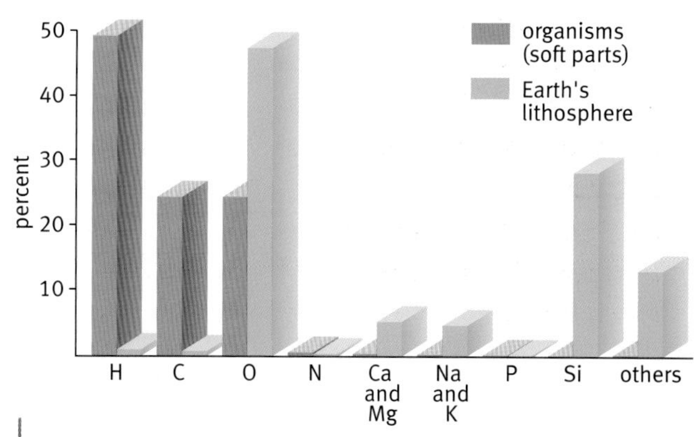

The percentage of different elements in living tissue and in the lithosphere compared.

The rocks of the lithosphere are made up mainly of silicates. These are compounds of silicon combined with oxygen and much smaller quantities of other elements.

The elements exist as different chemical species in the three spheres. In the lithosphere, carbon is combined with hydrogen to make the hydrocarbons in crude oil. Carbon is also hidden in the calcium carbonate of chalk and limestone. In the atmosphere, carbon is present as the gas carbon dioxide. Carbon is a vitally important element for life: most of the chemicals that make up living things are compounds of carbon combined with hydrogen, oxygen, and a few other elements.

Flowing between the spheres

Chemicals do not always stay in one sphere. They are constantly on the move between the spheres. Think of a carbon atom: it may start in the atmosphere; be taken into a plant; be washed into water in the hydrosphere; then buried in sediment of the lithosphere.

Water flows freely between the spheres. The obvious place for water is the hydrosphere. But think of clouds in the atmosphere; then rain sinking into the lithosphere, reappearing as a spring, from where it flows into a river, and then into the sea.

Questions

1 Look at the bar charts on this page.
   a What are the three most common elements in living tissue?
   b What are the three most common elements in the lithosphere?
   Comment on your answers.

2 'Living things inhabit the atmosphere, hydrosphere, and lithosphere.' Give some examples to illustrate this statement.

3 Make a list of four different things you use in a typical day that come from the lithosphere.

Find out about

- gases in the air
- weak attractions between molecules
- strong covalent bonding

nitrogen	N N
oxygen*	O O
argon**	Ar
carbon dioxide	O C O

The main of atmospheric gases.
* The element oxygen is unusual as it can exist as normal oxygen, O_2, or as ozone, O_3.
** The element argon is a noble gas. All the noble gases are made up of single atoms, so argon is represented by Ar.

The Earth is just the right size for its gravity to hold onto its atmosphere. It is also just the right distance from the Sun to have the right temperature for water to exist as a liquid. This water, together with the carbon dioxide and oxygen in the atmosphere, allows the Earth to support a great variety of plant and animal life.

The relative proportions of the main gases in the atmosphere are about 78% N_2, 21% O_2, 1% Ar, and 0.04% CO_2. The atmosphere also contains water vapour and small amounts of other gases.

The presence of argon (Ar) in the air was not discovered until the 1890s – over 100 years after the discovery of oxygen. Argon is a noble gas but it is not a rare gas. Every time you breathe in you take about 5 cm^3 of argon into your lungs.

The average temperature of the atmosphere at the Earth's surface is 15 °C. At a height of 80 km the temperature drops to –90 °C, its lowest point. Chemicals in the atmosphere have melting and boiling points low enough for them to exist as gases at temperatures between these extremes.

All the chemicals in the atmosphere are either non-metallic elements (eg, O_2, N_2, Ar) or compounds made from non-metallic elements (eg, CO_2).

Atoms and molecules in the air

Most of the chemicals in the atmosphere are made of **small molecules**. Only the noble gases exist as single atoms.

All molecules have a slight tendency to stick together. For example, there is an attraction between one O_2 and another O_2. But these **attractive forces** are very weak. This is why the chemicals that make up the atmosphere are gases with low melting and boiling points.

One way to picture this is that the molecules are moving so quickly that, when two O_2 molecules come close to each other, the attractive force between them is not strong enough to hold them together.

Strong bonds in molecules

The forces inside molecules that hold the atoms together are very strong, many times stronger than the weak attractions between molecules. Small molecules such as O_2 or H_2 do not split up into atoms except at very, very high temperatures.

Key words

- small molecules
- attractive forces
- molecular models
- electrostatic attraction
- covalent bonding

Chemists often use a single line to represent a single bond between atoms in a molecule. For example, the simple molecules in hydrogen, oxygen, and carbon dioxide can be represented by the molecular formulae H_2, O_2, and CO_2. But if you want to show their bonds, they can be represented by:

H—H   O=O   O=C=O

Some **molecular models** use the same idea. A coloured ball represents each atom, and a stick or a spring is used for each bond.

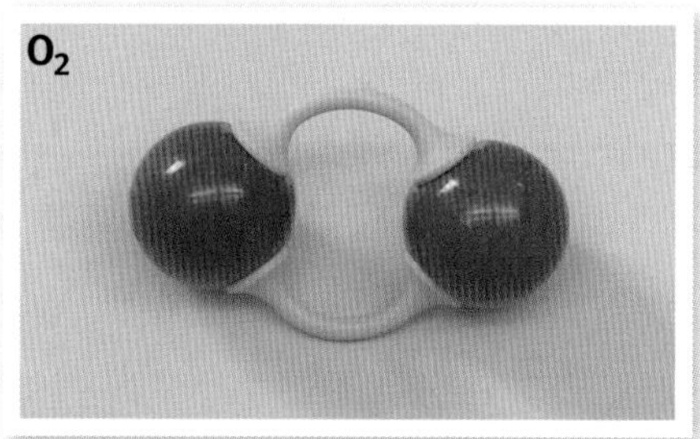

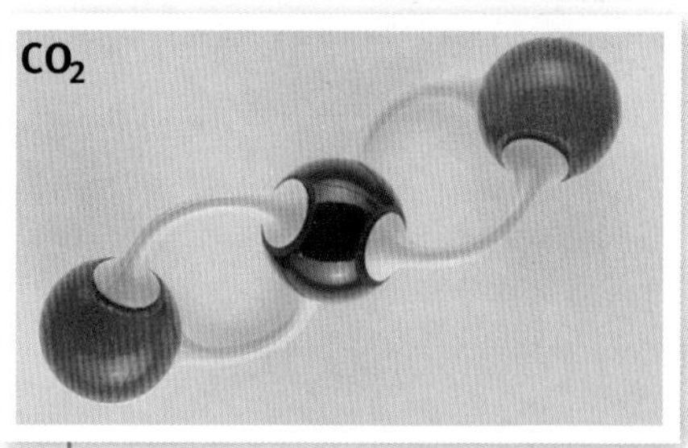

Ball-and-stick models of H_2, O_2, H_2O, and CO_2. Notice that the molecules have a definite shape. There are fixed angles between the bonds in molecules.

Electrons and bonding

A knowledge of atomic structure can help to explain the bonding in molecules (see C4, K: Ionic theory and chemical structure). When non-metal atoms combine to form molecules, they do so by sharing electrons in their outer shells.

The atoms are held strongly together by the attraction of their nuclei for the pair of electrons they share.

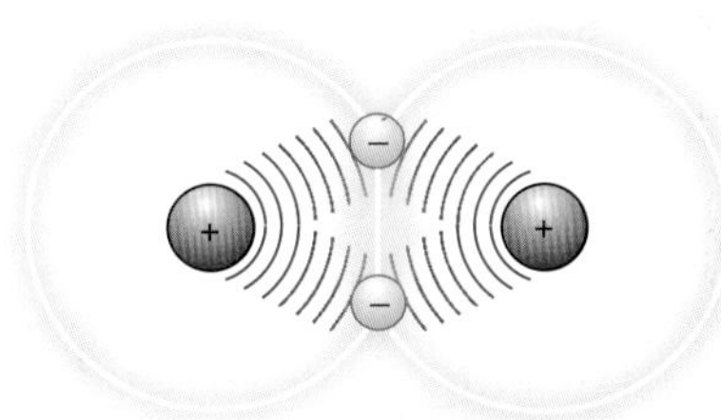

The atoms cannot move any closer together because the repulsion between the two positively charged nuclei will push them apart again.

The formation of a covalent bond between two hydrogen atoms.

The molecule of hydrogen is held together by the **electrostatic attraction** between the two nuclei and the shared pair of electrons. This is a single bond.

This type of strong bonding is called **covalent bonding**. 'Co' means 'together' or 'joint', while the Latin word *valentia* means strength. So we have strength by sharing.

The number of bonds that an element can form depends on the electrons in the outer shell of its atoms. The table gives the number of covalent bonds normally formed by atoms of some non-metal elements.

Atom	Usual number of covalent bonds
H, hydrogen	1
Cl, chlorine	1
O, oxygen	2
N, nitrogen	3
C, carbon	4
Si, silicon	4

Questions

1 Draw diagrams to show the covalent bonding in these molecules:
  a hydrogen chloride, HCl
  b ammonia, NH_3
  c methane, CH_4
  d ethene, C_2H_4

2 Estimate the volume of argon in the room in which you are sitting.

Find out about

- unusual properties of water
- bonding within and between water molecules
- ions in solution

On Earth, water exists mostly in the liquid state (the oceans and the clouds), with some in the solid state (ice) and a smaller amount as gas (water vapour).

Key words

- salts
- dissolve

Properties of water

We see water so often that we take its special chemical and physical properties for granted.

One of these special properties is that it is a liquid at room temperature. The H_2O molecule has a smaller mass than molecules of O_2, N_2, or CO_2, which are all gases at room temperature. So you might expect H_2O to be a gas at room temperature too. But H_2O melts at 0 °C and boils at 100 °C. So molecules of H_2O must have a greater tendency to stick to each other than the molecules of a gas like oxygen.

Strange things happen when water cools from room temperature. As it cools to 4 °C, the liquid contracts as expected. However, on cooling further to 0 °C, it starts to expand. This means that ice is less dense than the cold water surrounding it, so it floats. This is very important in nature. It means that winter ice does not sink to the bottom of lakes where it would be far from the warming rays of spring sunshine.

Another special property of water is that it is a good solvent for **salts**. Most solvents do not **dissolve** ionic compounds, but water does.

Pure water does not conduct electricity – in this way it resembles other liquids made up of small molecules. This shows that it does not contain charged particles that are free to move. Even though pure water does not conduct, you should never touch electrical devices with wet hands. This is because the water on your hands is not pure, so it will conduct electricity and increases the risk of you receiving an electric shock.

Water molecules

Knowledge of the structure of water molecules can help to explain its remarkable properties. The three atoms in a water molecule are not arranged in a straight line, but are at an angle.

The electrons are not evenly shared in the covalent bonds between atoms. The oxygen atom has more than its fair share. It also has four unshared electrons in its outer shell. These two effects give each water molecule a small negative charge on its oxygen side, and a small positive charge on its hydrogen side. Overall, each molecule is electrically neutral.

The small charges on opposite sides of the molecules cause slightly stronger attractive forces between the molecules. The small charges also help water dissolve ionic compounds by attracting the ions out of their crystals.

The attractions between water molecules and their angular shape mean that they line up in ice to create a very open structure. As a result, ice is less dense than liquid water.

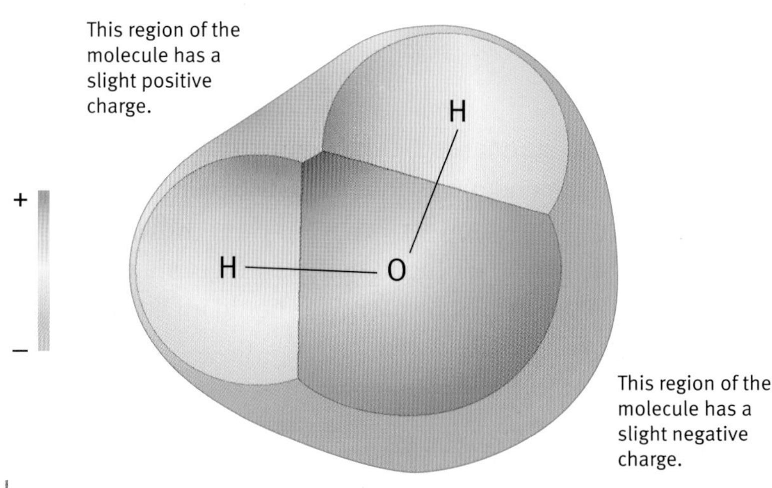

A water molecule, showing that oxygen has a slightly greater share of the pair of electrons in each bond.

Ions separate and can move about freely when dissolved in water. In seawater there is a mixture of positive ions and negative ions.

Why is seawater salty?

The diagram below shows how soluble chemicals get carried from rocks to the sea during part of the water cycle.

The main soluble chemical carried into the sea is sodium chloride. River water does not taste salty because the concentration is so low, but the concentration of salt in the sea has built up over millions of years, and so it does taste salty.

Other chemicals that dissolve include potassium chloride, potassium bromide, magnesium chloride, magnesium sulfate, and sodium sulfate. One litre of typical seawater contains about 40 g of dissolved chemicals from rocks. Most of the compounds that are dissolved in seawater are made up of positively charged metal ions and negatively charged non-metal ions. They are salts.

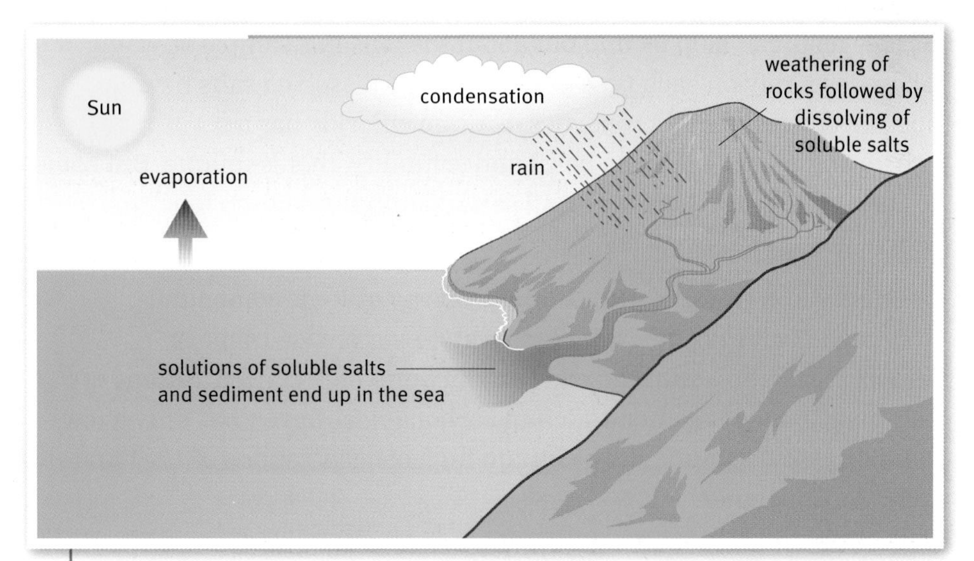

Weathering slowly breaks down rocks. This exposes the inside of the rocks to water. Soluble chemicals in the rocks dissolve in the water and get washed away.

Questions

1 Water has an unusually high boiling point compared with other small molecules. What other unusual properties does water have?

2 Write out the formulae of these salts with the help of the periodic table showing the ion charges in C4 Section K. The sulfate ion is SO_4^{2-}.
   a Potassium bromide.
   b Magnesium chloride.
   c Magnesium sulfate.
   d Sodium sulfate.

3 How would the possibility of life on Earth have been affected had ice been denser than water?

D Detecting ions in salts

Find out about

- properties of ions
- precipitation reactions
- ionic equations
- tests for ions

Why test water?

Government regulations set standards for the quality of water supplied to our homes, and control the chemicals that industrial companies can discharge into rivers. The Foundation for Environmental Education awards beaches all over the world a Blue Flag if they meet standards that include the quality of the local seawater. We should all be interested in the make-up of the Earth's hydrosphere.

A domestic water supply.

A Blue Flag beach in Devon.

Industrial discharge to a river.

Solubility

Simple tests can help us find out about the ions in a sample of water. These tests rely on each type of ion from the dissolved salts having special, characteristic properties. Sodium chloride has a set of properties, such as melting point and solubility, that is unique. But some of its properties are shared by all salts containing sodium ions, and others are shared by all salts containing chloride ions.

Solubility in water is an important property of ionic compounds. Sodium chloride is quite soluble in water over a wide range of temperatures. So a solution can contain quite high concentrations of Na^+ and Cl^- ions together. Calcium carbonate, however, has only a low solubility, so no solution can contain high concentrations of Ca^{2+} and CO_3^{2-} ions together.

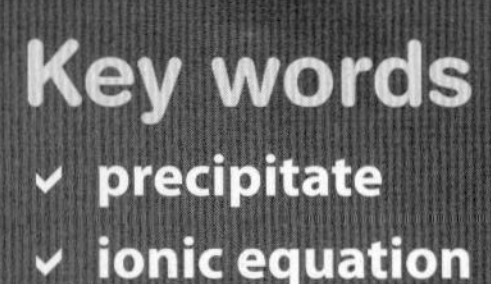

If you try to put high concentrations of Ca^{2+} and CO_3^{2-} ions together, you need two solutions: one where you have dissolved lots of a soluble *calcium* salt, and another where you have dissolved lots of a soluble *carbonate* salt. Calcium nitrate and sodium carbonate would be good examples to use. Then you mix the two solutions together. Very quickly, huge numbers of Ca^{2+} and CO_3^{2-} meet each other and cluster together to make solid crystals, which fall to the bottom of the solution as a solid white **precipitate**. The process is exactly the reverse of dissolving. The other two ions, sodium and nitrate, stay dissolved in the solution because sodium nitrate is a soluble salt.

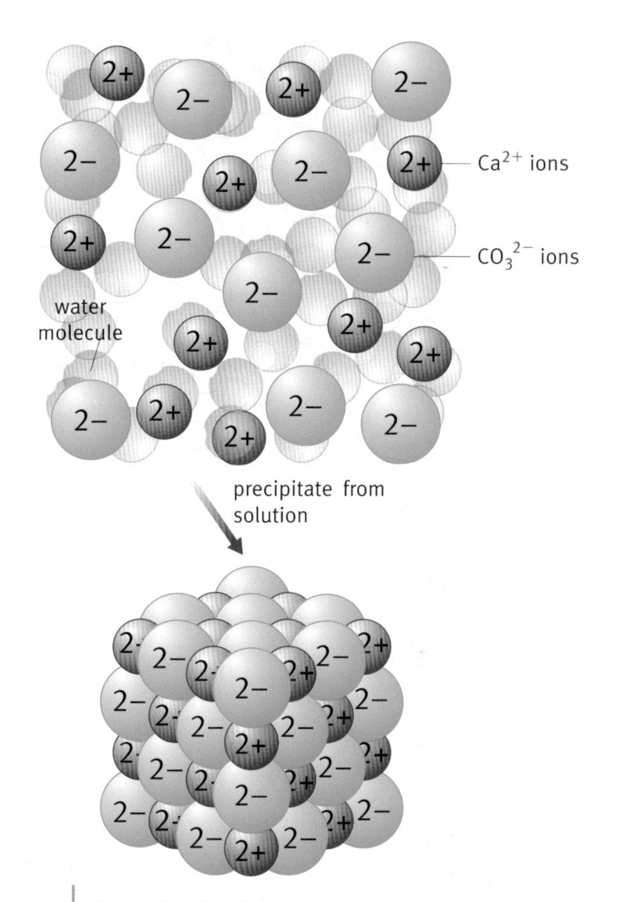

Ions of an insoluble salt cluster together and make a solid precipitate.

Ionic equations

When a solution of calcium nitrate is mixed with a solution of sodium carbonate, only two of the four ions in the mixture make a precipitate. Chemists use **ionic equations** to summarise reactions like this. Ionic equations allow you to focus attention just on the ions involved. For the precipitation of calcium carbonate:

$$Ca^{2+}(aq) + CO_3^{2-}(aq) \longrightarrow CaCO_3(s)$$

It is important to include the state symbols in these equations.

To write ionic equations correctly, use the following steps:

1. Write the correct formula of the precipitate, with the state symbol (s), on the right-hand side of the arrow.
2. Write the ions that make the precipitate separately, with their charges and state symbols (aq), on the left-hand side.
3. If necessary, put numbers in front of the ions on the left to show the ratio in which they stick together. You can work this out from the formula of the precipitate.

Questions

1. What would happen if you mixed solutions of potassium carbonate and calcium nitrate?
2. Write an ionic equation showing the precipitation of magnesium carbonate by mixing solutions of magnesium nitrate and sodium carbonate.

Shellfish use calcium carbonate to make their protective shells. At the surface of the sea, the concentrations of Ca^{2+} and CO_3^{2-} in the seawater are high enough for the calcium carbonate to stay as a solid precipitate and not dissolve. When shellfish die, their remains sink and build up as sedimentary limestone rocks on the seabed.

Salt or hydroxide	Solubility
all nitrates	soluble
all salts of sodium and potassium	soluble
silver chloride	insoluble
silver bromide	insoluble
silver iodide	insoluble
barium sulfate	insoluble
calcium carbonate	insoluble
hydroxides of metals not in Group 1	insoluble

The solubility of some salts and hydroxides.

Predicting when precipitates form

If you know the solubility of salts, you can predict when precipitation reactions will happen. This can help you to detect the ions present in a solution.

Looking for metal ions

Adding a solution of sodium carbonate and observing a white precipitate would be a good test for calcium ions in a solution – but there are other metal ions that would also form insoluble, white carbonate precipitates. So the results would not be conclusive.

Several metal ions form insoluble hydroxide compounds, too. The ions of transition metals are special, though, because they are coloured. This gives their hydroxide precipitates characteristic colours that can be used to identify the metal ion in the original solution. To carry out the test, you add a small volume of a dilute solution of sodium hydroxide to a solution of the substance to be investigated.

The ionic equation for the Fe^{3+} test is:

$$Fe^{3+}(aq) + 3OH^{-}(aq) \longrightarrow Fe(OH)_3(s)$$

Metal ion tested	Precipitate	Observations
copper, $Cu^{2+}(aq)$	$Cu(OH)_2(s)$	light blue (insoluble in excess NaOH (aq))
iron(II), $Fe^{2+}(aq)$	$Fe(OH)_2(s)$	green (insoluble in excess NaOH (aq))
iron(III), $Fe^{3+}(aq)$	$Fe(OH)_3(s)$	red-brown (insoluble in excess NaOH (aq))
calcium, $Ca^{2+}(aq)$	$Ca(OH)_2(s)$	white (insoluble in excess NaOH (aq))
zinc, $Zn^{2+}(aq)$	$Zn(OH)_2(s)$	white (soluble in excess NaOH (aq))

Tests for positively charged ions.

In the test for Zn^{2+} ions, if you add excess sodium hydroxide the precipitate dissolves, leaving a colourless solution. The other hydroxide precipitates in the table are insoluble in excess sodium hydroxide.

Water companies check the concentrations of transition metal ions in drinking water, and keep them below levels that would be toxic. They also keep the concentrations of Mg^{2+} and Ca^{2+} below levels that would cause too much limescale build-up in water pipes.

The hydroxide precipitates $Fe(OH)_2$, $Fe(OH)_3$, and $Cu(OH)_2$.

Looking for non-metal ions

Some tests for non-metal ions also use precipitation reactions.

The ions chloride (Cl^-), bromide (Br^-), and iodide (I^-) from the halogen group each make insoluble precipitates with silver ions (Ag^+). Each precipitate has a characteristic colour.

The ionic equation for the chloride ion, Cl^-, test is:

$$Ag^+(aq) + Cl^-(aq) \longrightarrow AgCl(s)$$

Sulfate ions (SO_4^{2-}) make a characteristic white precipitate with barium ions (Ba^{2+}).

The ionic equation for the sulfate ion, SO_4^{2-}, test is:

$$Ba^{2+}(aq) + SO_4^{2-}(aq) \longrightarrow BaSO_4(s)$$

The precipitates AgCl(s), AgBr(s), and AgI(s).

Ion tested	Test	Precipitate	Colour of precipitate
$Cl^-(aq)$	acidify with dilute nitric acid, then add silver nitrate solution	AgCl(s)	white
$Br^-(aq)$	acidify with dilute nitric acid, then add silver nitrate solution	AgBr(s)	cream
$I^-(aq)$	acidify with dilute nitric acid, then add silver nitrate solution	AgI(s)	yellow
$SO_4^{2-}(aq)$	acidify, then add barium chloride solution or barium nitrate solution	$BaSO_4(s)$	white

Tests for negatively charged ions.

Water companies test for sulfate ions and keep the concentration below about 1g/litre. Anything higher has a laxative effect.

Carbonate ions make carbon dioxide gas when you add a dilute acid. Bubbles form, creating effervescence and the gas can be tested to see if it makes limewater go cloudy. Carbon dioxide reacts with calcium hydroxide in limewater to make a milky white precipitate of calcium carbonate.

The test for carbonate ions.

Questions

3 What would you expect to see if you added a solution of barium nitrate to an acidified solution of sodium sulfate?

4 Which tests would you carry out to distinguish between solutions of sodium chloride and sodium iodide? What results would you expect?

5 Write ionic equations for the following:
  a Cu^{2+} ions reacting with OH^- ions to make a precipitate of $Cu(OH)_2$
  b Ag^+ ions reacting with acidified I^- ions to make a precipitate of AgI
  c Barium nitrate reacting with potassium sulfate.

6 Solution X gives a green precipitate when sodium hydroxide solution is added, and a white precipitate when acidified barium nitrate solution is added. What salt is dissolved in solution X?

7 Solution Y gives a white precipitate when sodium hydroxide solution is added, which dissolves in excess sodium hydroxide. Solution Y gives a cream precipitate when acidified and a solution of silver nitrate is added. What salt does Y contain?

Find out about

- the Earth's crust
- ionic bonding
- giant ionic structures

Granite is a mixture of minerals: glassy grains of quartz, black crystals of mica, and large, pink or white crystals of feldspar.

Key words

- rock
- mineral
- ore
- abundant
- ionic compound
- ionic bonding
- giant ionic lattice

Questions

1. Write a sentence that makes clear the differences between the words *rock*, *mineral*, and *lithosphere*.
2. Draw up a table to compare the properties of a molecular compound such as water and an ionic compound such as sodium chloride.

Rocks and minerals

The lithosphere is made of **rocks**. Rocks can be big (mountains), medium-sized (boulders), or small (stones and pebbles).

Rocks may contain one or more **minerals**. Minerals are naturally occurring chemicals; they may be elements or compounds.

Limestone rock is made mainly from the mineral calcite, with the chemical formula $CaCO_3$. Sandstone rock is made mainly from the mineral quartz, with the chemical formula SiO_2. Both calcite and quartz are compounds. Granite rock is a mixture of several minerals; quartz, feldspar, and mica, all of which are compounds.

Gold is found naturally on its own in the lithosphere. So a gold nugget is a rock made from one mineral that is an element.

A metal **ore** (see Section G) is a rock that contains minerals from which a metal can be extracted economically.

Haematite, Fe_3O_4. Crystals of this mineral range from metallic black to dull red.

Calcite, $CaCO_3$.

The non-metals oxygen and silicon are the two most **abundant,** or common, elements in the lithosphere. They form silica and silicate minerals, which make up 95% of the Earth's continental crust. The feldspars are silicates, which also contain ions of metals including aluminium, the third most abundant element in the lithosphere.

Quartz is a form of silica, with silicon and oxygen atoms joined by covalent bonds. Haematite and calcite are examples of minerals made up of ions.

Evaporite minerals

Seawater contains abundant dissolved chemicals. When the water evaporates, **ionic compounds** crystallise. Sodium chloride, NaCl, or rock salt, is a common example. The mineral is halite.

Minerals that are formed in this way are called evaporites. Roughly 200 million years ago, in the Triassic era, vast salt deposits were laid down as seawater evaporated off the northwest coast of England. The salt was later covered with other sediments and is now under the county of Cheshire in England. People have been extracting and trading this salt since before Roman times.

The Great Basin: a hot, dry area in Western USA. Salty water evaporates quickly, leaving salt flats.

The structure and properties of salts

Crystals of sodium chloride are cubes. They are made up of sodium and chloride ions.

In every crystal of sodium chloride, the ions are arranged in the same regular pattern. All crystals of the compound have the same cubic shape.

As a sodium chloride crystal forms, millions of Na^+ ions and millions of Cl^- ions pack closely together. The ions are held together very strongly by the attraction between their opposite charges. This is called **ionic bonding**, and the structure is called a **giant ionic lattice**. Unlike compounds such as water, which is made up of individual molecules of H_2O, there is *not* an individual NaCl molecule.

Because of the very strong attractive forces, it takes a lot of energy to break down the regular arrangement of ions. So NaCl has to be heated to 801°C before it melts, and to 1413°C before it boils. Compare this with melting ice (melting point 0°C) and boiling water (boiling point 100°C), where you only need to supply sufficient energy to overcome the relatively weak attractive forces *between* the H_2O molecules.

Sodium and chloride ions are charged particles; however, solid ionic compounds do not conduct electricity because the ions are not free to move. When an ionic compound is molten or dissolved in water, it can conduct electricity because its ions separate and can move around.

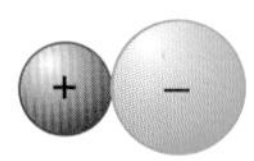

1 The Na^+ attracts other Cl^- ions that are close to it. The Cl^- attracts other Na^+ ions that are close to it.

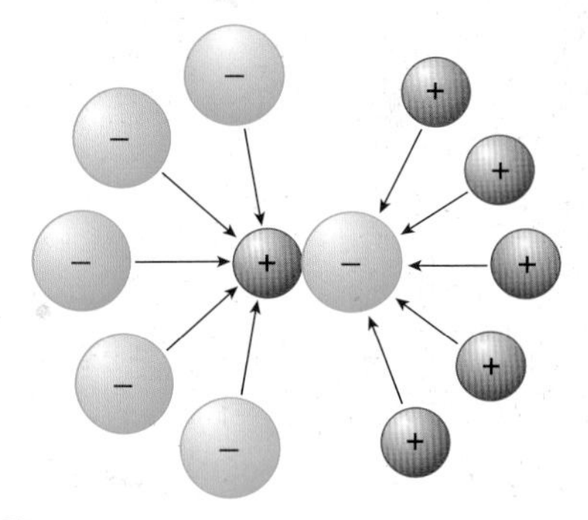

2 Another five Cl^- ions can fit around the Na^+, making six in total, and another five Na^+ ions can fit around the Cl^- ion, making six in total.

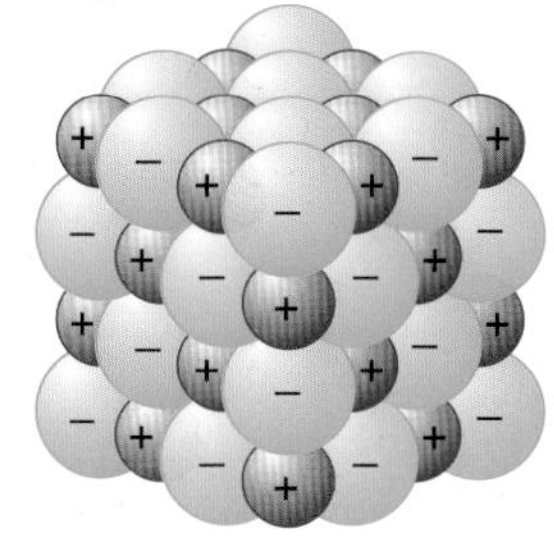

3 Each of these ions then attracts other ions of the opposite charge, and the process continues until millions and millions of oppositely charged ions are all packed closely together.

How a sodium chloride crystal forms.

Carbon minerals – hard and soft

Find out about

- diamond, graphite, and quartz
- giant covalent structures

Diamonds are transparent, and are cut so they reflect light and sparkle.

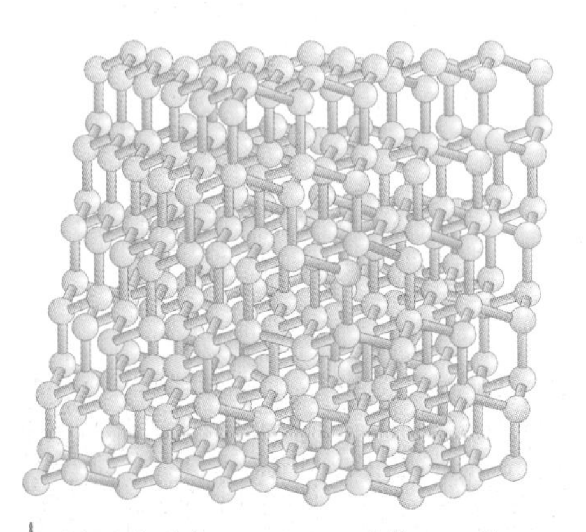

Model of the structure of diamond. Each carbon atom inside the structure is covalently bonded to four others in a huge 3D arrangement.

Diamond and graphite – non-identical twins

If you heat a **diamond** strongly, it will eventually burn and make carbon dioxide. **Graphite** will do the same. This is because both these minerals are made from carbon and nothing else. They both burn, but in many ways they could not be more different. How the carbon atoms are arranged and stuck together in diamond and in graphite is very different, but is easy to picture. Diamond and graphite are good examples of how the behaviour of a solid depends on its structure and bonding.

Geologists think diamonds formed about 3 billion years ago, at very high temperatures and pressures, in the Earth's mantle. Volcanic eruptions forced them up to the surface. Graphite was made in the Earth's crust, where the decayed remains of living things were squashed in metamorphic rocks. The two different sets of conditions have produced two very different minerals.

Diamond and graphite head-to-head

Property	Diamond	Graphite
hardness (1 = softest, 10 = hardest)	10	1–2
melting point (°C)	3560	3650
boiling point (°C)	4830	4830
solubility in water	insoluble	insoluble
electrical conductivity	low	high

Some properties of diamond and graphite.

Carbon atoms in diamond and graphite

In diamond, covalent bonds join each carbon atom to its four nearest neighbours. The four bonds are evenly spaced in *three dimensions* around each carbon atom, making pyramid (tetrahedron) shapes. The arrangement is called a **giant covalent structure**, as it repeats over and over again until you have billions and billions of atoms covalently bonded together.

Graphite has a giant covalent structure, too. Covalent bonds join each carbon atom to its three nearest neighbours. The three bonds are evenly spaced in *two dimensions*, making flat sheets of hexagons that go on and on to include billions and billions of atoms. Each atom has one outer-shell electron that has not been used to make a covalent bond. These electrons drift around freely in the gaps between the layers of atoms. They help stick the layers together, but only weakly.

Different structures – different properties – different uses

Diamond and graphite both have high melting and boiling points. This is because many strong covalent bonds would need to be broken for them to melt. They are also both insoluble in water. This is because of their many strong covalent bonds, and also because, unlike ionic compounds, they are not made up of ions. Ionic compounds dissolve because the ions are attracted to the water molecules.

Strong covalent bonds in three dimensions make diamond the hardest known material. Drills used in mining have bits with diamond tips.

In graphite, weak forces between the layers make it easy for them to slide over one another. This slipperiness makes graphite useful as a lubricant. The electrons drifting between the layers are free to move and take their charge with them. This allows graphite to conduct electricity – this is very unusual for a giant covalent structure. In contrast, diamond has no charged particles that are free to move, so it is an insulator.

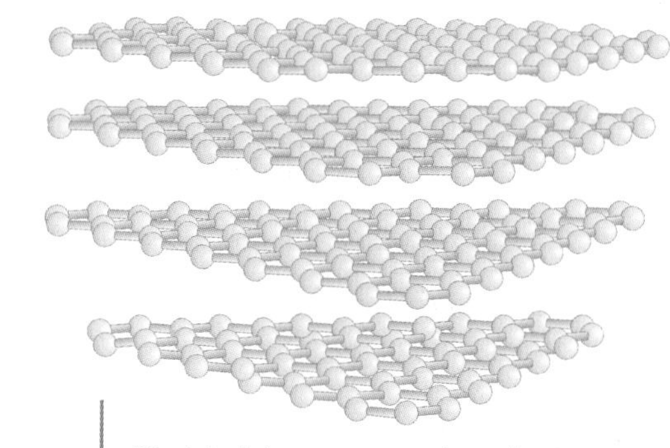

Model of the structure of graphite. Each carbon atom inside the structure is covalently bonded to three others, making flat sheets of hexagons with weak forces between them.

A diamond-tipped drill. The extreme hardness of diamond allows it to cut through anything.

Quartz – a diamond-like mineral

Sandstone is mostly made up of the mineral quartz. Quartz is a common crystalline form of silicon dioxide (SiO_2). It has a giant covalent structure similar to diamond. The strong bonds between the silicon and oxygen atoms make it a strong and rigid mineral. It, too, is hard, so it is suitable for use as an abrasive in sandpaper, for example. Sandstone is a traditional building material – its insoluble quartz helps it to stay weatherproof.

Pure quartz crystals are transparent and very hard.

Pencil lead is made of graphite and clay. As you slide the pencil across the paper, layers of graphite flake off and leave a mark.

Questions

1. Look back at C4 Section J. Draw a table to compare the properties of giant ionic substances with those of giant covalent substances. You should include melting and boiling points, solubility, and electrical conductivity.
2. Explain why graphite is soft but diamond is hard.

Key words

- graphite
- diamond
- giant covalent structure

Metals from the lithosphere

Find out about

- metal ores
- extracting metals
- chemical quantities
- electrolysis

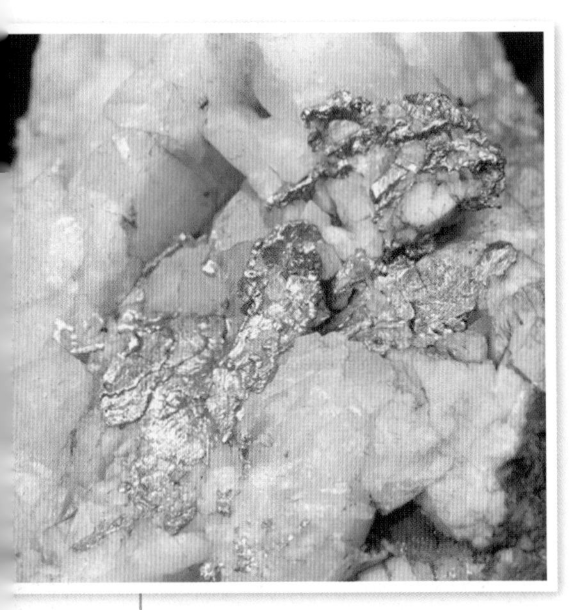

Gold is so unreactive that it occurs uncombined in the lithosphere. Most metals occur as compounds.

An open-pit copper mine in Utah, USA. Mining on this scale has a big impact on the environment.

Metal ores

The wealth of societies has often depended on their ability to extract and use metals. Mining and quarrying for metal ores takes place on a large scale and can have a major impact on the environment.

All metals come from the lithosphere, but most metals are too reactive to exist on their own in the ground. Instead, they exist combined with other elements as compounds. Like other compounds found in the lithosphere, they are called minerals.

Rocks that contain useful minerals are called ores. The valuable minerals are very often the **oxides** or sulfides of metals.

Metal	Name of the ore	Chemical in the mineral
aluminium	bauxite	aluminium oxide, Al_2O_3
copper	copper pyrite	copper iron sulfide, $CuFeS_2$
gold	gold	gold, Au
iron	haematite	iron oxide, Fe_2O_3
sodium	rock salt	sodium chloride, NaCl

Because it occurs in an uncombined state, gold has been used by humans for more than 5000 years. More **reactive metals** like iron were not used by humans until methods for **extracting** them from their ores had been developed. Extraction methods include **reduction** reactions that separate the metal from minerals where it is combined with other elements.

Mineral processing

Over hundreds of millions of years, rich deposits of ores have built up in certain parts of the Earth's crust. But even the richest deposits do not contain pure mineral. The valuable mineral is mixed with lots of useless dirt and rock, which have to be separated off as much as possible. This is called concentrating the ore.

Some ores are already fairly concentrated when they are dug up – iron ore is often over 85% pure Fe_2O_3. But other ores are much less concentrated – copper ore usually contains less than 1% of the pure copper mineral.

Extracting metals: some of the issues

There is a range of factors to weigh up when thinking about the method for extracting a metal.

- **How can the ore be reduced?**
  The more reactive the metal, the harder it is to reduce its ore. The table on the right compares the methods used to reduce different ores.
- **Is there a good supply of ore?**
  Metals ores are mined in different parts of the world. If ore is not very pure, it may not be worth using – the cost of concentrating the ore may be too great. The more valuable the metal, the lower the quality of ore that can be used.
- **What are the energy costs?**
  It takes energy to extract metals, as well as a good supply of ore. This is especially true if the metal is extracted by electrolysis. For example, a quarter of the cost of making aluminium is the cost of electricity.
- **What is the impact on the environment?**
  Metals like iron and aluminium are produced on a huge scale. Millions of tonnes of ore are needed. Mining this ore can have a big environmental impact. This is why it is important to recycle metals. It takes about 250 kg of copper ore to make 1 kg of copper. So recycling 1 kg of copper means that 250 kg of ore need not be dug up.

	Metal	Method
MORE REACTIVE ↑	potassium sodium calcium magnesium aluminium	electrolysis of molten ores
	zinc iron tin lead copper	reduction of ores using carbon
LESS REACTIVE	silver gold	metals occur uncombined

Hot metal can be poured when molten.

Question

1 Suggest explanations for these facts:
   a The Romans used copper, iron, and gold, but not aluminium.
   b Iron is cheap compared with many other metals.
   c Gold is expensive, even though it is found uncombined in nature.
   d About half the iron we use is recycled, but nearly all the gold is recycled.
   e The tin mines in Cornwall have closed, even though there is still some tin ore left in the ground.

Key words

- ✓ **oxides**
- ✓ **reactive metals**
- ✓ **extracting (a metal)**
- ✓ **reduction**

Key words

- **reducing agent**
- **oxidised**

Extracting metals from ores

Zinc is a metal that can be extracted from its oxide. Zinc is found in the lithosphere as ZnS, called zinc blende. This can be easily turned to ZnO by heating it in air.

The task is then to remove the oxygen from the zinc oxide, to convert ZnO to Zn. Removing oxygen in this way is called reduction. The process needs a **reducing agent** that will remove oxygen. In this case, the reducing agent is carbon. In removing oxygen, the reducing agent is **oxidised**.

zinc oxide + carbon $\longrightarrow$ zinc + carbon monoxide

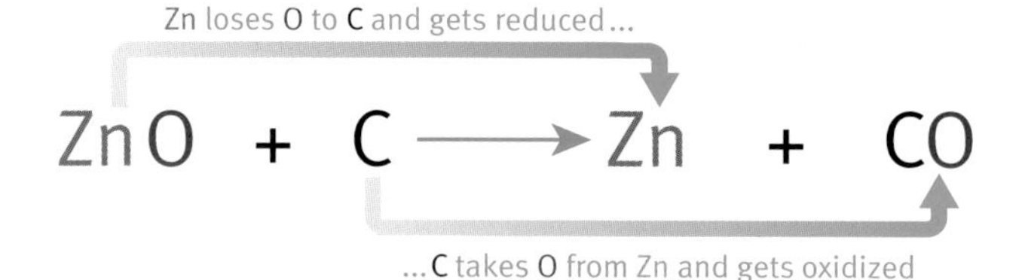

Reducing zinc oxide to zinc using carbon.

Further oxidation of the carbon forms carbon dioxide. Carbon is often used as a reducing agent to extract metals. Carbon, in the form of coke, can be made cheaply from coal. At high temperatures, carbon has a strong tendency to react with oxygen, so it is a good reducing agent. What is more, the carbon monoxide formed is a gas, so it is not left behind to make the zinc impure.

Carbon is also used to extract iron and copper. These reactions can be summarised as:
iron oxide + carbon $\longrightarrow$ iron + carbon dioxide
copper oxide + carbon $\longrightarrow$ copper + carbon dioxide

How much metal?

Chemists often ask the 'How much?' question. It's useful to know, say, how much iron it's possible to get from 100 kg of pure iron ore, Fe_2O_3.

Questions

2 a Write an equation for the reaction of zinc sulfide with oxygen to make zinc oxide and sulfur dioxide.
  b What problems might arise from the formation of sulfur dioxide on a large scale?
  c What might be done to deal with this problem?

3 Why do oxidation and reduction always go together when carbon extracts a metal from a metal oxide?

Relative atomic masses

Chemists need to know the relative masses of the atoms involved to answer questions such as 'How much iron, Fe, could you get from 100 kg of iron oxide, Fe_2O_3?'

Atoms are far too small to weigh directly. For example, it would take around a million million million million hydrogen atoms to make one gram.

Instead of working in grams, chemists find the mass of atoms relative to one another. Chemists can do this using an instrument called a mass spectrometer.

Values for the **relative atomic masses** of elements are shown in the periodic table on page 47. The relative mass of the lightest atom, hydrogen, is 1.

One Mg atom weighs twice as much as one C atom.

Key words

- **relative atomic mass**
- **relative formula mass**

Formula masses

If you know the formula of a compound, you can work out its **relative formula mass** by adding up the relative atomic masses of all the atoms in the formula:

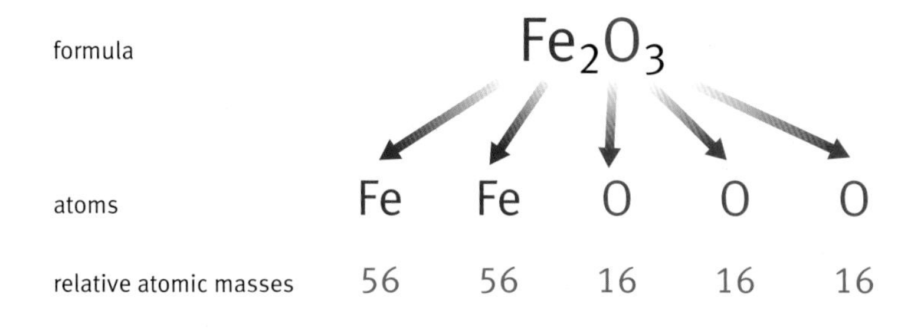

Finding the formula mass of Fe_2O_3.

Worked example

How much Fe could you get from 100 kg of Fe_2O_3?

Fe_2O_3 has a relative formula mass of **160**.

In this formula, there are two atoms of Fe.
2Fe has relative mass $2 \times 56 =$ **112**.

This means that, in 160 kg of Fe_2O_3, there must be 112 kg of Fe.

So 1 kg of Fe_2O_3 would contain 112/160 kg of Fe.

So 100 kg of Fe_2O_3 would contain 100 kg × 112/160 of Fe = 70 kg.

Another way of saying this is that the percentage of Fe in Fe_2O_3 is 70%.

Questions

Look up relative atomic masses in the periodic table.

4 What is the relative formula mass of carbon dioxide?

5 What mass of Al could be made from 1 tonne of Al_2O_3?

6 What mass of Na could be made from 2 tonnes of NaCl?

7 The main ore of chromium is $FeCr_2O_4$. What is the mass of Cr in 100 kg of $FeCr_2O_4$?

8 1000 tonnes of copper ore are dug out of the ground. Only 1% of this is the pure mineral, $CuFeS_2$. What mass of the Cu could be made from 1000 tonnes of the ore?

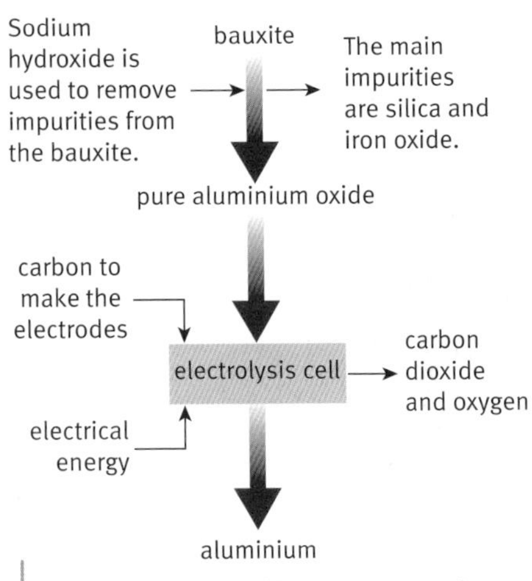

The processing of bauxite to produce aluminium.

Extracting aluminium

Some reactive metals, such as aluminium, hold on to oxygen so strongly that they cannot be extracted using carbon as a reducing agent. To extract these metals, the industry has to use **electrolysis**. Electrolysis is a process that uses an electric current to split a chemical up into its elements.

Aluminium is the most abundant metal in the lithosphere. Much of the metal is in aluminosilicates. It is very hard to separate the aluminium from these minerals.

The main ore of aluminium is bauxite. This consists mainly of aluminium oxide, Al_2O_3. There is some iron oxide in bauxite, which has to be removed before extraction of aluminium.

The diagram below shows the equipment used to extract aluminium by electrolysis. The process takes place in steel tanks lined with carbon. The carbon lining is the negative **electrode** and conducts electricity.

The **electrolyte** is hot, molten Al_2O_3, which contains Al^{3+} and O^{2-} ions.

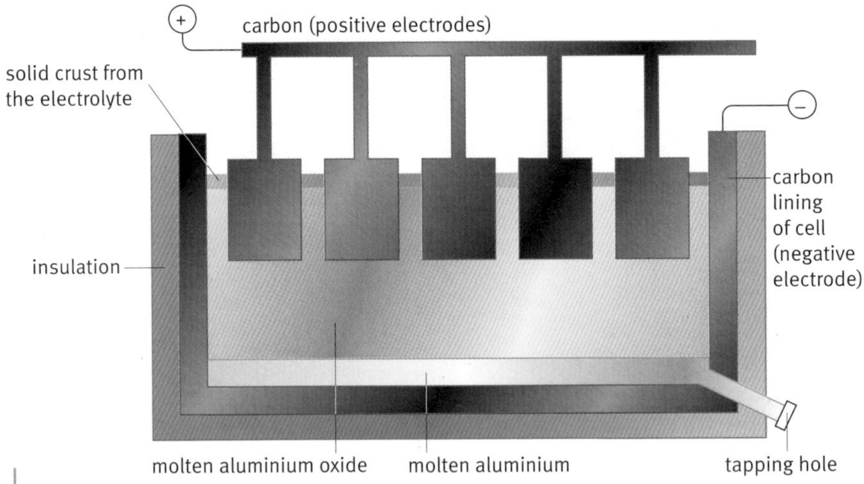

Equipment for extracting aluminium from its oxide by electrolysis.

When an ionic compound, such as Al_2O_3, melts, the ions become free to move around independently, and an electric current can pass through it. The electric current is used to decompose the electrolyte.

Aluminium forms at the negative electrode lining the tank. Because it is very hot in the tank, the aluminium is a liquid and forms a pool of molten metal at the bottom of the tank.

The positive electrodes are blocks of carbon dipping into the molten aluminium oxide. Oxygen forms at the positive electrodes. Some of this combines with the carbon to make carbon dioxide.

Key words

- electrolysis
- electrode
- electrolyte

Ions into atoms and molecules

Electrolysis turns ions back into atoms. Metal ions are positively charged, so they are attracted to the negative electrode. It is a flow of electrons from the power supply into this electrode that makes it negative.

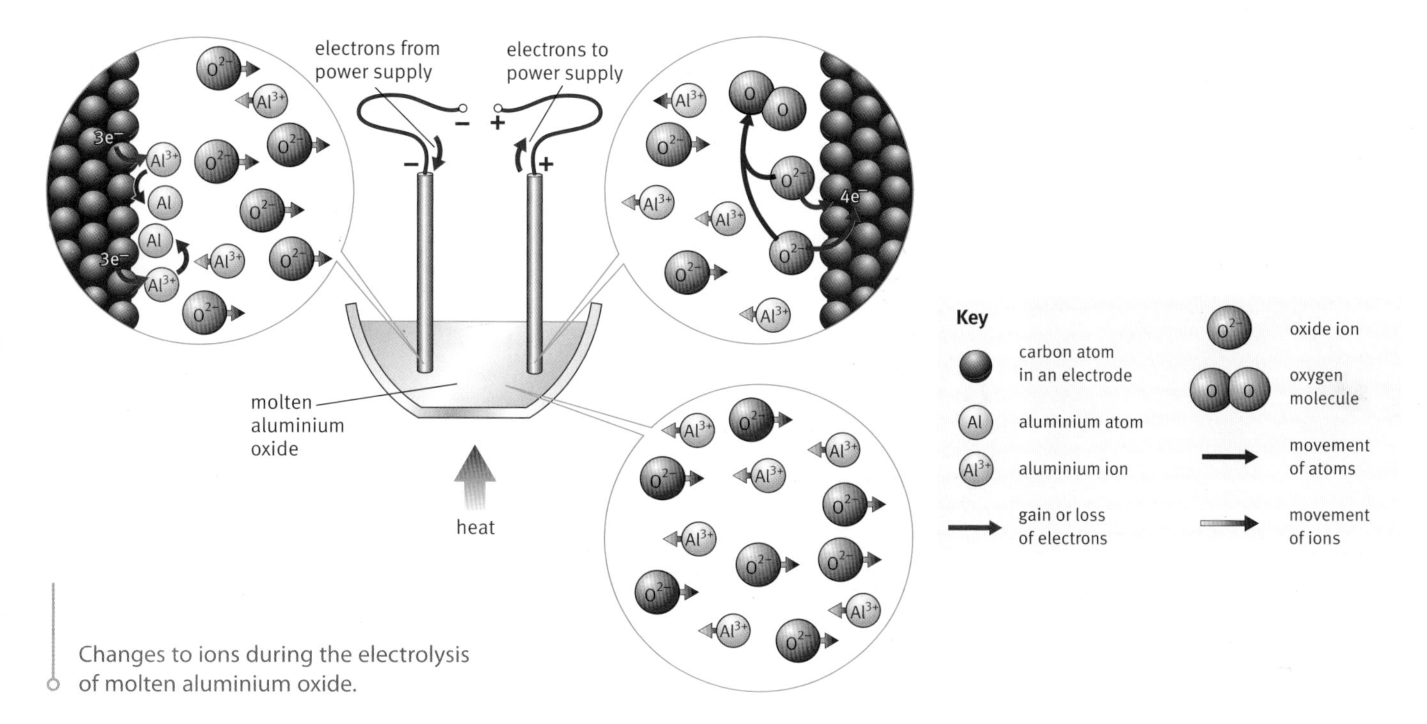

Changes to ions during the electrolysis of molten aluminium oxide.

Positive metal ions gain electrons from the negative electrode and turn back into atoms. During the electrolysis of molten aluminium oxide, the aluminium ions turn into aluminium atoms:

$$Al^{3+} + 3e^- \longrightarrow Al$$

ion + electrons supplied by the negative electrode ⟶ atom

Non-metal ions are negatively charged, so they are attracted to the positive electrode. This electrode is positive because the power supply pulls electrons away from it.

Negative ions give up electrons to the positive electrode and turn back into atoms. During the electrolysis of molten aluminium oxide, the oxide ions turn into oxygen atoms, which pair up to make oxygen molecules:

$$O^{2-} \longrightarrow O + 2e^-$$

ion ⟶ atom + electrons removed by the positive electrode

$$O + O \longrightarrow O_2$$

atom + atom ⟶ molecule

Questions

9 Draw a diagram to show the electron arrangements in:
  a an aluminium atom
  b an aluminium ion.

10 Sodium is extracted by the electrolysis of molten sodium chloride.
  a What are the two products of the process?
  b Use words and symbols to describe the changes at the electrodes during this process.

Structure and bonding in metals

Find out about

- the properties of metals
- the structure of metals
- bonding in metals

Metal properties

Metals have been part of human history for thousands of years. Our lives still depend on metals, despite the development of new materials, including all the different plastics. The varied uses of metals reflect their properties.

Technologists have learnt to use new metals and mixtures of metals called **alloys** so that, as well as steel and aluminium, other metals such as titanium and magnesium can now be used in engineering.

Most metals have high melting points.

Many metals are strong. The titanium hull of this research submarine is strong enough to withstand the pressure at a depth of 6 km. Titanium is also used to make hip joints and racing cars.

Metals can be bent or pressed into shape. They bend without breaking. They are malleable. Aluminium sheet can be moulded under pressure to make cans.

Metals conduct electricity. Copper and aluminium are commonly used as conductors.

Metallic structures

Data can be collected about the properties of a metal, for example, the melting point or tensile strength. Most metals have high melting points and are strong, but to explain *why* they have these properties, materials scientists need to know something about their structure.

Scientists use models to describe what they have discovered about the structures of metals. The models show that metals are made up of atoms that are:

- tiny spheres
- arranged in a regular pattern
- packed closely together in each crystal as a giant structure

This model was thought up creatively to explain the properties of metals. A model is a good one if it explains all the data.

Key words

- metallic bonding
- alloys

The diagram on the right shows the arrangement of atoms in copper, a typical metal. You can see how closely together the atoms of copper are packed – as close together as it is possible to be. Every atom inside the structure has 12 others touching it, the maximum number possible. The atoms are held to each other by strong metallic bonding. Because the bonding is strong, copper is strong and difficult to melt.

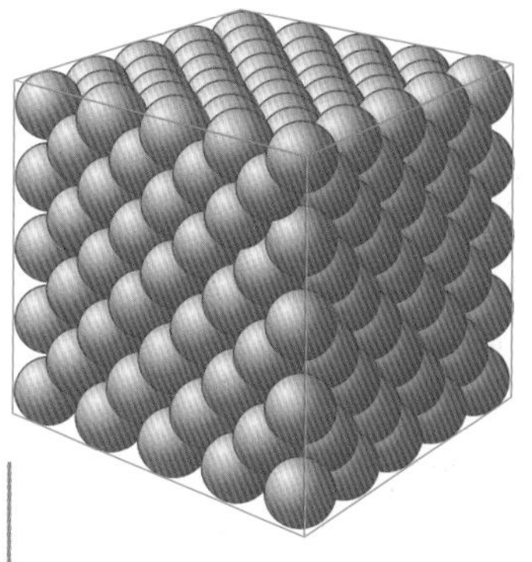

The arrangement of atoms in copper. Because the metallic bonding is strong, copper is strong and difficult to melt. Because the bonding is flexible, copper is malleable – the atoms can be moved around without shattering the structure.

Metallic bonding

Metals have a special kind of bonding – not ionic, nor covalent, but metallic. **Metallic bonding** is strong, but flexible, allowing the atoms to slide into new positions.

Metal atoms tend to share the electrons in their outer shell easily. In the solid metal, the atoms lose these shared electrons and become positively charged. The electrons, no longer held by the atomic nuclei, drift freely between the metal atoms. The attraction between the 'sea' of negative electrons and the positively charged metal atoms holds the structure together.

Overall, a metal crystal is not charged. This is because the total negative charge on the electrons balances the total positive charge on the metal atoms. The charged atoms are often described as metal ions but in their chemical reactions metals behave as collections of atoms.

The electrons can move freely through the giant structure, which explains why metals conduct electricity well. When an electric current flows through a metal wire, the free electrons drift from one end of the wire towards the other. Although the electrons are free, the metal atoms themselves are packed closely together in a regular lattice.

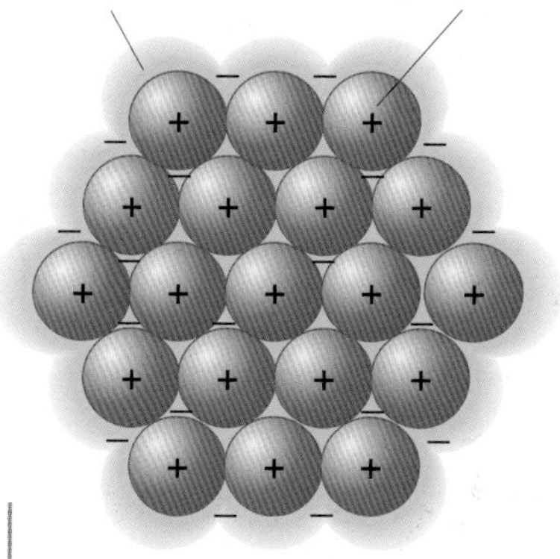

A model of metallic bonding. This model pictures the atoms sharing outer electrons, which hold the charged atoms (metal ions) together.

Questions

1. Give five examples of metals used for their strength. For each metal, give an example of a use that depends on the strength of the metal.
2. Someone looking at a model showing the arrangement of atoms in a copper crystal might think that the following statements are true. Which of these ideas are correct? Which ideas are false? What would you say to put someone right who believed the false ideas?
   a. Copper crystals are shaped like cubes because the atoms are packed in a cubic pattern.
   b. There is air between the atoms in a crystal of copper.
   c. Copper is dense because the atoms are closely packed.
   d. The atoms in a copper crystal are not moving at room temperature.
   e. Copper has a high melting point because the atoms are strongly bonded in a giant structure.
   f. Copper melts when strongly heated because the atoms melt.
3. There are positive metal ions in a metal crystal, but a metal is not an ionic compound. Explain.

I The life cycle of metals

Find out about

- impacts of extracting, using, and disposing of metals

Mining, mineral processing, and metal extraction produce many valuable metal products, but these activities can be hazardous and can also have a serious impact on the environment. There can be a conflict between those who want to build up profitable industries and create jobs and those whose aim is to protect the natural world.

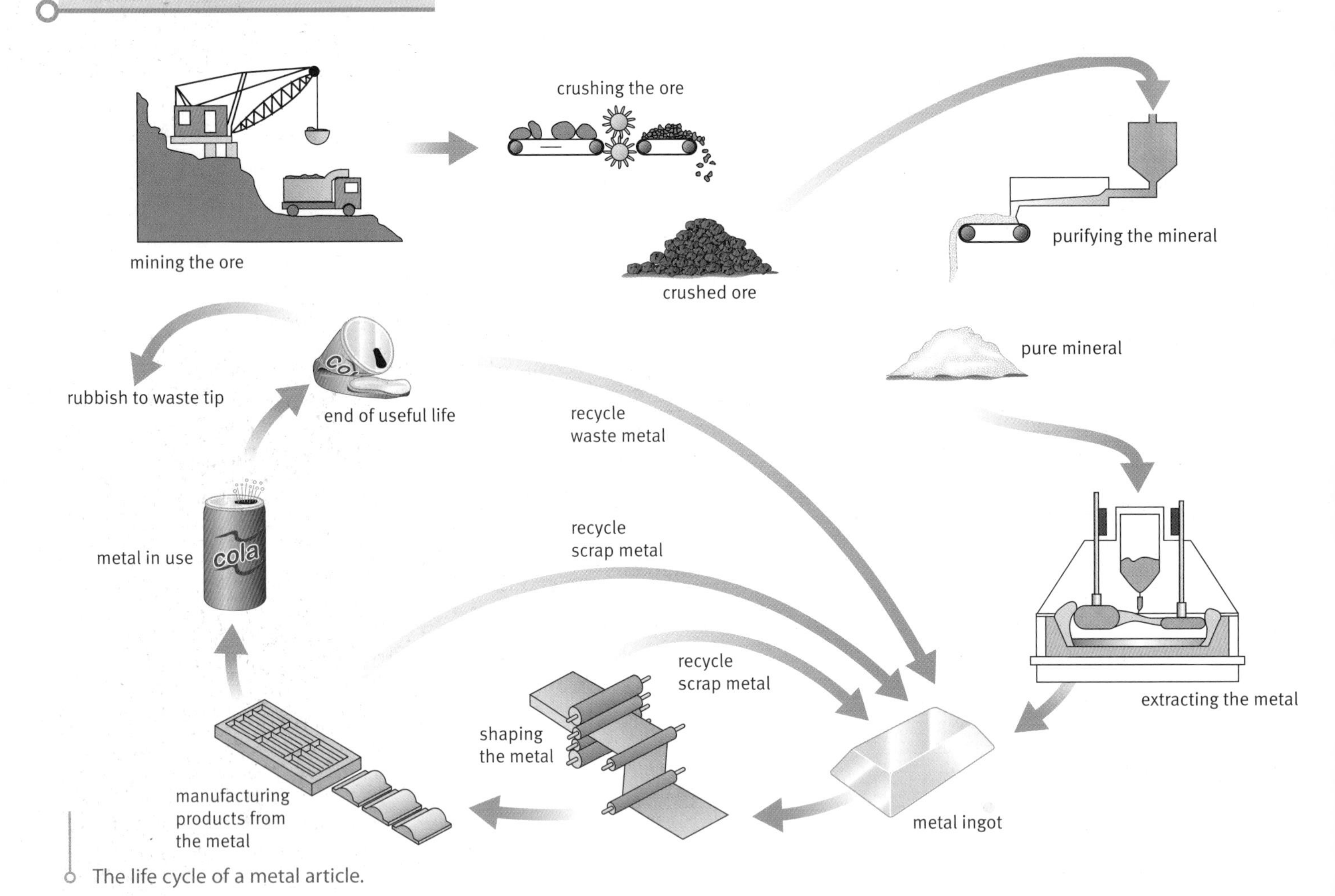

The life cycle of a metal article.

Mining

Mines can be on the surface (open-cast) or underground. Both types of mine produce large volumes of waste rock and can leave very large holes in the ground. Miners use explosives to blast the rock. This is noisy and produces dust.

Miners working in underground mines are at risk from dust, heat, and the possibility of the mine caving in. Modern mines have ventilation shafts and fans to provide fresh air and sensors to monitor the air quality. Walls and ceilings are braced with strong boards to prevent cave-ins.

Mining is much safer than it has been in the past but accidents still happen it is impossible to make mining completely safe, but most countries now have regulations to ensure that mining companies reduce risks.

Processing ores

Many metal ores are high value but low grade. The ore in an open-cast copper mine may contain as little as 0.4% of the metal and still be profitable. This means that 99.6% of the rock dug from the ground becomes waste. Near any mine there are waste tips, which might be large ponds. These can be hazardous if they contain traces of toxic metals such as lead or mercury.

Metal extraction

All the stages of metal extraction and metal fabrication need energy, use large volumes of water, and give off air pollutants. Higher expectations from society and tighter regulation mean that industries have to do more to prevent harmful chemicals escaping into the environment. Economic pressures favour the development of equipment and procedures that minimise the use of energy, water, and other resources.

Metals in use

Careful choice of metals can reduce the environmental impact of our life style. In transport, for example, lighter cars, trucks, and trains mean less fuel consumption and fewer emissions, as well as less wear and tear on roads and tracks. Vehicles can be designed to be lighter by replacing steel with lighter metals, such as aluminium, or with plastics.

Recycling

Recycling is well established in the metal industries. Scrap metal from all stages of production is routinely recycled. Much metal is also recycled at the end of the useful life of metal products.

Recycled steel is as good as new after reprocessing. The scrap is fed to a furnace and melted with fresh metal to make new steel. For every tonne of steel recycled, there is a saving of 1.5 tonnes of iron ore and half a tonne of coal. There is also a big reduction in the total volume of water needed, since large quantities of water are used in mineral processing.

Recycling aluminium is particularly cost-effective because so much energy is needed to extract the metal from its oxide by electrolysis. Recycling also reduces the impact on the environment by cutting the use of raw materials, and the associated mining and processing.

A large pond in Jamaica used to contain the waste from a bauxite mine. Bauxite is impure aluminium oxide. The main impurity is iron oxide, which ends up in the rusty-looking waste.

Questions

1. Why are recycling rates for metal waste from manufacturing higher than recycling rates of metal after use?
2. Draw up a table to list three groups of people affected by metal mining, production, and use. Show the benefits and costs to each group.
3. The price of metals can go up and down. A mining company is considering opening a new copper mine. The price of copper has just fallen – how might this affect the company's decision? Explain your answer.
4. Known bauxite reserves will last for hundreds of years. The aluminium industry claims this makes aluminium a sustainable material. Do you agree?

Science Explanations

Theories of structure and bonding can help to explain the physical properties and chemical reactions of the chemicals we find in the atmosphere, hydrosphere, and lithosphere.

You should know:

- that the elements and compounds in the Earth's atmosphere consist of small molecules
- that the Earth's hydrosphere consists mainly of water with some ionic salts in solution
- that silicon, oxygen, and aluminium are very abundant in the Earth's lithosphere
- why chemicals that are molecular are often liquids or gases at room temperature, and do not conduct electricity
- why ionic compounds have high melting and boiling points and conduct electricity when molten or dissolved in water, but not when solid
- that the formulae of salts show that the total charge on the metal positive ions is balanced by the total charge on the negative ions
- how chemists use precipitation reactions to detect which ions are present in an ionic compound
- how to use ionic equations to describe precipitation reactions
- why chemicals such as silicon dioxide and diamond have very high melting points, do not dissolve in water, and do not conduct electricity
- why metals are strong and conduct electricity
- why the methods used to extract metals from their ores are related to their reactivity
- how to calculate formula masses and reacting masses based on equations, to find the mass of the metal that can be extracted from a metal compound
- why electrolysis turns ions back into atoms and splits ionic compounds into their elements
- how to use equations to show what happens to ions when electrolysis is used to extract metals.

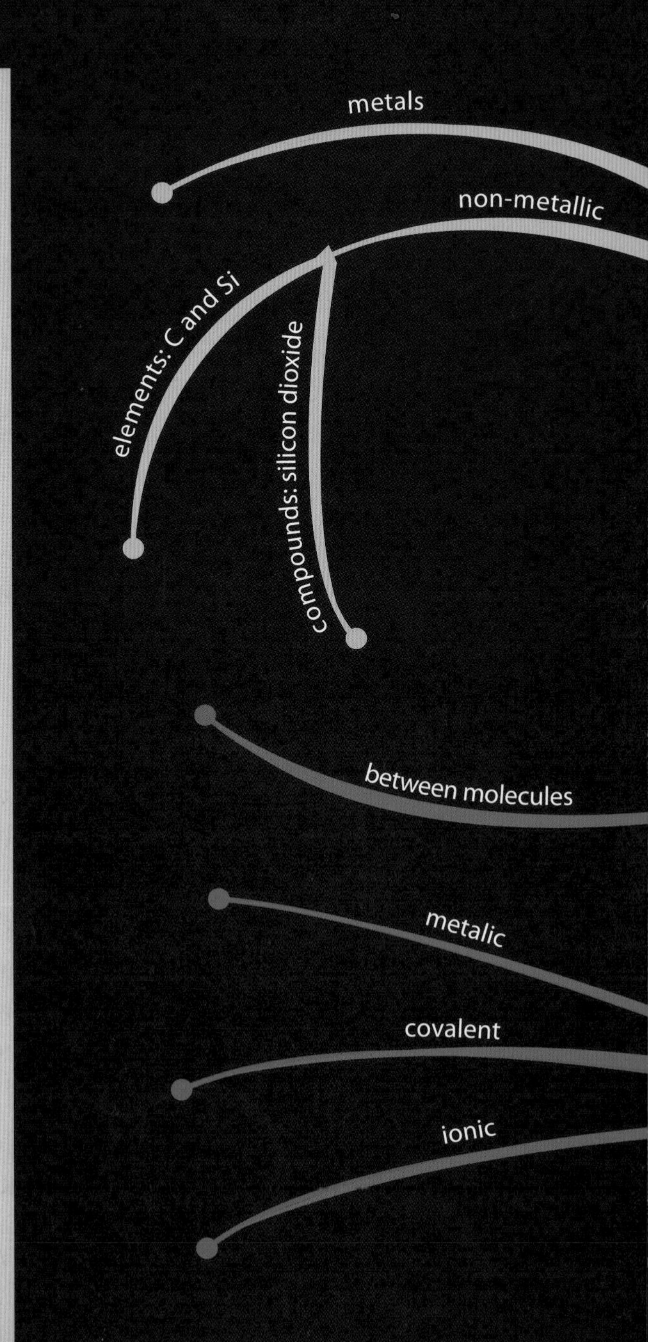

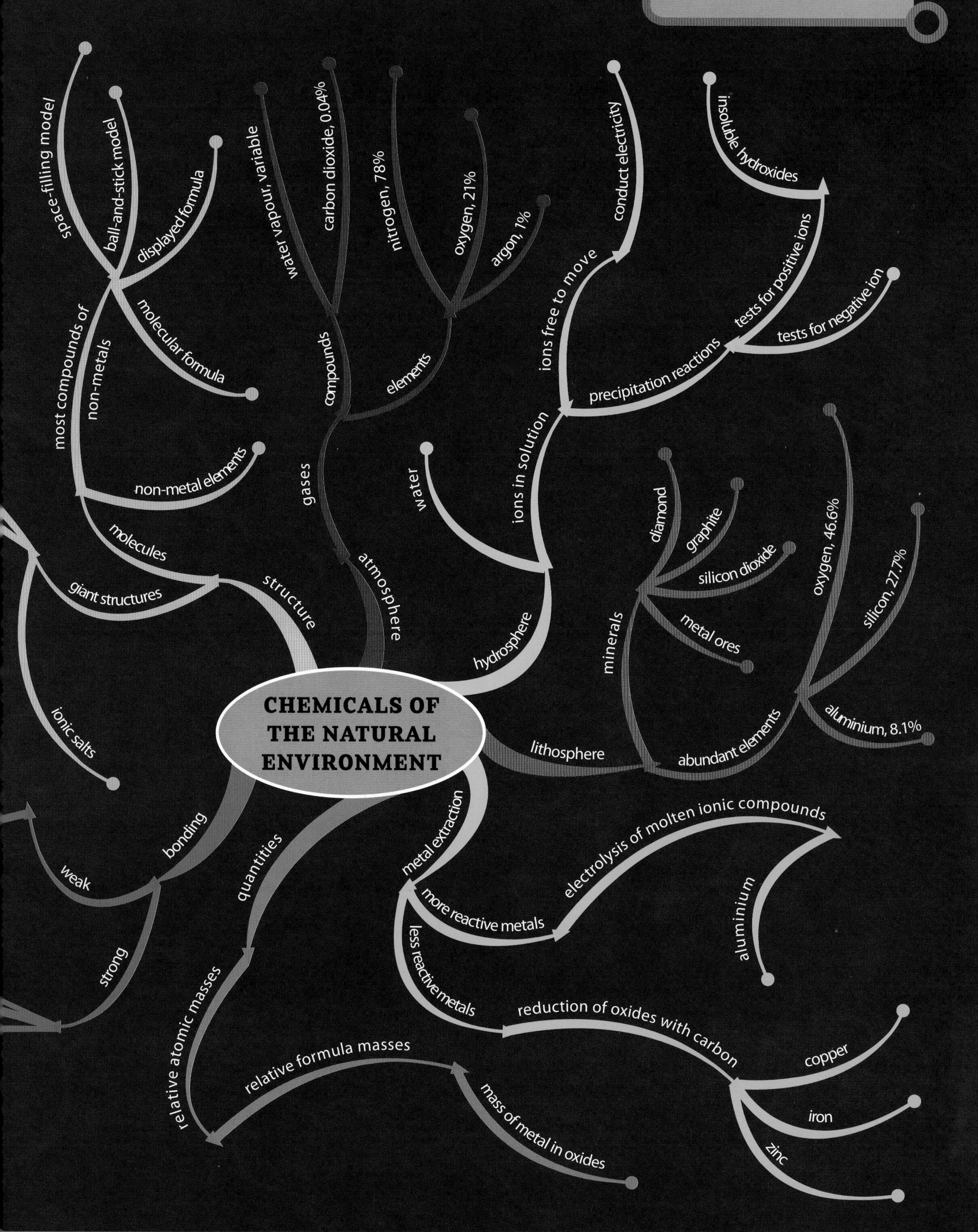

CHEMICALS OF THE NATURAL ENVIRONMENT
structure
molecules
giant structures
most compounds of non-metals
non-metal elements
space-filling model
ball-and-stick model
displayed formula
molecular formula
ionic salts
atmosphere
gases
compounds
elements
water vapour, variable
carbon dioxide, 0.04%
nitrogen, 78%
oxygen, 21%
argon, 1%
hydrosphere
water
ions in solution
ions free to move
conduct electricity
precipitation reactions
tests for positive ions
insoluble hydroxides
tests for negative ion
lithosphere
minerals
diamond
graphite
silicon dioxide
metal ores
abundant elements
oxygen, 46.6%
silicon, 27.7%
aluminium, 8.1%
bonding
weak
strong
quantities
relative atomic masses
relative formula masses
mass of metal in oxides
metal extraction
more reactive metals
electrolysis of molten ionic compounds
aluminium
less reactive metals
reduction of oxides with carbon
copper
iron
zinc

Ideas about Science

Scientific explanations are based on data but they go beyond the data and are distinct from them. An explanation has to be thought up creatively to account for the data. In the context of the theories of structure and bonding you should be able to:

- give an account of scientific work and distinguish statements that report data from statements of explanatory ideas (hypotheses, explanations, and theories)
- recognise data, such as measures of the properties of elements and compounds, that is accounted for by explanations based on theories of structure and bonding.

New technologies and processes based on scientific advances sometimes introduce new risks. Some people are worried about the effects arising from the extraction and use of metals.

You should be able to:

- explain why nothing is completely safe
- identify examples of risks that arise from mining and metal extraction
- suggest ways of reducing a given risk.

Some applications of science, such as the extraction and use of metals, can have unintended and undesirable impacts on the quality of life or the environment. Benefits need to be weighed against costs. You should be able to:

- identify the groups affected and the main benefits and costs of a course of action for each group
- suggest reasons why different decisions on the same issue might be appropriate in view of differences in social and economic context
- identify and suggest examples of unintended impacts of human activity on the environment, such as the production of large volumes of waste by mining, mineral processing, and metal extraction based on low-grade ores
- explain the idea of sustainability, and apply it to the methods used to obtain, use, recycle, and dispose of metals.

Some forms of scientific work have ethical implications that some people will agree with and others will not. When an ethical issue is involved, you need to be able to:

- state clearly what the issue is
- summarise the different views that people might hold.

When discussing ethical issues, common arguments are that:

- the right decision is the one that leads to the best outcome for the majority of the people involved
- certain actions are right or wrong whatever the consequences.

Review Questions

1 A chemist tests a solution of an impure solid. His results are in the table. Use pages 146 and 147 to help you answer the questions.

Test number	Test	Observation
1	acidify then add silver nitrate solution	white precipitate
2	acidify then add barium nitrate solution	white precipitate
3	add dilute sodium hydroxide solution	light-blue precipitate

a Which test shows that the powder contains copper ions?

b Which test shows that the powder contains chloride ions?

c What other ion was found to be present?

2 Zinc metal can be extracted from its oxide by heating with carbon. The equation for the reaction is:

$ZnO + C \longrightarrow Zn + CO$

a In the equation above:

i name the element that is oxidised

ii name the reducing agent.

b Calculate:

i the mass of zinc oxide needed to make 1 kg of zinc (relative atomic masses: Zn=65, O=16, C=12)

ii the mass of carbon monoxide produced when 1 kg of zinc is made.

c Explain why recycling helps to make the use of zinc metal more sustainable.

3 Aluminium metal is extracted from its oxide, Al_2O_3, by electrolysis.

a Explain why aluminium cannot be extracted from its oxide by heating with carbon.

b Explain why aluminium oxide conducts electricity when molten.

c Write an equation to summarise what happens at the negative electrode during the electrolysis.

4 The table gives some boiling points.

Chemical	Boiling point (°C)	Solubility in water
nitrogen	–210	insoluble
potassium chloride	770	soluble
silicon dioxide (quartz)	1610	insoluble

a Use ideas about structure and bonding to give reasons for the differences in melting point.

b Why is silicon dioxide found in the lithosphere and not in the hydrosphere or the atmosphere?

C6 Chemical synthesis

Why study chemical synthesis?

We use chemicals to preserve food, treat disease, and decorate our homes. Many of these chemicals do not occur naturally: they are synthetic. Developing new products, such as drugs to treat disease, depends on chemists who synthesise and test new chemicals.

What you already know

- Atoms are rearranged during chemical reactions.
- The number of atoms of each element stays the same in a chemical reaction.
- Raw materials can be used to make synthetic materials.
- Alkalis neutralise acids to make salts.
- Chemical reactions can be represented by word equations and balanced symbol equations.
- Some substances are made up of electrically charged particles called ions.
- Data is more reliable if it can be repeated.

Find out about

- the importance of the chemical industry
- a theory to explain acids and alkalis
- reactions that give out and take in energy
- techniques for controlling the rate of chemical change
- the steps involved in the synthesis of a new chemical
- ways to measure the efficiency of chemical synthesis.

The Science

Chemists who synthesise new chemicals need practical skills and an understanding of science explanations. They must control reactions so that they are neither too slow nor too fast. They must calculate how much of the reactants to use to make the amount of product required. Chemists also take into account any energy changes. Acids are important reactants in synthesis. Ionic theory explains the characteristic behaviours of these chemicals.

Ideas about Science

Chemists make sure they use the right grade of chemical for a reaction. Technical chemists test chemicals from suppliers to check the purity. They take measurements and make sure the data they collect is as accurate and reliable as possible. They can then make the best estimate of the true value of the purity.

A The chemical industry

Find out about

- the chemical industry
- bulk and fine chemicals
- the importance of chemical synthesis

The **chemical industry** converts raw materials, such as crude oil, natural gas, minerals, air and water, into useful products. The products include chemicals for use as food additives, fertilisers, pigments, dyes, paints, and pharmaceutical drugs.

The industry makes **bulk chemicals** on a scale of thousands or even millions of tonnes per year. Examples are ammonia, sulfuric acid, sodium hydroxide, chlorine, and ethene.

On a much smaller scale, the industry makes **fine chemicals** such as drugs, herbicides, and pesticides. It also makes small quantities of speciality chemicals needed by other manufacturers for particular purposes. These include such things as flame retardants, food additives, and the liquid crystals for flat-screen televisions and computer displays.

The chemical industry converts raw materials into pure chemicals, which are then used in synthesis to make a wide range of products.

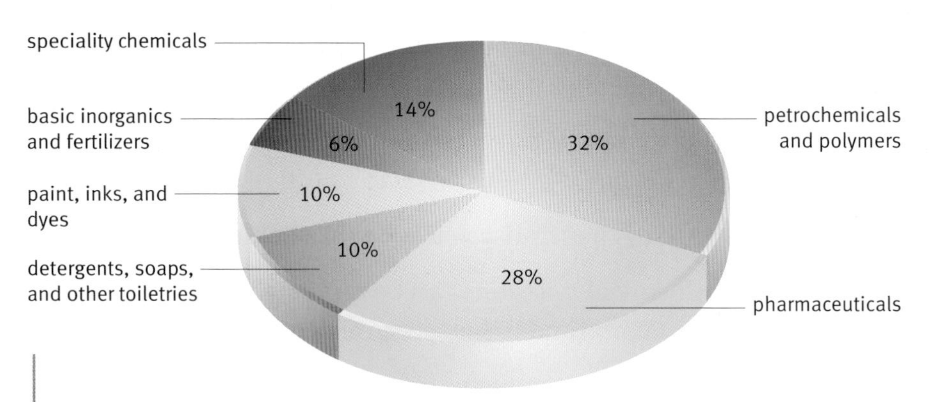

The range of products made by the chemical industry in the UK by value of sales.

The part of a chemical works that produces a chemical is called a **plant**. Some of the chemical reactions occur at a high temperature, so a source of energy is needed. Also, a lot of electrical power is needed for pumps to move reactants and products from one part of the plant to another. Sometimes the energy to produce this power can be supplied by chemical reactions that give out energy.

Sensors monitor the conditions, such as temperature and pressure, in all areas of the plant. The data is fed to computers in the control centre, where the technical team controls the plant.

Industrial chemists work in the plant and the laboratory.

Key words

- chemical industry
- bulk chemicals
- fine chemicals
- plant
- pilot plant
- scale up

People in the chemical industry

People with many different skills are needed in the chemical industry. Research chemists work in laboratories to find new processes and develop new products.

The industry needs new processes so that it can be more competitive and more sustainable. The aim is to use smaller amounts of raw materials and energy while creating less waste.

People devising new products have to work closely with people in the marketing and sales department. They are able to say if the novel product is wanted. If the new product is promising, it may first be tried out by making small amounts of it in a **pilot plant**.

As part of the market research, possible new products are given to customers for trial. At the same time, financial experts estimate the value of the new product in the market. They then compare this with the cost of making the product to check that the new process will be profitable.

Chemical engineers have to **scale up** the process and design a full-scale plant. This can cost hundreds of millions of pounds.

Some chemicals from the industry go directly on sale to the public, but most of them are used to make other products. Transport workers carry the chemicals to the industry's customers.

Every chemical plant needs managers and administrators to control the whole operation. There are also people in service departments who look after the needs of the people working in the plant. These include medical and catering staff, and training and safety officers.

Questions

1. Classify the following as raw materials or products of the chemical industry: air, ammonia, aspirin, water, crude oil, polythene.
2. Give the name and chemical formula of a bulk chemical.
3. List these chemicals under two headings: 'bulk chemical' and 'fine chemical'.
   - the drug aspirin
   - the hydrocarbon ethene
   - the perfume chemical citral
   - the acid sulfuric acid
   - the herbicide glyphosate
   - the alkali sodium hydroxide
   - the food dye carotene

Plant operators monitor the processes from a control room.

Maintenance workers help to keep the plant running.

Find out about

- acids and alkalis
- the pH scale
- reactions of acids

Key words

- acid
- alkali

Acids

The word **acid** sounds dangerous. Nitric, sulfuric, and hydrochloric acids are very dangerous when they are concentrated. You must handle them with great care. These acids are less of a hazard when diluted with water. Dilute hydrochloric acid, for example, does not hurt the skin if you wash it away quickly, but it stings in a cut and rots clothing. It is in fact present in our stomachs where it helps to break down food and kill bacteria. Our stomachs are lined with a protective layer of mucus.

Not all acids are dangerous to life. Many acids are part of life itself. Biochemists have discovered the citric acid cycle. This is a series of reactions in all cells. The cycle harnesses the energy from respiration for movement and growth in living things.

Organic acids

Organic acids are molecular. They are made of groups of atoms. Their molecules consist of carbon, hydrogen, and oxygen atoms. The acidity of these acids arises from the hydrogen in the —COOH group of atoms.

Citric and tartaric acids are examples of solid organic acids. Ethanoic acid is a liquid organic acid.

Acetic acid (chemical name: ethanoic acid) is a liquid. It is the acid in vinegar. Most white vinegar is just a dilute solution of acetic acid. Brown vinegars have other chemicals in the solution that give the vinegar its colour and flavour. Most microorganisms cannot survive in acid, so vinegar is used as a preservative in pickles (E260).

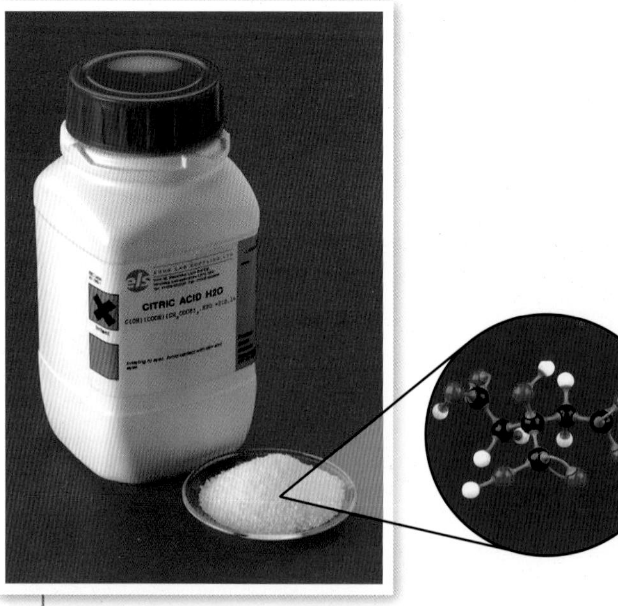

Citric acid (a solid acid) is found in citrus fruits like oranges and lemons. The human body processes about 2 kg of citric acid a day during respiration. Citric acid and its salts are added to food to prevent them reacting with oxygen in the air, and to give a tart taste to drinks and sweets.

Mineral acids

Sulfuric, hydrochloric, and nitric acids come from inorganic or mineral sources. The pure acids are all molecular. Sulfuric acid and nitric acid are liquids at room temperature. Hydrogen chloride is a gas, and becomes hydrochloric acid when it dissolves in water.

Sulfuric acid, H_2SO_4, is manufactured from sulfur, oxygen, and water. The pure, concentrated acid is an oily liquid. The chemical industry in the UK makes about 2 million tonnes of the acid each year. The acid is essential for the manufacture of other chemicals, including detergents, pigments, dyes, plastics, and fertilisers.

Alkalis

Pharmacists sell antacids in tablets to control heartburn and indigestion. The chemicals in these medicines are the chemical opposites of acids. They are designed to neutralise excess acid produced in the stomach – hence the name 'antacids'.

Antacids in medicines are usually insoluble in water. Other chemical antacids are soluble in water and give a solution with a pH above 7. Chemists call them **alkalis**. Common alkalis are sodium hydroxide, NaOH; potassium hydroxide, KOH; and calcium hydroxide, $Ca(OH)_2$.

The traditional name for sodium hydroxide is caustic soda. The word caustic means that the chemical attacks living tissue, including skin. Alkalis can do more damage to delicate tissues than dilute acids. Caustic alkalis are used in the strongest oven and drain cleaners. They have to be used with great care.

Hydrogen chloride forms when concentrated sulfuric acid is added to salt (sodium chloride) crystals. Hydrogen chloride, HCl, is a gas that fumes in moist air and is very soluble in water.

Oven cleaners often contain caustic alkalis.

Questions

1 From the pictures of molecules work out the formulae of:
   a acetic acid
   b citric acid.

2 What are the formulae of these antacids?
   a Magnesium hydroxide made up of magnesium ions, Mg^{2+}, and hydroxide ions, OH^-.
   b Aluminium hydroxide made up of aluminium ions, Al^{3+}, and hydroxide ions, OH^-.

pH

14 — dilute sodium hydroxide
13
12 — limewater
11
10 — some brands of toothpaste
9
8
7 — blood — pure water — fresh cows' milk
6 — distilled water
5
4
3 — vinegar
2 — lemon juice
1 — digestive fluids in the stomach
0 — dilute hydrochloric acid

alkaline
neutral
acidic

The pH scale.

A pH meter can be used to measure pH values.

Using a feather to brush away hydrogen bubbles while etching a metal plate with acid.

Indicators and the pH scale

Indicators change colour to show whether a solution is acidic or alkaline. Blue litmus turns red in acid solution and red litmus turns blue in alkalis. Special mixed indicators, such as universal indicator, show a range of colours and can be used to estimate pH values.

pH values can also be measured electronically using a pH meter with an electrode that dips into the solution. The meter can be read directly from the display or it may be connected to a datalogger or computer.

The term pH appears on many cosmetic, shampoo, and food labels. It is a measure of acidity. The **pH scale** is a number scale that shows the acidity or alkalinity of a solution in water. Most laboratory solutions have a pH in the range 1–14.

Hydrangea flowers contain natural indicators – they are blue if grown on acid soil and pink on alkaline soil. Note that this is the opposite of the litmus colours.

Reaction of acids

Acids with metals

Acids react with **metals** to produce **salts**. The other product is hydrogen gas.

$$\text{acid} + \text{metal} \longrightarrow \text{salt} + \text{hydrogen}$$

For example: $2HCl(aq) + Mg(s) \longrightarrow MgCl_2(aq) + H_2(g)$

Not all metals will react in this way. You may remember the list of metals in order of reactivity in C5, G. Metals below lead in the list do not react with acids, and even with lead it is hard to detect any change in a short time.

Acids with metal oxides or hydroxides

An acid reacts with a **metal oxide** or **hydroxide** to form a salt and water. No gas forms.

acid + metal oxide (or hydroxide) ⟶ salt + water

For example: $2HCl(aq) + MgO(s) \longrightarrow MgCl_2(aq) + H_2O(l)$

The reaction between an acid and a metal oxide is often a vital step in making useful chemicals from ores.

Acids with carbonates

Acids react with **carbonates** to form a salt, water, and bubbles of carbon dioxide gas.

acid + metal carbonate ⟶ salt + water + carbon dioxide

Geologists can test for carbonates by dripping hydrochloric acid onto rocks. If they see any fizzing, the rocks contain a carbonate. This is likely to be calcium carbonate or magnesium carbonate.

The word equation is:

hydrochloric acid + calcium carbonate ⟶ calcium chloride + water + carbon dioxide

The balanced equation is:

$$2HCl(aq) + CaCO_3(s) \longrightarrow CaCl_2(aq) + H_2O(l) + CO_2(g)$$

This is a foolproof test for the carbonate ion. So the term 'the acid test' has come to be used to describe any way of providing definite proof.

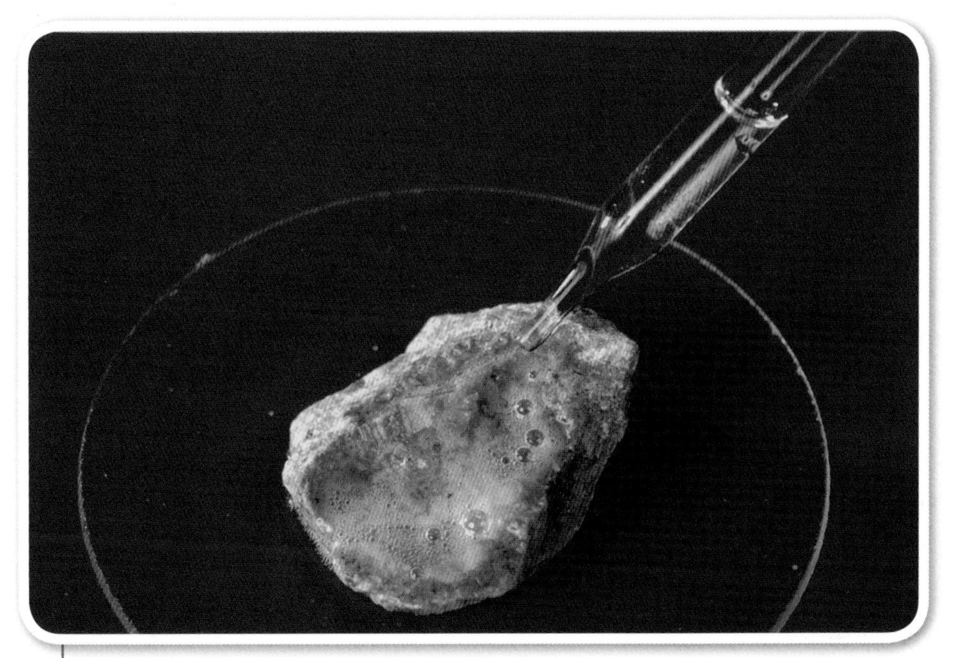

Testing for carbonate using hydrochloric acid.

Key words

- pH scale
- indicators
- metals
- salts
- metal oxide
- metal hydroxide
- carbonates

Questions

3 A pattern can be etched onto a zinc plate using hydrochloric acid to react with the zinc, forming soluble zinc chloride, $ZnCl_2$. Write a word equation and a balanced symbol equation for the reaction.

4 Magnesium hydroxide, $Mg(OH)_2$, is an antacid used to neutralise excess stomach acid, HCl. Write a word equation and a balanced symbol equation for the reaction.

5 There is a volcano in Tanzania, Africa, whose lava contains sodium carbonate, Na_2CO_3. The cooled lava fizzes with hydrochloric acid. Write a word equation and a balanced symbol equation for the reaction.

6 Limescale forms in kettles where hard water is heated. Limescale consists of calcium carbonate. Three acids are often used to remove limescale: citric acid, acetic acid (in vinegar), and dilute hydrochloric acid. Which acid would you use to de-scale an electric kettle and why?

Find out about

- an ionic explanation for neutralisation reactions
- salts and their formulae

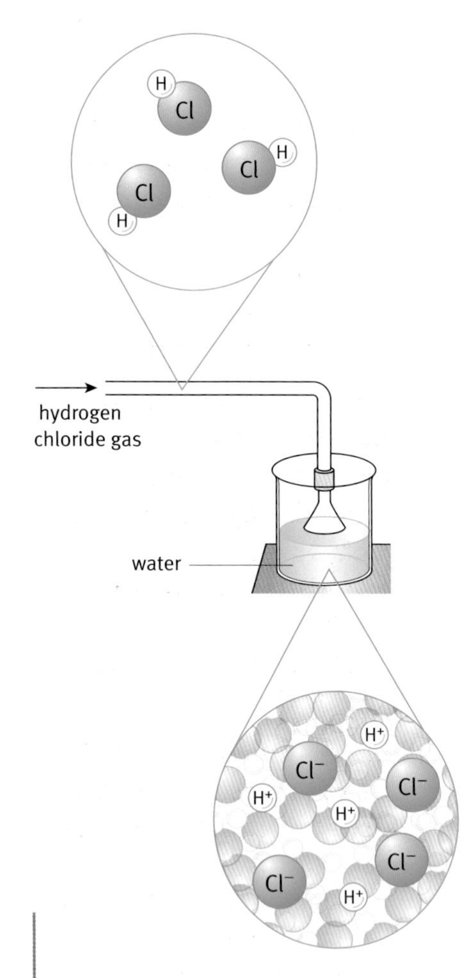

Hydrogen chloride dissolves in water to make hydrochloric acid. The HCl molecules react with water to form ions.

Key words

- hydrogen ions
- hydroxide ions
- neutralisation reaction

What makes an acid an acid?

Chemists have a theory to explain why all the different compounds that are acids behave in a similar way when they react with indicators, metals, carbonates, metal oxides, and metal hydroxides.

It turns out that acids do not simply mix with water when they dissolve. They react, and when they react with water they produce hydrogen ions (H^+). For example, hydrochloric acid is a solution of hydrogen chloride in water. The HCl molecules react with the water to produce **hydrogen ions** and chloride ions.

$$HCl(g) \xrightarrow{water} H^+(aq) + Cl^-(aq)$$

The theory of acids is an ionic theory. Any compound is an acid if it produces hydrogen ions when it dissolves in water.

All acids contain hydrogen in their formula. Nitric acid, HNO_3, and phosphoric acid, H_3PO_4, both contain hydrogen. But not all chemicals that contain hydrogen are acids. Ethane, C_2H_6, and ethanol, C_2H_5OH, are not acids.

In an organic acid it is only the hydrogen atom in the —COOH group that can ionise when the acid dissolves in water.

What makes a solution alkaline?

Alkalis such as the soluble metal hydroxides are ionic compounds. They consist of metal ions and **hydroxide ions** (OH^-). When they dissolve, they add hydroxide ions to water. It is these ions that make the solution alkaline.

$$NaOH(s) \xrightarrow{water} Na^+(aq) + OH^-(aq)$$

Neutralisation

Sodium hydroxide and hydrochloric acid react to produce a salt (sodium chloride) and water.

$$Na^+(aq) + OH^-(aq) + H^+(aq) + Cl^-(aq) \longrightarrow Na^+(aq) + Cl^-(aq) + H_2O(l)$$

During a **neutralisation reaction** the hydrogen ions from an acid react with hydroxide ions from the alkali to make water.

$$H^+(aq) + OH^-(aq) \longrightarrow H_2O(l)$$

The remaining ions in the solution make a salt.

Salts

Salts form when a metal oxide, or hydroxide, neutralises an acid. So every salt can be thought of as having two parents. Salts are related to a parent metal oxide or hydroxide and to a parent acid.

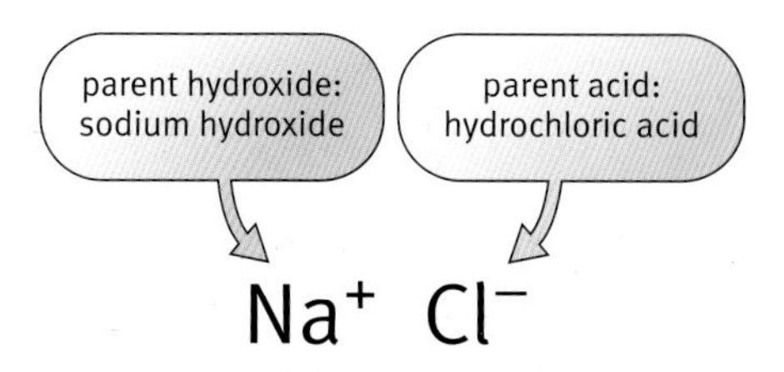

Salts are ionic (see C4, J: Ionic theory). Most salts consist of a positive metal ion combined with a negative non-metal ion. The metal ion comes from the parent metal oxide or hydroxide. The non-metal ion comes from the parent acid.

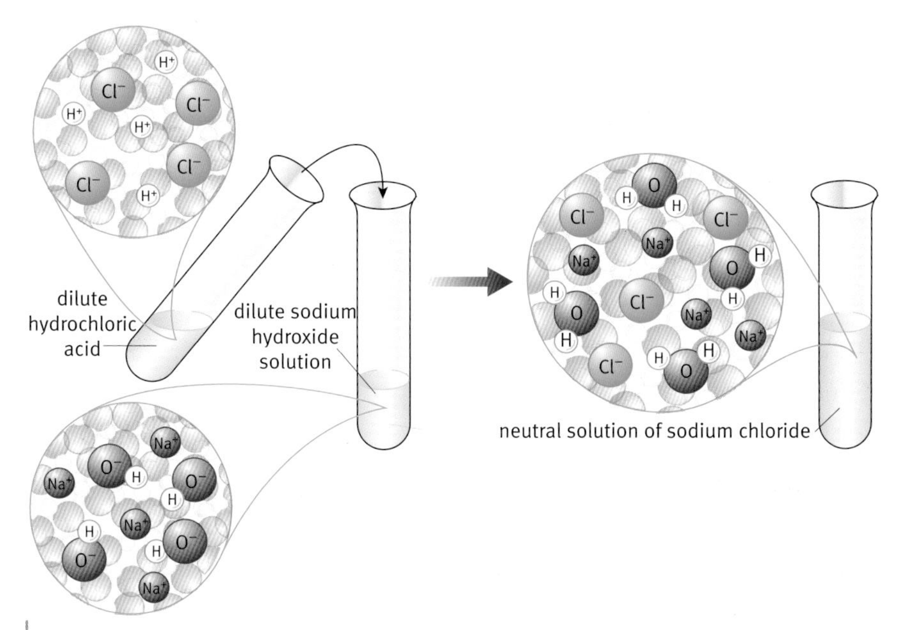

Dilute sodium hydroxide solution neutralises dilute hydrochloric acid, forming a neutral solution of sodium chloride. Water molecules not involved in the reaction are shown in a paler colour.

It is possible to work out the formulae of salts knowing the charges on the ions. Remember that all compounds are overall electrically neutral (see C4, K: Ionic theory and atomic structure). Some non-metal ions consist of more than one atom. The table below includes some examples. In the formula for magnesium nitrate, $Mg(NO_3)_2$, the brackets around the NO_3 show that two complete nitrate ions appear in the formula.

Non-metal ions that consist of more than one atom	Symbols
carbonate	CO_3^{2-}
hydroxide	OH^-
nitrate	NO_3^-
sulfate	SO_4^{2-}

Questions

1 Write equations to show what happens when these compounds dissolve in water:
  a nitric acid
  b sulfuric acid
  c calcium hydroxide.

2 Write down the name of the salt produced in the following reactions:
  a lithium hydroxide with hydrochloric acid
  b calcium carbonate with nitric acid
  c magnesium oxide with sulfuric acid.

3 Use the tables of ions in C4 Section K and on this page to write down the formulae of these salts:
  a potassium nitrate
  b magnesium carbonate
  c sodium sulfate
  d calcium nitrate.

4 Use the tables of ions in C4 Section K and on this page to write down the charge on the metal ion in each of these salts:
  a $CuCO_3$
  b $PbBr_2$
  c Fe_2O_3.

Purity of chemicals

Find out about

- purity
- titrations for testing purity

CALCIUM CARBONATE PRECIPITATED CP
QTY: 1kg BNO: C1042/R6 - 708717

Assay	99%
Chloride (Cl)	0.005%
sulphate (SO_4)	0.05%
Iron (Fe)	0.002%
Lead (Pb)	0.002%

Label on a bottle of laboratory grade calcium carbonate. The term 'assay' tells you how pure the chemical is. The calcium carbonate is 99% pure with the small amounts of impurities shown.

Key words

- titration
- burette
- end point

Questions

1 Uses of sodium chloride (salt) include:
   i flavouring food
   ii melting ice on roads
   iii saline drips in hospitals.
   Put these in order of the grade of sodium chloride required, with the purest first.

2 From 'steps involved in a titration' in which step is:
   a a solution made?
   b a pipette used, and what is it used for?
   c the end point reached, and how does the technician know?

Grades of purity

The reactions of acids with metals, oxides, hydroxides, and carbonates can be used to make valuable salts. For uses such as food or medicines, these salts have to be made pure so that they are safe to swallow.

Chemicals do not always have to be pure. Calcium carbonate, for example, is used in a blast furnace to extract iron from its ores. The iron industry can use limestone straight from a quarry. Limestone has some impurities but they do not stop it from doing its job in a blast furnace.

Suppliers of chemicals offer a range of grades of chemicals. In a school laboratory you might use one of these grades: technical, general laboratory, and analytical. The purest grade is the analytical grade.

Purifying a chemical is done in stages. Each stage takes time and money, and becomes more difficult. So the higher the purity, the more expensive the chemical. Manufacturers therefore buy the grade most suitable for their purpose.

When deciding what grade of chemical to use for a particular purpose, it is important to know:

- the amount of impurities
- what the impurities are
- how they can affect the process
- whether they will end up in the product, and whether it matters if they do.

Testing purity

Medicines contain an active ingredient. Other ingredients are included to make the medicine pleasant to taste and easy to take. This means that the pharmaceutical companies that make medicines need sweeteners, food flavours, and other additives.

The companies buy in many of their ingredients. Technical chemists working for the companies have to make sure that the suppliers are delivering the right grade of chemical.

Citric acid is often added to syrups, such as cough medicines, to control their pH. Technicians can check the purity of the acid using a procedure called a **titration**, which measures the volume of alkali that it can neutralise. The technician has to know the accurate concentration of the alkali.

Steps involved in a titration

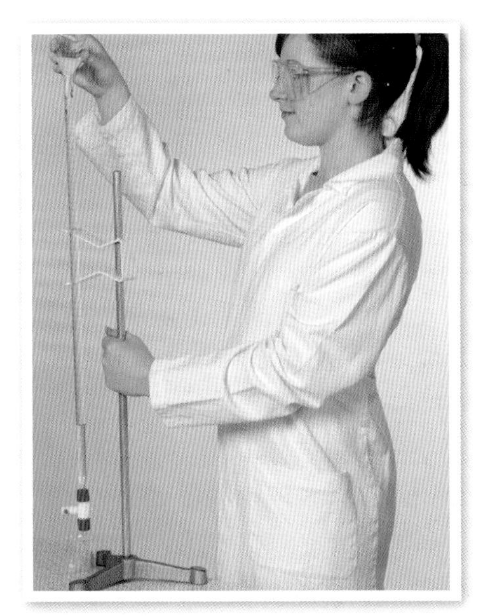

1 The technician fills a **burette** with a solution of sodium hydroxide. She knows the concentration of the alkali.

2 The technician weighs out a sample of citric acid accurately.

3 The technician dissolves the acid in pure water. Then she adds a few drops of phenolphthalein indicator. The indicator is colourless in the acid solution.

4 The technician adds alkali from the burette. She swirls the contents of the flask as the alkali runs in. Near the end she adds the alkali drop by drop. At the **end point** all the citric acid is just neutralised. The indicator is now permanently pink.

The technician will repeat the titration several times. If there are any results that differ greatly from the rest, and there is reason to doubt their accuracy, they will be discarded. The remaining values will be averaged to find the mean.

Questions

3 A 1.35 g sample of impure citric acid was dissolved in water and titrated with sodium hydroxide solution of concentration 40 g/dm^3. The average titre was found to be 20.6 cm^3.

a Use the following equation to find the mass of citric acid in the sample:
mass of citric acid (g) = average titre (cm^3) × 0.064.

b Use the following equation to calculate the percentage purity of the sample:
percentage purity =

$$\frac{\text{mass of citric acid}}{\text{mass of sample}} \times 100\ \%$$

4 A technical chemist measures the purity of tartaric acid by titration, and obtains these results: 98.7%, 99.0%. 105.4%, 80.0%, 98.8%, 98.5%

a Suggest reasons why the results are not exactly the same.

b Which two values should be checked?

c Calculate the mean value of purity after discarding the two outlying values.

d Suggest the limits between which the true value is likely to lie.

Energy changes in chemical reactions

Find out about

- reactions that give out energy and reactions that take in (absorb) energy

Exothermic and endothermic reactions

Most chemical reactions need a supply of energy to get them started. Some also need energy to keep them going. That is why Bunsen burners and electric heating mantles are so common in laboratories.

Most reactions give out energy once they are going – combustion (burning), and neutralisation of acids to make salts are common examples. Sulfuric acid is an important chemical in industry. One of the key reactions in making this acid has to be carefully controlled because it gives out a great deal of energy.

Reactions that give out energy to the surroundings (the surroundings get hotter) are called **exothermic**. There are also reactions that absorb energy from the surroundings (the surroundings get cooler). These are called **endothermic**.

It is possible to tell whether a chemical reaction is exothermic or endothermic by measuring the temperature of the reactants before the reaction starts, and the temperature of the products after it has finished. If the temperature has risen, the reaction is exothermic. If it has dropped, the reaction is endothermic.

Both exothermic and endothermic reactions have practical uses. An exothermic reaction provides the energy for welding. A cold pack absorbs energy from an injured muscle using an endothermic reaction.

Energy changes in the chemical industry

Scientists working in the chemical industry need to know whether a reaction is exothermic or endothermic when it is part of a synthesis. There are several reasons for this:

- It takes fuel, which costs money, to provide the energy input for endothermic reactions.
- The energy given out by exothermic reactions can be used elsewhere in the plant, to produce electricity, for example.
- A temperature increase makes chemical reactions go faster (see Section F) – a reaction that gives out heat energy will tend to get faster and faster, and may 'run away', possibly causing an explosion.

Such a 'runaway' reaction may have been the cause of a major disaster at a chemical plant in Bhopal, India, in 1984. Between 2000 and 15 000 people may have died and over 25 years later, the health of many people is still being affected by the incident. Proper understanding and control of the energy changes in the reactions concerned might have prevented the disaster.

Energy-level diagrams

All chemical reactions give out or take in energy. Understanding energy changes helps chemists to control reactions.

- In an exothermic reaction (where energy is given out), the products must have less energy than the reactants.
- In an endothermic reaction (where energy is absorbed), the products must have more energy than the reactants.

Chemists keep track of the changes in energy in chemical reactions using **energy-level diagrams**. These show the energies of the reactants and products. Notice that energy is plotted vertically up the diagram. Reactants are on the left and products on the right. These are usually shown using a balanced equation.

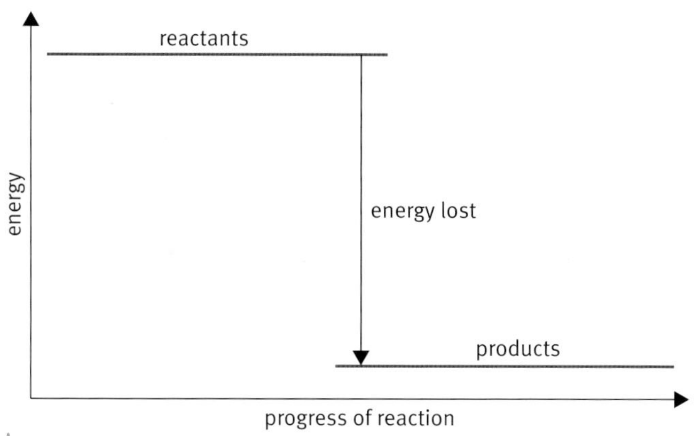

An exothermic reaction.

The energy-level diagram for the reaction of magnesium and hydrochloric acid, which is exothermic. Energy is given out in this reaction, so the products have less energy than the reactants.

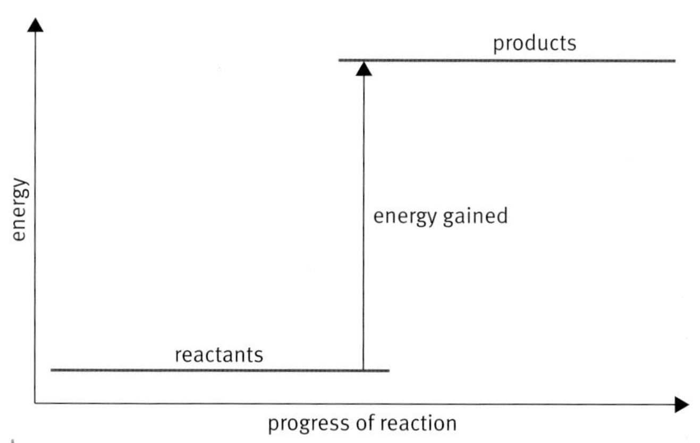

An endothermic reaction.

Energy-level diagram for the reaction of citric acid and sodium hydrogencarbonate, which is endothermic. Energy is absorbed in this reaction, so the products have more energy than the reactants.

Key words

- exothermic
- endothermic
- energy-level diagram

An electric heating mantle, used to heat up reactions without a naked flame.

Question

1. If you have ever had a plaster cast for a broken bone, apart from the pain of the injury, you might remember the pleasant feeling of warmth as the wet plaster begins to set. If you put a sherbet sweet on your tongue you will feel your mouth getting cold. Both of these are chemical reactions.
   - **a** Which reaction is exothermic and which is endothermic?
   - **b** Draw an energy-level diagram for each reaction. Use 'reactants' and 'products' for the names of the chemicals involved.

F Rates of reaction

Find out about

- measuring rates of reaction
- factors affecting rates of reaction
- catalysts in industry
- collision theory

Controlling reaction rates

Some chemical reactions seem to happen in an instant. An explosion is an example of a very fast reaction.

Other reactions take time – seconds, minutes, hours, or even years. Rusting is a slow reaction and so is the rotting of food.

It is the chemist's job to work out the most efficient way to synthesise a chemical. It is important that the chosen reactions happen at a convenient speed. A reaction that occurs too quickly can be hazardous. A reaction that takes several days to complete is not practical because it ties up equipment and people's time for too long, which costs money.

An explosion is an example of a very fast chemical reaction.

Measuring rates of reaction

Your pulse rate is the number of times your heart beats every minute. The production rate in a factory is a measure of how many articles are made in a particular time. Similar ideas apply to chemical reactions.

Chemists measure the **rate of a reaction** by finding the quantity of product produced or the quantity of reactant used up in a fixed time.

For the reaction

$$Mg(s) + 2HCl(aq) \longrightarrow MgCl_2(aq) + H_2(g)$$

the rate can be measured quite easily by collecting and measuring the hydrogen gas produced.

$$\text{average rate} = \frac{\text{change in the volume of hydrogen}}{\text{time for the change to happen}}$$

In most chemical reactions, the rate changes with time. The graph on the left is a plot of the volume of hydrogen formed against time for the reaction of magnesium with acid. The graph is steepest at the start, showing that the rate of reaction was greatest at that point. As the reaction continues the rate decreases until the reaction finally stops. The steepness of the line is a measure of the rate of reaction.

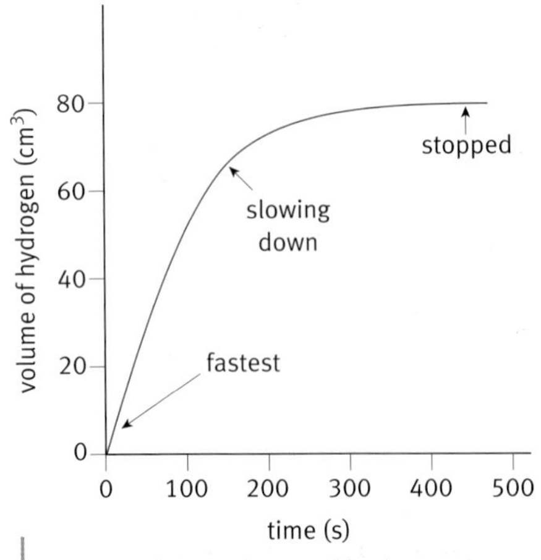

A plot of the volume of hydrogen formed against time for a reaction of magnesium with hydrochloric acid.

Question

1 For each of these reactions, pick a method from the opposite page that could be used to measure the rate of reaction: (Hint: Look at the states of the reactants and products.)

a $CaCO_3(s) + 2HCl(aq) \longrightarrow CaCl_2(aq) + CO_2(g) + H_2O(l)$

b $Zn(s) + H_2SO_4(aq) \longrightarrow ZnSO_4(aq) + H_2(g)$

c $Na_2S_2O_3(aq) + 2HCl(aq) \longrightarrow 2NaCl(aq) + SO_2(g) + S(s) + H_2O(l)$

Key word

- rate of reaction

Methods of measuring rates of reaction

Collecting and measuring a gas product

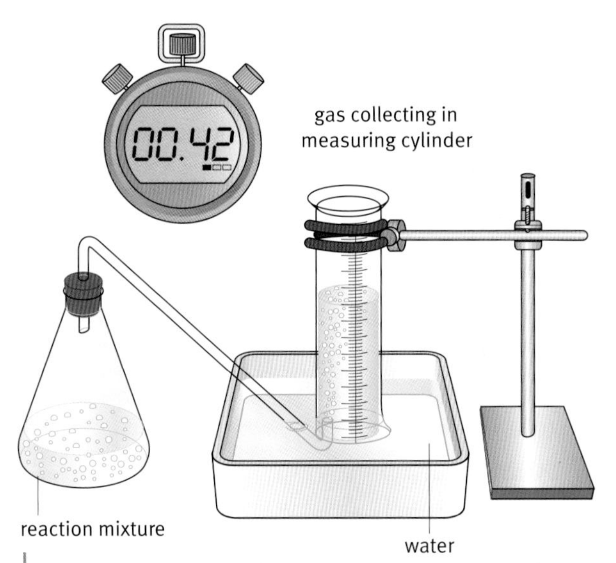

Record the volume at regular intervals, such as every 30 or 60 seconds. A gas syringe could be used to collect the gas, instead of a measuring cylinder – see the next page.

Measuring the loss of mass as a gas forms

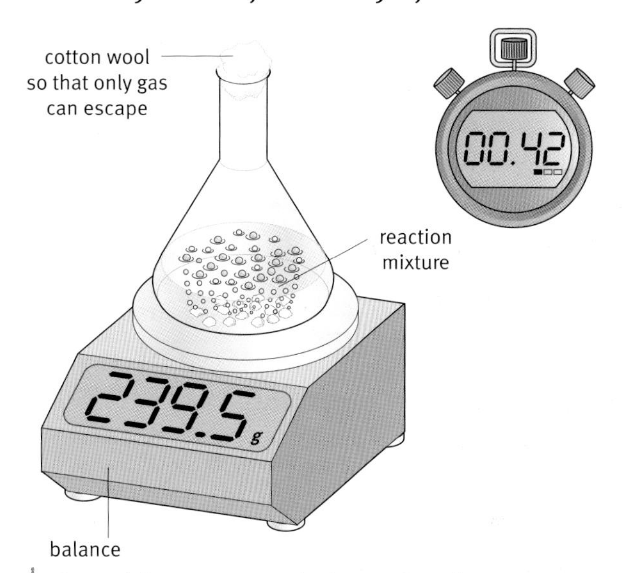

Record the mass at regular intervals such as every 30 or 60 seconds.

Timing how long it takes for a small amount of solid reactant to disappear

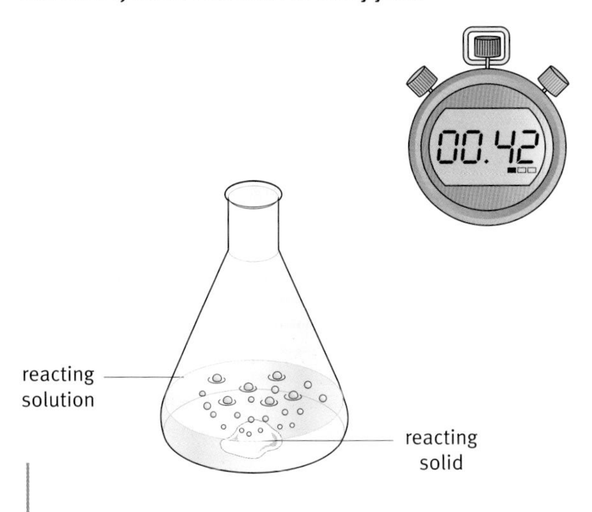

Mix the solid and solution in the flask and start the timer. Stop it when you can no longer see any solid.

Timing how long it takes for a solution to turn cloudy

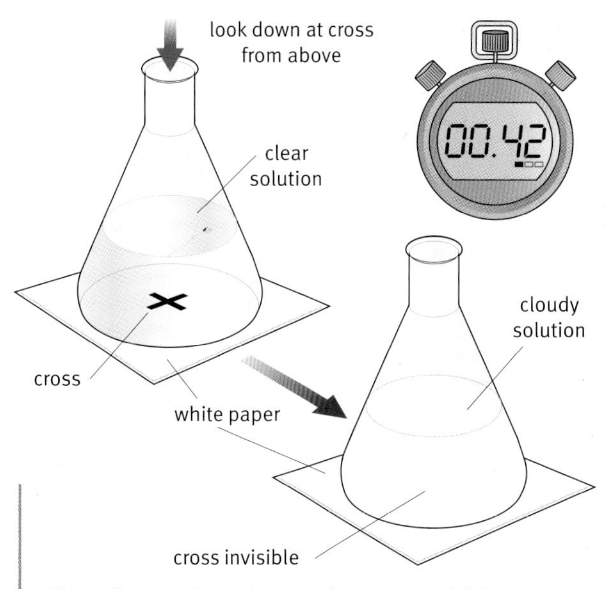

This is for reactions that produce an insoluble solid. Mix the solutions in the flask and start the timer. Stop it when you can no longer see the cross on the paper through the solution.

Manual timing of reactions is only suitable for reactions that are over in a few minutes (or more). Modern methods using lasers and dataloggers allow chemists to measure the rates of reactions that are over in less than one thousand billion billionths of a second (one femtosecond).

Key words

- concentration
- surface area
- catalyst

Factors affecting reaction rates

Powdered Alka-Seltzer reacts faster in water than a tablet. Milk standing in a warm kitchen goes sour more quickly than milk kept in a refrigerator. Changing the conditions alters the rate of these processes and many others.

Factors that affect the rate of chemical reactions are:

- the *concentration* of reactants in solution – the higher the concentration, the faster the reaction
- the *surface area* of solids – powdering a solid increases the surface area in contact with a liquid, solution, or gas, and so speeds up the reaction.
- the *temperature* – typically a 10 °C rise in temperature can roughly double the rate of many reactions
- *catalysts* – these are chemicals that speed up a chemical reaction without being used up in the process.

The factors in action

The apparatus in the diagram was used in an investigation into the effect of changing the conditions on the reaction of zinc metal with sulfuric acid. The graph below shows the results.

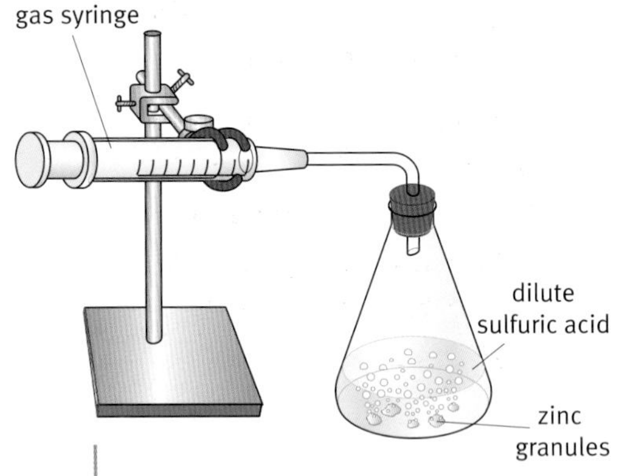

Apparatus used to investigate the factors affecting the rate of reaction of zinc with sulfuric acid.

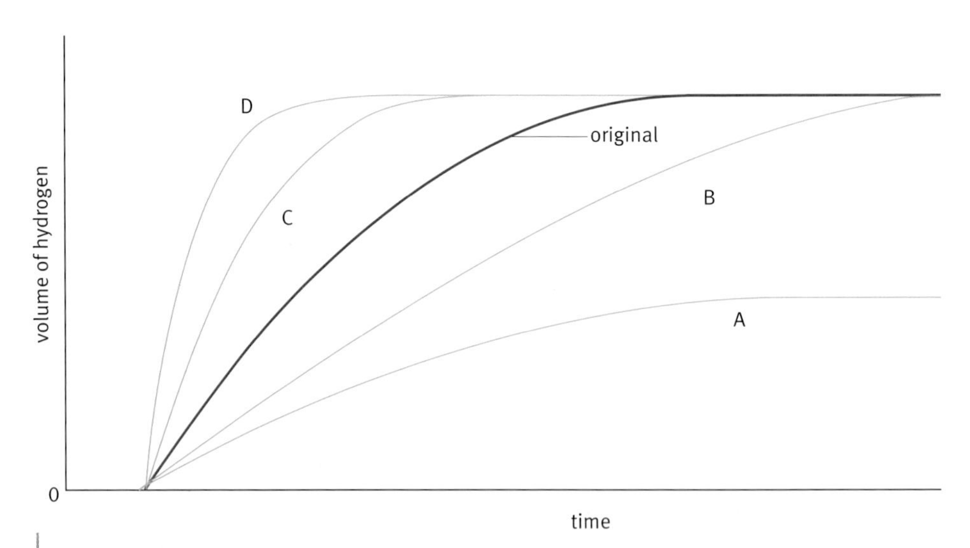

The volume of hydrogen formed over time during an investigation of the factors affecting the rate of reaction of zinc with sulfuric acid. The investigator used the same mass of zinc each time. There was more than enough metal to react with all the acid.

$$Zn(s) + H_2SO_4(aq) \longrightarrow ZnSO_4(aq) + H_2(g)$$

The red line on the graph plots the volume of hydrogen gas against time using zinc granules and 50 cm^3 of dilute sulfuric acid at 20 °C. The reaction gradually slows down and stops because the acid concentration falls to zero. There is more than enough metal to react with all the acid. The zinc is in excess.

The effect of concentration

Line A on the graph shows the result of using acid that was half as concentrated while leaving all the other conditions the same as in the original set up.

The investigator added 50 cm^3 of this more dilute acid. Halving the acid **concentration** lowers the rate at the start. The final volume of gas is cut by half because there was only half as much acid in the 50 cm^3 of solution to start with.

The effect of surface area

Line B on the graph shows the result of using the same excess of zinc metal in fewer larger pieces. All other conditions were the same as in the original setup. Fewer larger lumps of metal have a smaller total **surface area** so the reaction starts more slowly. The amount of acid is unchanged and the metal is still in excess so that the final volume of hydrogen is the same.

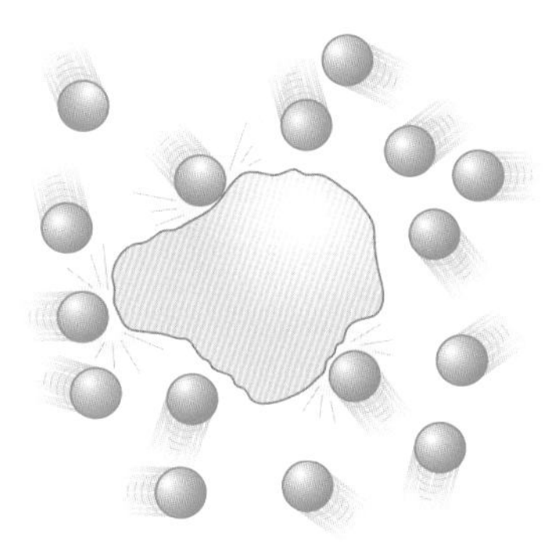

one big lump (slow reaction)

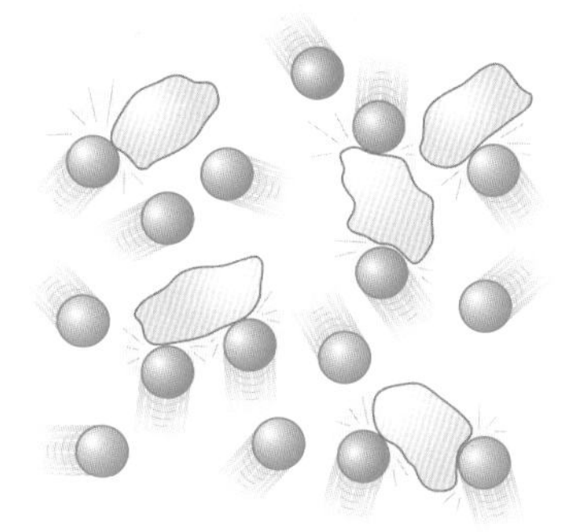

several small lumps (fast reaction)

Breaking up a solid into smaller pieces increases the total surface area. This increases the amount of contact between the solid and the solution, making it possible for the reaction to go faster.

The effect of temperature

Line C on the graph shows the result of carrying out the reaction at 30 °C while leaving all the other conditions the same as in the original set up. This more or less doubles the rate at the start. The quantities of chemicals are the same so the final volume of gas collected is the same as it was originally.

The effect of adding a catalyst

Line D on the graph shows what happens when the investigation is repeated with everything the same as in the original set up but with a **catalyst** added. The reaction starts more quickly. Catalysts do not change the final amount of product, so the volume of gas at the end is the same as before.

Questions

2 When a lump of calcium carbonate is placed into a beaker of acid it reacts and carbon dioxide is given off. Suggest three ways to speed up the reaction.

3 How is it possible to control conditions to speed up these changes:
   a the setting of an epoxy glue
   b the cooking of an egg
   c the conversion of oxides of nitrogen in car exhausts to nitrogen.

4 When investigating the effect of temperature on a chemical reaction, why is it important to keep all other conditions the same?

5 The effect of concentration on the rate of reaction between zinc and sulfuric acid was investigated. The results were plotted on a graph.

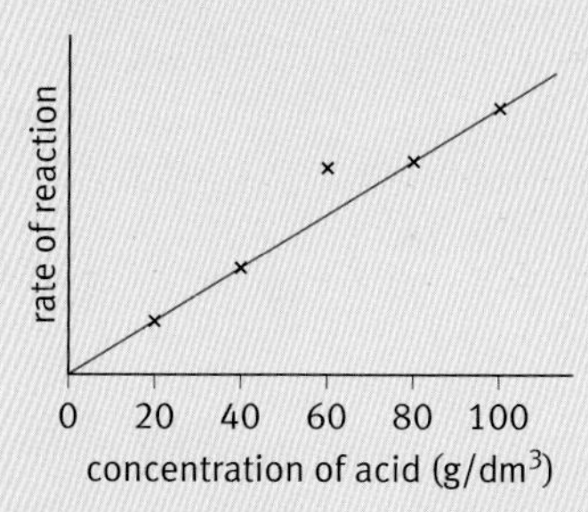

   a Is there a correlation? If so, describe it.
   b Which result is an outlier? Suggest a reason why this result is different from the expected value.

Catalysts in industry

What is a catalyst?

A catalyst is a chemical that speeds up a chemical reaction. It takes part in the reaction, but is not used up.

Modern catalysts can be highly selective. This is important when reactants can undergo more than one chemical reaction to give a mixture of products. With a suitable catalyst it can be possible to speed up the reaction that gives the required product, but not speed up other possible reactions that create unwanted by-products.

The manufacture of ethanoic acid from methanol and carbon monoxide uses a catalyst to speed up the reaction.

methanol + carbon monoxide ⟶ ethanoic acid

$CH_3OH(g) + CO(g) \longrightarrow CH_3COOH(g)$

Better catalysts

Catalysts are essential in many industrial processes. They make many processes economically viable. This means that chemical products can be made at a reasonable cost and sold at affordable prices.

Research into new catalysts is an important area of scientific work. This is shown by the industrial manufacture of ethanoic acid (see Section B) from methanol and carbon monoxide. This process was first developed by the company BASF in 1960 using a cobalt compound as the catalyst at 300 °C and at a pressure 700 times atmospheric pressure.

About six years later the company Monsanto developed a process using the same reaction, but a new catalyst system based on rhodium compounds. This ran under much milder conditions: 200 °C and 30–60 times atmospheric pressure.

In 1986, the petrochemical company BP bought the technology for making ethanoic acid from Monsanto. It has since devised a new catalyst based on compounds of iridium. This process is faster and more efficient. Iridium is cheaper, and less of the catalyst is needed. Iridium is even more selective so the amount of ethanoic acid produced (yield) is greater and there are fewer by-products. This makes it easier and cheaper to make pure ethanoic acid and there is less waste.

Collision theory

Chemists have a theory to explain how the various factors affect reaction rates.

The basic idea is that particles, such as molecules, atoms, and ions can only react if they bump into each other. Imagining these particles colliding with each other leads to a theory that can account for the effects of concentration, temperature, and catalysts on reaction rates.

According to **collision theory**, when molecules collide some bonds between atoms can break while new bonds form. This creates new molecules.

Molecules are in constant motion in gases, liquids, and solutions. There are millions upon millions of collisions every second. Most reactions would be explosive if every collision led to a reaction. It turns out that only a very small proportion of all the collisions are successful and actually lead to a reaction. These are the collisions in which the molecules are moving with enough energy to break bonds between atoms.

Any change that increases the number of successful collisions per second has the effect of increasing the rate of reaction. Increasing the concentration of solutions of dissolved chemicals increases the frequency of collisions. The same small proportion of these collisions will be successful, but now there are more of them. This means that there will also be more successful collisions.

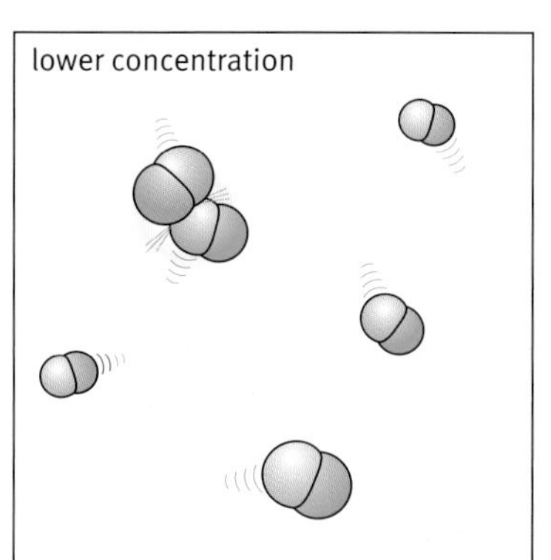

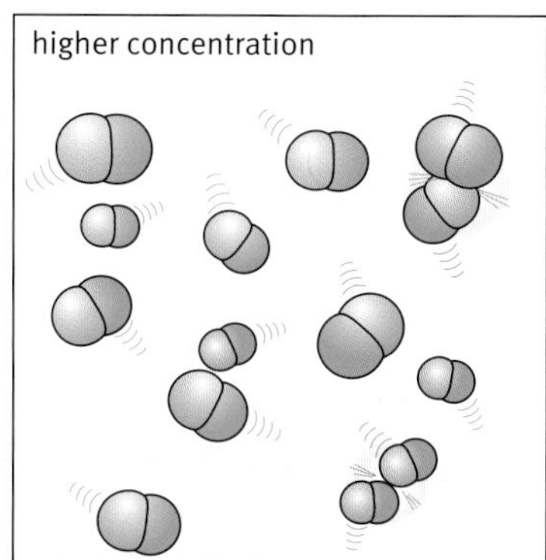

Molecules have a greater chance of colliding in a more concentrated solution. More frequent collisions means – faster reactions. Reactions get faster if the reactants are more concentrated.

Breaking up a solid into smaller pieces increases its surface area. Increased surface area means that there are more atoms, molecules, or ions of the solid available to react. This speeds up the reaction by increasing the frequency of successful collisions with particles in liquid, solution, or gas.

Key word

- collision theory

Questions

6 a Where do cobalt, rhodium and iridium appear in the periodic table?

  b Why is not surprising that these three metals can be used to make catalysts for the same process?

7 Suggest reasons why it is important to develop industrial processes that:

  a run at lower temperatures and pressures

  b produce less waste.

Find out about

- steps in synthesis
- making a soluble salt

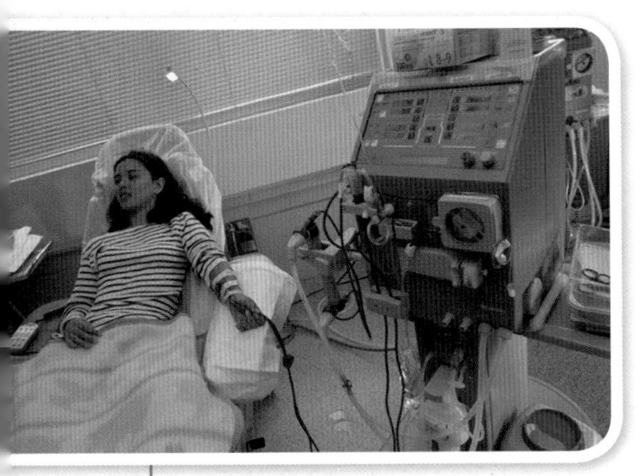

Kidney dialysis.

Making calcium chloride for dialysis

Chemical synthesis is a way of making new compounds. Synthesis puts things together to make something new. It is the opposite of analysis, which takes things apart to see what they are made of.

The kidneys remove toxic chemicals from the blood. In cases of kidney failure, patients are put on dialysis machines that do the job outside the body while they await a transplant. Blood passes out of the body through a tube into the dialysis machine.

Inside the machine, the blood flows past a special membrane. On the other side of the membrane is a solution containing a mixture of salts at the same concentrations as the same salts in the blood. The toxic chemicals pass from the blood through the membrane into the solution and are carried away. It is also possible for useful salts to pass back into the blood.

One of the salts in the dialysis solution is calcium chloride. It is a particularly important salt as the level of calcium in the blood has to be maintained at a particular level. Just a little bit too much or too little and the patient could become very ill indeed. The calcium chloride therefore has to be very pure, and the quantity added to the solution has to be measured accurately.

The process for making calcium chloride is shown in the flow chart.

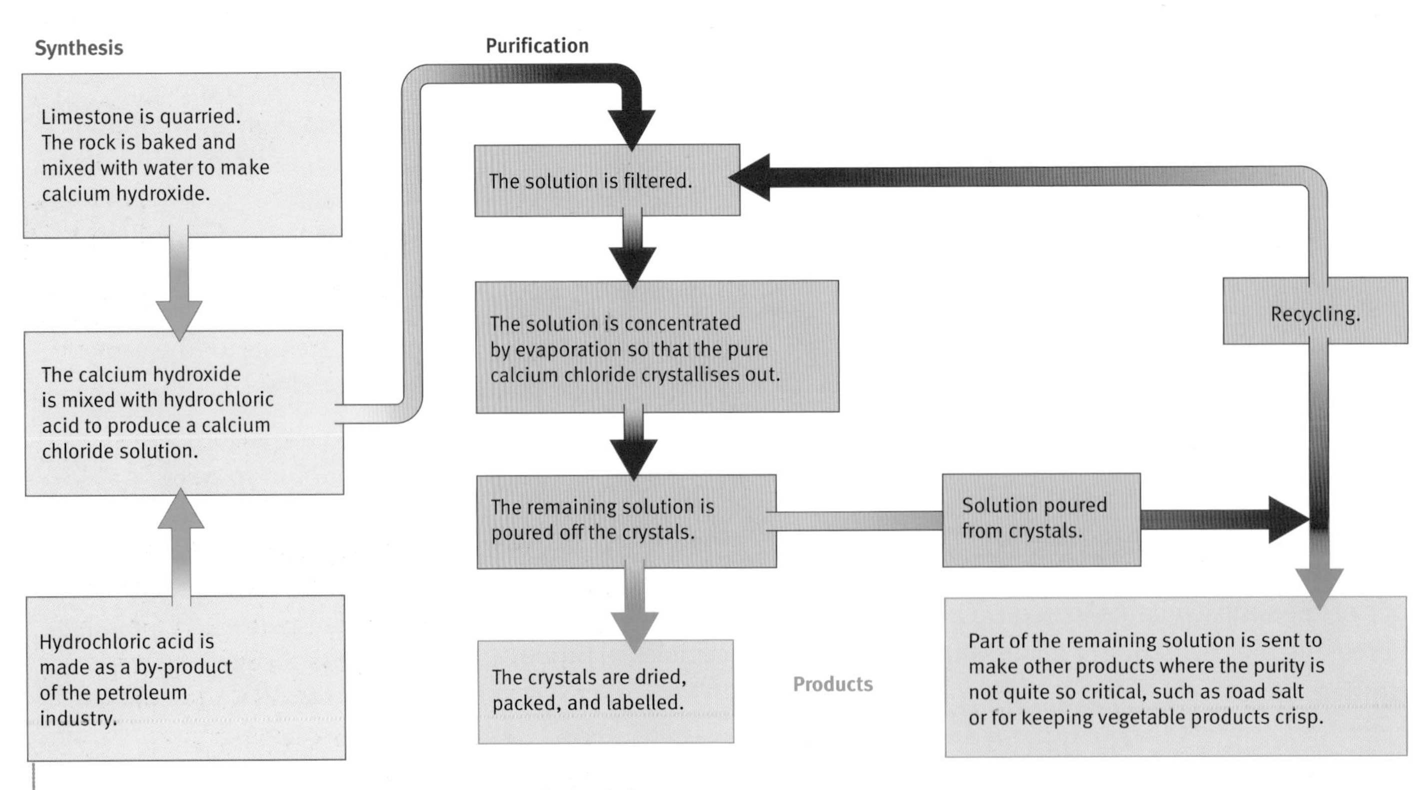

The process for making very pure calcium chloride for dialysis.

Making a sample of magnesium sulfate

Magnesium sulfate is another salt produced by the chemical industry. It has many uses including as a nutrient of plants.

The process of making magnesium sulfate (or any other soluble salt) on a laboratory scale illustrates the stages in a chemical synthesis.

In the following method an excess of solid is added to make sure that all the acid is used up. This method is only suitable if the solid added to the acid is either insoluble in water or does not react with water.

Choosing the reaction

Any of the characteristic reactions of acids can all be used to make salts:

- acid + metal ⟶ salt + hydrogen
- acid + metal oxide or hydroxide ⟶ salt + water
- acid + metal carbonate ⟶ salt + carbon dioxide + water

Magnesium metal is relatively expensive because it has to be extracted from one of its compounds. So it makes sense to use either magnesium oxide or carbonate as the starting point for making magnesium sulfate from sulfuric acid.

Carrying out a risk assessment

It is always important to minimise exposures to risk. You should take care to identify hazardous chemicals. You should also look for hazards arising from equipment or procedures. This is a **risk assessment**.

In this preparation the magnesium compounds are not hazardous. The dilute sulfuric acid is an irritant, which means that you should keep it off your skin and especially protect your eyes. You should always wear eye protection when handling chemicals, for example.

An operator emptying magnesium sulfate into the tank of a sprayer on a farm. He is wearing protective clothing because magnesium sulfate is harmful. As well as being a micronutrient needed for healthy plant growth, the salt is a:

- raw material in soaps and detergents
- laxative in medicine
- refreshing additive in bath water
- raw material in the manufacture of other magnesium compounds
- supplement in feed for poultry and cattle
- coagulant in the manufacture of some plastics.

wear eye protection

harmful

Questions

1 Refer to the flow diagram for the process of making calcium chloride.

  a Write the word and balanced symbol equations for the reaction used to make the salt.

  b Identify steps taken to make the yield of the pure salt as large as possible.

2 Epsom salts consist of magnesium sulfate. Magnesium sulfate is soluble in water. Produce a flow diagram to show how you could remove impurities that are insoluble in water from a sample of Epsom salts. Processes you might use include: crystallisation, dissolving, drying, evaporation, filtration.

Questions

3 Write the balanced equation for the reaction of magnesium carbonate with sulfuric acid.

4 Why does the mixture of magnesium carbonate and sulfuric acid froth up?

5 What is the advantage of:
  a using powdered magnesium carbonate?
  b warming when most of the acid has been used up?
  c adding a slight excess of the solid to the acid?

6 Why is it impossible to use this method to make a pure metal sulfate by the reaction of dilute sulfuric acid with:
  a lithium metal?
  b sodium hydroxide?
  c potassium carbonate?

7 Look at the procedure on this and the following page for making magnesium sulfate, and identify the step when risks might arise from:
  a chemicals that react vigorously and spill over
  b chemicals that might spit or splash on heating
  c hot apparatus that might cause burns
  d apparatus that might crack and form sharp edges.

Key word

✓ **risk assessment**

Working out the quantities to use

Reacting masses can be used to work out the amount of reactants needed to produce a particular amount of product (see Section H). In this procedure the solid is added in excess. This means that the amount of product is determined by the volume and concentration of the sulfuric acid. The concentration of dilute sulfuric acid is 98 g/litre, and a volume of 50 cm^3 dilute sulfuric acid is used. This contains 4.9 g of the acid.

Carrying out the reaction in suitable apparatus under the right conditions

The reaction is fast enough at room temperature, especially if the magnesium carbonate is supplied as a fine powder.

This reaction can be safely carried out in a beaker. Stirring with a glass rod makes sure that the magnesium oxide or carbonate and acid mix well. Stirring also helps to prevent the mixture frothing up and out of the beaker.

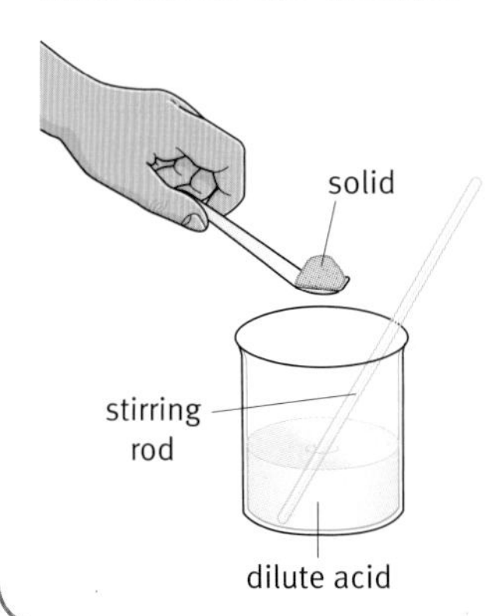

Measure the required volume of acid into a beaker. Add the metal oxide or carbonate bit by bit until no more dissolves in the acid. Warm gently until all of the acid has been used up. Make sure that there is a slight excess of solid before moving on to the next stage.

Separating the product from the reaction mixture

Filtering is a quick and easy way of separating the solution of the product from the excess solid. The mixture filters more quickly if the mixture is warm.

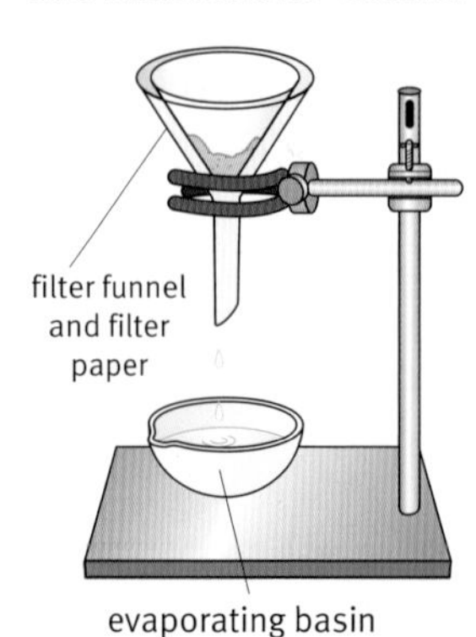

Filter off the excess solid, collecting the solution of the salt in an evaporating basin. The residue on the filter paper is the excess solid.

Purifying the product

After the mixture has been filtered, the filtrate contains the pure salt dissolved in water. Evaporating much of the water speeds up crystallisation. This is conveniently carried out in an evaporating basin. The concentrated solution can then be left to cool and crystallise. The drying can be completed in a warm oven. The crystals can then be transferred to a desiccator. This is a closed container that contains a solid that absorbs water strongly.

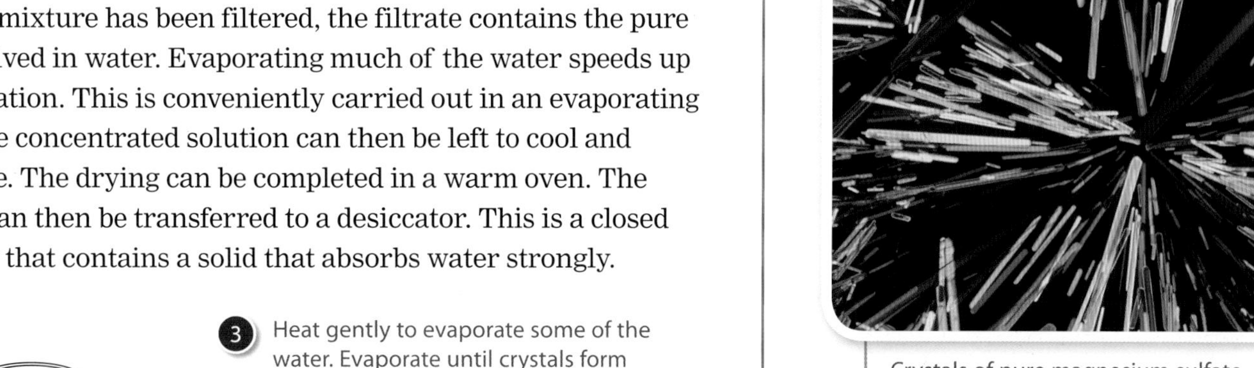

Crystals of pure magnesium sulfate seen through a Polaroid filter (×60).

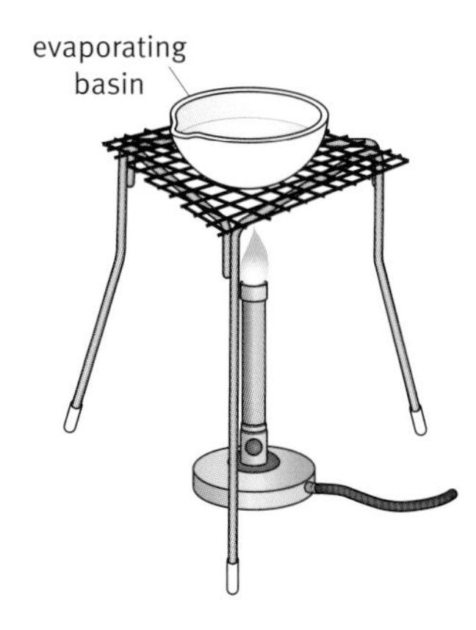

3 Heat gently to evaporate some of the water. Evaporate until crystals form when a droplet of solution picked up on a glass rod crystallises on cooling.

Petri dish

salt crystals

4 Pour the concentrated solution into a labelled glass Petri dish and set it aside to cool slowly.

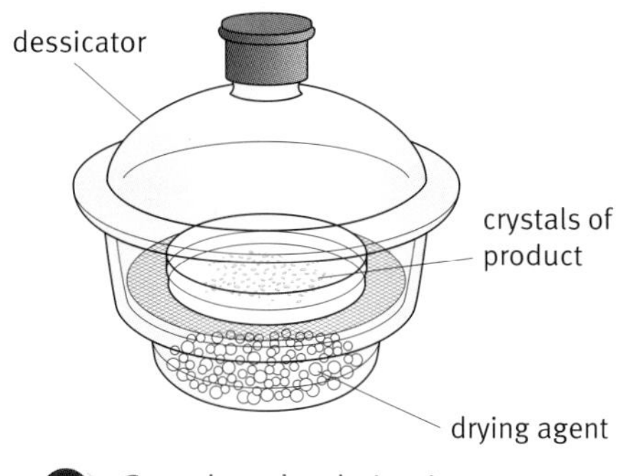

5 Complete the drying in an oven and then store in a dessicator.

Measuring the yield and checking the purity of the product

The final step is to transfer the dry crystals to a weighed sample tube and re-weigh it to find the actual yield of crystals. Often it is important to carry out tests to check that the product is pure.

The appearance of the crystals can give a clue to the purity of the product. A microscope can help if the crystals are small. The crystals of a pure product are often well-formed and even in shape.

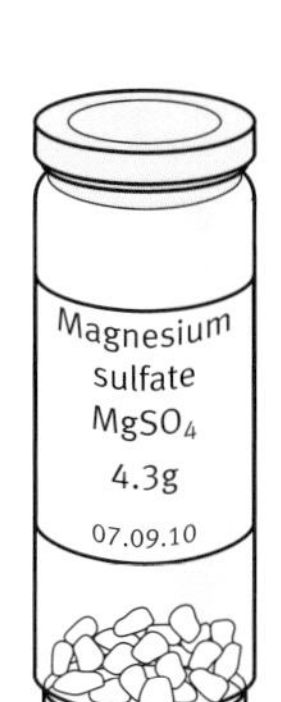

6 The weighed sample of product showing the name and formula of the chemical, the mass of product, and the date it was made.

Questions

8 Identify the impurities removed during the purification stages of the magnesium sulfate synthesis.

9 Why is it important that the magnesium carbonate is added in excess to the sulfuric acid?

H Chemical quantities

Find out about

- reacting masses
- yields from chemical reactions

Key words

- relative formula mass
- reacting mass
- actual yield
- theoretical yield
- percentage yield

Question

1 What mass of:
   a HCl in hydrochloric acid reacts with 100 g calcium carbonate?
   b HNO_3 in dilute nitric acid neutralises 56 g of potassium hydroxide?
   c copper(II) oxide (CuO) would you need to react with sulfuric acid (H_2SO_4) to make 319 g of copper sulfate ($CuSO_4$)?

Chemists wanting to make a certain quantity of product need to work out how much of the starting materials to order. Getting the sums right matters – especially in industry, where a higher yield for a lower price can mean better profits.

The trick is to turn the symbols in the balanced chemical equation into masses in grams or tonnes. This is possible given the relative masses of the atoms in the periodic table.

Reacting masses

Adding up the relative atomic masses for all the atoms in the formula of a compound gives the **relative formula mass** of chemicals . Given the relative formula masses, it is then possible to work out the masses of reactants and products in a balanced equation. These are the **reacting masses**.

RULES FOR WORKING OUT REACTING MASSES

STEP 1 Write down the balanced symbol equation.

STEP 2 Work out the relative formula mass of each reactant and product.

STEP 3 Write the relative reacting masses under the balanced equation, taking into account the numbers used to balance the equation.

STEP 4 Convert to reacting masses by adding the units (g, kg, or tonnes).

STEP 5 Scale the quantities to amounts actually used in the synthesis or experiment.

Worked example

What are the masses of reactants and products when sulfuric acid reacts with sodium hydroxide?

Step 1 $2NaOH + H_2SO_4 \longrightarrow Na_2SO_4 + 2H_2O$

Step 2 relative formula mass of NaOH = 23 + 16 + 1 = 40

relative formula mass of H_2SO_4 = (2 × 1) + 32 + (4 × 16) = 98

relative formula mass of Na_2SO_4 = (2 × 23) + 32 + (4 × 16) = 142

relative formula mass of H_2O = (2 × 1) + 16 = 18

Steps 3 & 4	2NaOH	+	H_2SO_4	$\longrightarrow$	Na_2SO_4	+	$2H_2O$
	2 × 40 = 80		98		142		2 × 18 = 36
	80 g		98 g		142 g		36 g

Yields

The yield of any synthesis is the quantity of product obtained from known amounts of starting materials. The **actual yield** is the mass of product after it is separated from the mixture, purified, and dried.

Theoretical yield

The **theoretical yield** is the mass of product expected if the reaction goes exactly as shown in the balanced equation. This is what could be obtained in theory if there are no by-products and no losses while chemicals are transferred from one container to another. The actual yield is always less than the theoretical yield.

Worked example

What is the theoretical yield of ethanoic acid made from 8 tonnes of methanol?

Step 1 *Write down the balanced equation*

methanol	+ carbon monoxide	⟶	ethanoic acid
$CH_3OH(g)$	+ $CO(g)$	⟶	$CH_3COOH(g)$
32			60

Step 2 *Work out the relative formula masses*

methanol: $12 + 4 + 16 = 32$

ethanoic acid: $24 + 4 + 32 = 60$

Steps 3 & 4 *Write down the relative reacting masses and convert to reacting masses by adding the units*

Theoretically, 32 tonnes of methanol should give 60 tonnes of ethanoic acid.

Step 5 *Scale to the quantities actually used*

If the theoretical yield of ethanoic acid = x tonnes, then

$$\frac{\text{mass of ethanoic acid}}{\text{mass of methanol}} = \frac{60\text{ tonnes}}{32\text{ tonnes}} = \frac{x\text{ tonnes}}{8\text{ tonnes}}$$

So, the yield of ethanoic acid from 8 tonnes of methanol should be

$$8\text{ tonnes} \times \frac{60\text{ tonnes}}{32\text{ tonnes}} = 15\text{ tonnes}$$

Percentage yield

The **percentage yield** is the percentage of the theoretical yield that is actually obtained. It is always less than 100%.

Questions

2 What is the mass of salt that forms in solution when:
   a hydrochloric acid neutralises 4 g sodium hydroxide?
   b 12.5 g zinc carbonate, $ZnCO_3$, reacts with excess sulfuric acid?

3 A preparation of sodium sulfate began with 8.0 g of sodium hydroxide.
   a Calculate the theoretical yield of sodium sulfate from 8.0 g sodium hydroxide.
   b Calculate the percentage yield, given that the actual yield was 12.0 g.

Worked example

What is the percentage yield if 8 tonnes of methanol produces 14.7 tonnes of ethanoic acid?

From the previous example:

theoretical yield = 15 tonnes
actual yield = 14.7 tonnes

$$\text{percentage yield} = \frac{\text{actual yield}}{\text{theoretical yield}} \times 100$$

$$= \frac{14.7\text{ tonnes}}{15\text{ tonnes}} \times 100$$

$$= 98\%$$

Science Explanations

Chemists use their knowledge of chemical reactions to plan and carry out the synthesis of new compounds.

You should know:

- that the chemical industry provides useful products such as food additives, fertilisers, dyestuffs, paints, pigments, and pharmaceuticals
- how chemists use indicators and pH meters to detect acids and alkalis and to measure pH
- that common alkalis include the hydroxides of sodium, potassium, and calcium
- that there are characteristic reactions of acids with metals, metal oxides, metal hydroxides, and metal carbonates that produce salts
- how chemists use ionic theory to explain why acids have similar properties, and how alkalis neutralise acids to form salts
- that safety precautions are important when working with hazardous chemicals such as corrosive acids and alkalis
- how to follow the rate of a change by measuring the disappearance of a reactant or the formation of a product and then to analyse the results graphically
- that the concentrations of reactants, the particle size of solid reactants, the temperature, and the presence of catalysts are factors that affect the rates of reaction
- how collision theory can explain why changing the concentration of reactants, or the particle size of solids, affects the rate of a reaction
- that reactions are exothermic if they give out energy, and endothermic if they take in energy
- that a chemical synthesis involves a number of stages and a range of practical techniques, which are important to achieving a good yield of a pure product in a safe way
- that a titration is a procedure that can be used to check the purity of chemicals used in synthesis
- how to use the balanced equation for a reaction to work out the quantities of chemicals to use in a synthesis, and to calculate the theoretical yield.

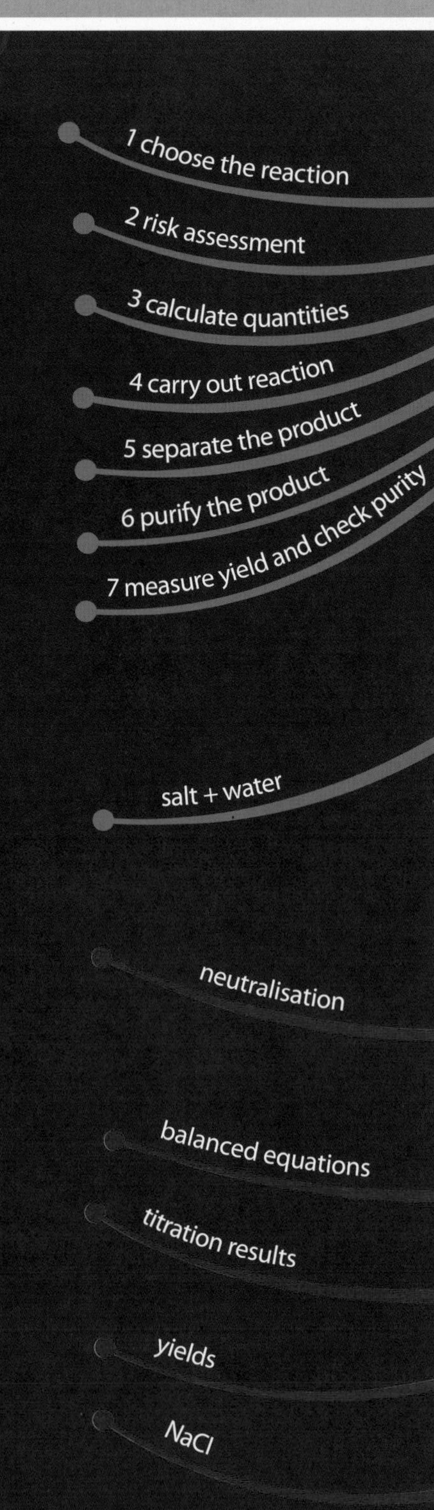

CHEMICAL SYNTHESIS
temperature
effect of concentration
collision theory
to explain
effect of surface area of solids
catalysts
rates
energy
changes
exothermic
endothermic
control
stages
synthesis
solids eg citric acid
liquids eg sulfuric acid
gases eg hydrogen chloride
hydrochloric acid, HCl(aq)
hydrogen ions in solution
ions react to form water
hydroxide ions in solution
neutralisation
acids
alkalis
hydroxides of Na, K, and Ca
chemicals
pH
measurement
universal indicator
pH meter
scale
pH 1–6 acidic
pH 7 neutral
pH 8–14 alkaline
making salts
metals
salt + hydrogen
oxides and hydroxides
carbonates
salt + carbon dioxide + water
chemical industry
products
fertilisers
food additives
pigments
dyes
drugs and medicines
raw materials
crude oil
natural gas
chemicals from plants
formulae
equations
ionic equations
formula masses
charges on ions
reacting masses
ionic formulae
techniques
drying
evaporating
filtering
crystallising
dissolving

Ideas about Science

Scientists can never be sure that a measurement tells them the true value of the quantity being measured. Data is more reliable if it can be repeated. If you take several measurements of the same quantity, for example, in a titration, the results are likely to vary. This may be because:

- the quantity you are measuring is varying, for example, the purity of separate solid samples of a product may not be uniform
- there are variations in the judgement of when the indicator colour corresponds to the end-point, or of the position of a meniscus
- there are limitations in the measuring equipment, such as burettes.

Usually the best estimate of the true value of a quantity is the mean (or average) of several repeat measurements. You should:

- be able to calculate the mean from a set of repeat measurements
- know that a measurement may be an outlier if it lies well outside the range of the other values in a set of repeat measurements
- be able to give reasons to explain why an outlier should be retained as part of the data or rejected when calculating the mean.

When comparing sets of titration results you should know that:

- a difference between their means is real if their ranges do not overlap.

To investigate the relationship between a factor and an outcome, it is important to control all the other factors that might affect the outcome, such as temperature and the concentrations of other reactants. When investigating the rates of chemical reactions you should be able to:

- identify the outcome and factors that may affect it
- suggest how an outcome might alter when a factor is changed.

In a plan for an investigation of the effect of a factor on an outcome, you should:

- be able to explain why it is necessary to control all the factors that might affect the outcome
- recognize that the fact that other factors are controlled is a positive design feature, and the fact that they are not is a design flaw.

If an outcome occurs when a specific factor is present but does not when it is absent, or if an outcome variable increases (or decreases) steadily as an input variable increases, we say that there is a correlation between the two. In the context of studying reaction rates you should:

- understand that a correlation shows a link between a factor and an outcome
- be able to identify where a correlation exists when data is presented as text, as a graph, or in a table
- understand that a correlation does not always mean that the factor causes the outcome.
- identify that where there is a machanism to explain a correlation, scientists are likely to accept that the factor causes the outcome.

Review Questions

1 A student is developing a new sports pack. She needs to find two chemicals that will cool an injury when mixed inside the pack.

a Should the reaction be endothermic or exothermic?

b Which of the energy-level diagrams below represents the reaction that is likely to cool injuries more effectively?

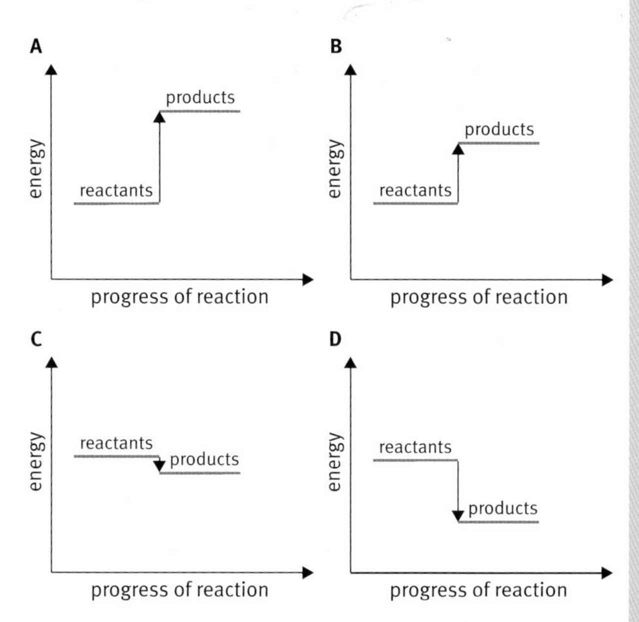

2 A chemist tested the purity of a sample of tartaric acid by titrating with dilute sodium hydroxide. The table shows his results.

	Run 1	Run 2	Run 3	Run 4
Initial burette reading (cm^3)	0.1	0.1	25.0	21.9
Final burette reading (cm^3)	25.2	25.0	50.0	46.9
Volume of acid added (cm^3)	25.1	24.9	25.0	

a Explain why the chemist repeated the titration four times.

b Calculate the volume of alkali added in run 4.

c The chemist calculates the mean of his titration results. Why does he do this?

d Suggest the range within which the true value probably lies.

3 A teacher adds 3 g of zinc granules to dilute hydrochloric acid in a flask. She uses a gas syringe to measure the volume of hydrogen gas given off.

Time (min)	Volume of gas in syringe (cm^3)
1	32
2	56
3	74
4	87
5	95
6	95

a Draw a line graph of the results.

b Calculate the average rate of reaction between three and four minutes, in cm^3 per minute.

c The teacher decides to repeat the experiment with 3 g of zinc powder. She keeps all other factors the same. Draw and label a line on the graph to show the expected results. Use collision theory to explain the effect of this change.

d Explain why it was important for the teacher to keep all other factors the same.

C7 Chemistry for a sustainable world

Why study chemistry?

Chemistry is the science that helps us to understand matter on an atomic scale. It is the central science. Knowledge of chemistry informs materials science and engineering as well as biochemistry, genetics, and environmental sciences.

What you already know

- The chemical industry produces bulk and fine chemicals.
- Manufactured chemicals bring both benefits and risks to society and the environment.
- New materials are made through chemical synthesis.
- Crude oil is a mixture of hydrocarbons.
- Hydrocarbons are the feedstock for many synthetic materials.
- Exothermic and endothermic reactions involve energy changes.
- Paper chromatography is used to separate mixtures.
- How to write chemical formulae.
- How to write word equations and balanced equations.

Find out about

- the 'greening' of the chemical industry
- the chemistry of carbon compounds (organic chemistry)
- energy changes in chemistry
- catalysts and the rates of reactions
- reversible reactions and equilibria
- chemical analysis by chromatography and titrations.

The Science

There are so many carbon compounds that chemists study them in families such as the alcohols, carboxylic acids, and esters. When chemists synthesise new chemicals they use ideas about energy changes and equilibrium to control reactions and maximise yield. Analytical techniques are used to protect us by checking that food and water are safe, diagnosing disease, and solving crimes.

Ideas about Science

Chemists have theories to explain observed data and answer key questions about chemical reactions – how much? how fast? how far? Science-based technologies often improve our quality of life, but may also harm the environment. Benefits must be weighed against costs. The chemical industry is changing to become more sustainable.

Topic 1: Green chemistry

The chemical industry takes crude oil, air, seawater, and other raw materials and converts them to pure chemicals such as acids, salts, solvents, compressed gases, and carbon compounds.

The chemical industry

The chemical industry converts raw materials into useful products. The products include chemicals for use as drugs, fertilisers, detergents, paints, and dyes.

The industry makes **bulk chemicals** on a scale of thousands or even millions of tonnes per year. Examples are ammonia, sulfuric acid, sodium hydroxide, and phosphoric acid.

On a much smaller scale the industry makes **fine chemicals** such as drugs, food additives, and fragrances. It also makes small quantities of speciality chemicals needed by other manufacturers for particular purposes. These include such things as flame retardants, and the liquid crystals for flat-screen televisions and computer displays.

Greener industry

The chemical industry is reinventing many of the processes it uses. The industry seeks to become 'greener' by:

- turning to renewable resources
- devising new processes that convert a high proportion of the atoms in the reactants into the product molecules
- cutting down on the use of hazardous chemicals
- making efficient use of energy
- reducing waste
- preventing pollution of the environment.

Key words

- bulk chemicals
- fine chemicals

Harvesting a natural resource. Lavender is distilled to extract chemicals for the perfume industry.

The work of the chemical industry

Find out about

- feedstocks for the chemical industry
- products from the chemical industry
- people who work in the chemical industry

Key words

- feedstocks
- synthesis
- by-products

Transport workers bring materials in and out of the chemical plant.

Raw materials

The basic raw materials of the chemical industry are:

- crude oil
- air
- water
- vegetable materials
- rocks and minerals such as metal ores, salt, limestone, and gypsum.

The first step in any process is to take the raw materials and convert them into a chemical, or mixture of chemicals, that can be fed into a process. Crude oil, for example, is a complex mixture of chemicals. An oil refinery distills the oil and then processes chemicals from the distillation column to produce purified **feedstocks** for chemical synthesis.

Chemical plants

At the centre of the plant is the reactor. This is where **synthesis** takes place and reactants are converted into products. The feedstock may have to be heated or compressed before it is fed to the reactor. The reactor often contains a catalyst.

Generally, a mixture of chemicals leaves the reactor. The mixture includes the desired product, but there may also be **by-products** and unchanged starting materials. So the chemical plant has to include equipment to separate the main product and by-products and to recycle unchanged reactants.

After separation, samples of the product are analysed to monitor the purity. By-products and waste are either fed back into production processes, sold on for other uses, or disposed of carefully.

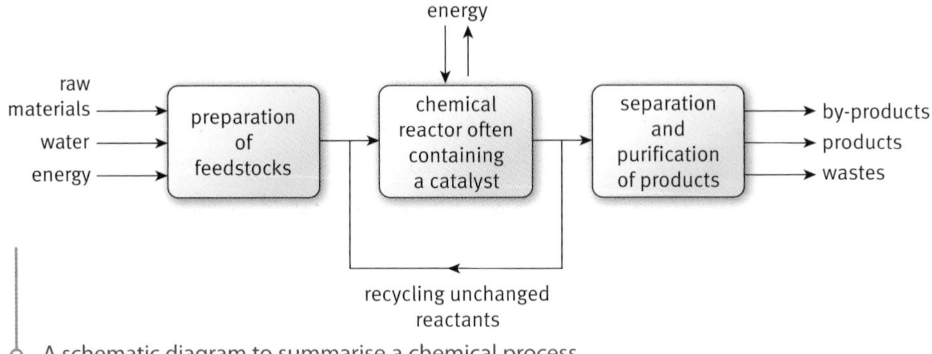

A schematic diagram to summarise a chemical process.

Chemical plants need energy. Some chemical reactions occur at high temperature, so energy is often needed for heating. Also, a lot of electric power is needed for pumps to move reactants and products from one part of the plant to another.

Sensors monitor the conditions at all the key points in the plant. The data is fed to computers in the control centre where the technical team controls the plant.

Products from the chemical industry

The chemical industry produces five main types of product. Some of these are made in huge quantities. Other chemicals have high value but are made in much smaller amounts.

Basic inorganics including fertilisers

The industry makes large amounts of these chemicals. Chlorine, sodium hydroxide, sulfuric acid, and fertilisers are all bulk chemicals.

Petrochemicals and polymers

Petrochemical plants use hydrocarbons from crude oil to make a great variety of products, including polymers.

Dyes, paints, and pigments

Modern dyes are made from petrochemicals, but in the past coal-tar was the main source of carbon compounds for chemicals such as dyes.

Pharmaceuticals

The pharmaceutical industry grew from the dyestuffs industry. The industry produces drugs and medicines.

Speciality chemicals

Speciality chemicals are used to make other products. They include food flavourings and the liquid-crystal chemicals in flat-screen displays.

People who work in the chemical industry

The chemical industry employs a wide range of people. Research chemists carry out investigations to find the best methods for making new products. Production chemists scale up these methods to see if they will work on a large scale. Members of the technical team monitor the data collected from computer sensors to make sure everything is working correctly. In quality assurance, analytical chemists test the purity of samples to make sure they are up to standard.

Research chemists play an important role in developing new materials.

A maintenance worker inside a large chemical reactor, checking to make sure the pipes are not leaking and that the rotating paddles can move freely.

Questions

1. Why do you think it is important for the chemical industry to employ
   - a a technical team?
   - b maintenance workers?
   - c analytical chemists?
2. When the tank in the picture above is in use, it is filled with a reaction mixture. Suggest the purpose of
   - a the rotating paddle in the centre of the tank
   - b the network of pipes round the edge of the tank.

Innovations in green chemistry

Find out about

- new sources of chemicals
- measures of efficiency
- safer ways to make chemicals
- energy efficiency
- reducing waste and recycling

In the second half of the 20th century, the chemical industry found that its reputation with the public was falling. Many people were worried about synthetic chemicals and their impact on health and the environment. Politicians responded by passing laws to regulate the industry. This included controls on the storage and transport of chemicals. At first, these laws were aimed at dealing with the industry as it was then by treating pollution and minimising its effects.

Crops can be grown to supply feedstocks for the chemical industry rather than for their food value. In Europe this includes growing wheat, maize, sugar beet, and potatoes as sources of sugars or starch.

Controlling the flow of oil at a refinery. Crude oil is the main source of organic chemicals today.

More recently, legislation has set out to prevent pollution by changing the industry. New laws encourage companies to reduce the formation of pollutants through changes in production methods, and using raw materials that are cost effective and renewable.

In the early stages the aim was to cut risks by controlling people's exposure to hazardous chemicals. Now green chemistry attempts to eliminate the hazard altogether. If the industry can avoid using or producing hazardous chemicals, then the risk is avoided.

Green chemistry has the potential to bring both social and economic benefits. Innovative green chemistry can benefit industry by increasing efficiency and cutting costs, and benefit society by helping to avoid the dangers of hazardous chemicals.

Renewable feedstocks

One of the aims of green chemistry is to use renewable raw materials. At the moment, crude oil is the main source of chemical feedstocks. Less than 3% of crude oil is used to make chemicals. All the rest is burnt as fuel or used to make lubricants. Even so, reducing our reliance on petrochemicals would help to make the industry more **sustainable**.

Currently, most polymers are made from petrochemicals. This includes the polyester fibres that are widely used in clothing.

DuPont has developed a way of making a new type of polyester by fermenting renewable plant materials. The company calls this polymer Sorona. Manufacturers convert Sorona into fibres for clothing, upholstery, and carpets.

The chemical starting point for the synthesis of Sorona is malonic acid. DuPont has found that it can produce this acid by fermenting corn starch with bacteria. Other feedstocks include soybeans, sugar cane, and wheat.

Plants can be grown year after year. They are a **renewable resource**. However, growing plants for chemicals takes up land that could be used to grow food. Energy is needed to make fertilisers and for harvesting crops, but the production of Sorona uses 40% less energy than that required to make the same amount of nylon. This reduces emissions of greenhouse gases and saves crude oil.

Manufacturers have to weigh up both sides of the argument and decide which is the most sustainable process.

Key words

- sustainable
- renewable resource

Fabrics made from Sorona are used for clothing. It is made using renewable resources, reducing the use of raw materials based on crude oil. Sorona can be recycled. This helps reduce waste.

Questions

1 Classify these raw materials as 'renewable' or 'non-renewable':
   a salt (sodium chloride)
   b crude oil
   c wood chippings
   d limestone
   e sugar beet.

2 Which of the ways of making industry 'greener' are illustrated by the development of Sorona?

Thermal decomposition of calcium carbonate.

RULES FOR WORKING OUT REACTING MASSES

STEP 1 Write down the balanced symbol equation.

STEP 2 Work out the relative formula mass of each reactant and product.

STEP 3 Write the relative reacting masses under the balanced equation, taking into account the numbers used to balance the equation.

STEP 4 Convert to reacting masses by adding the units (g, kg, or tonnes).

STEP 5 Scale the quantities to amounts actually used in the synthesis or experiment.

Spreading lime on a field. Lime contains calcium hydroxide, which is formed by reacting calcium oxide with water. Lime is used to reduce soil acidity.

Converting more of the reactants to products

Yields

Chemists calculate the **percentage yield** from a production process to measure its efficiency. This compares the quantity of product with the amount predicted by the balanced chemical equation.

A high yield is a good thing, but it is not necessarily an indicator that the process is 'green'. This is illustrated on a small scale by the thermal decomposition of calcium carbonate to produce calcium oxide.

Suppose that 10 g of calcium carbonate is heated strongly for 20 minutes and the product is 4.8 g of calcium oxide.

To work out the percentage yield, you must first find the theoretical yield. This is found from the reacting masses.

Step 1: $CaCO_3(s) \longrightarrow CaO(s) + CO_2(g)$

Step 2: relative formula mass of $CaCO_3 = 40 + 12 + (16 \times 3) = 100$

relative formula mass of $CaO = 40 + 16 = 56$

Steps 3 & 4: $CaCO_3(s) \longrightarrow CaO(s) + CO_2(g)$

100 g 56 g

So, 100 g of calcium carbonate could theoretically produce 56 g of calcium oxide.

Step 5: If the theoretical yield of calcium oxide = x g, then

$$\frac{\text{mass of calcium oxide}}{\text{mass of calcium carbonate}} = \frac{56\,\text{g}}{100\,\text{g}} = \frac{x\,\text{g}}{10\,\text{g}}$$

So the theoretical yield from 10 g of calcium carbonate

$$= 10\,\text{g} \times \frac{56\,\text{g}}{100\,\text{g}} = 5.6\,\text{g calcium oxide}$$

The percentage yield from the laboratory process is

$$\frac{\text{actual yield}}{\text{theoretical yield}} \times 100$$

$$= \frac{4.8\,\text{g}}{5.6\,\text{g}} \times 100 = 85.7\%$$

This would be a good yield. However, the actual result is that only 4.8 g of product is made from 10 g of reactant.

Even with 100% yield, only 5.6 g of calcium oxide could be produced from 10 g of calcium carbonate.

Atom economy

In 1998, Barry Trost of Stanford University, USA, was awarded a prize for his work in green chemistry. He introduced the term **atom economy** as a measure of the efficiency with which a reaction uses its reactant atoms.

$$\text{atom economy} = \left(\frac{\text{mass of atoms in the product}}{\text{mass of atoms in the reactants}}\right) \times 100\%$$

In an ideal chemical reaction, all the atoms in the reactants would end up in the products, and no atoms would be wasted. If this was the case the atom economy for the reaction would be 100%.

$CaCO_3(s) \longrightarrow CaO(s) + CO_2(g)$

Total of atoms in the reactants: 1Ca, 1C, 3O
(total relative atomic mass = 100)

Total of green atoms ending in the product: 1Ca, 1O
(total relative atomic mass = 56)

Total of brown atoms ending up as waste: 1C, 2O
(total relative atomic mass = 44)

$$\text{Atom economy} = \frac{56\,g}{100\,g} \times 100 = 56\%$$

Calculating the atom economy for the thermal decomposition of calcium carbonate. The atoms in the reactants that end up in the product are described as 'green'. The atoms that end up as waste are described as 'brown'.

At the very best for this thermal decomposition reaction, just over half of the mass of starting materials can end up as product. So this is not a green process.

This approach does not take yield into account and does not allow for the fact that many real-world processes use a deliberate excess of reactants.

For example, in many neutralisation reactions, such as the reaction between magnesium carbonate and sulfuric acid, the carbonate is in excess. Reactants are also used in excess in the preparation of many carbon-based compounds, such as the preparation of bromoethane from ethanol. This preparation needs an excess of sulfuric acid and sodium bromide. You will learn more about reactions like these later in this module.

Key words

- percentage yield
- atom economy

Questions

3 16 g of methane (CH_4) was burned in the air. 32 g of carbon dioxide was collected during the reaction. During the reaction, some sooty deposits were noticed. The equation for the combustion of methane is:

$CH_4 + 2O_2 \longrightarrow CO_2 + 2H_2O$

a What was the percentage yield of carbon dioxide?
b Why do you think the percentage yield was so low?
c Calculate the atom economy for the reaction.

4 Heating with a catalyst converts cyclohexanol, $C_6H_{11}OH$, to cyclohexene, C_6H_{10}.
a What is the percentage yield if 20 g of cyclohexanol gives 14.5 g of cyclohexene?
b What is the atom economy, assuming that the catalyst is recovered and reused?

Key words

- exothermic
- catalyst
- enzyme

This weedkiller can now be made by a method that does not involve the use of toxic cyanide compounds.

Avoiding chemicals that are hazardous to health

The chemical industry produces a large number of synthetic chemicals. Some of these are reactive intermediates that are only used in manufacturing processes.

One aim of green chemistry is to replace reactants that are highly toxic with alternative chemicals that are not a threat to human health or the environment.

The aim is to protect the health of people working in the industry and also people who live near industrial plants. It is important to avoid chemical accidents, including accidental release of chemicals through explosions and fires.

Originally the company Monsanto used hydrogen cyanide in the process to produce the weedkiller that the company markets as 'Roundup'. Hydrogen cyanide is extremely toxic. The company has now developed a new route for making the herbicide. The new method has a different starting material and runs under milder conditions because of a copper catalyst.

Similarly, a new process for making polycarbonate plastics has replaced the gas phosgene with safer starting materials: methanol and carbon monoxide. Carbon monoxide is poisonous but it is not as dangerous as phosgene, which is so nasty that it has been used as a poison gas in warfare.

Energy efficiency

All manufacturing processes need energy to convert raw materials into useful products. In the chemical industry, energy is used in several ways, such as:

- to raise the temperature of reactants so a reaction begins or continues
- to heat mixtures of liquids to separate and purify products by distillation
- to dry product material
- to process waste.

The energy used in separation, drying, and waste management may be more than that used in the reaction stages.

Burning natural gas or other fossil fuels is the usual source of energy. Often the energy from burning is used to produce super-heated steam, which can then be used for heating around the chemical plant.

The most direct way of reducing the use of energy is to prevent losses of steam from leaking valves on steam pipes and by installing efficient insulation on reaction vessels or pipes.

Some of the reactions in the chemical process may be so **exothermic** that they provide the energy to raise steam and generate electricity. The energy is transferred using a heat exchanger. The first step in the manufacture of sulfuric acid is to burn sulfur. This is so exothermic that a sulfuric acid plant has no fuel bills and can raise enough steam to generate sufficient electric power to contribute significantly to the income of the operation.

Chemical production in general has become much more energy efficient than in the past. The average energy required per tonne of chemical product is less than half that needed 50 years ago.

The development of efficient **catalysts** has made a significant contribution. With a suitable catalyst, it is possible to speed up the reaction that gives the desired products, without speeding up other reactions that give unwanted by-products. This reduces waste, and reactions work at lower temperatures, saving energy.

One aim of green chemistry is not only to make processes more energy efficient but also to lower their energy demand. New processes are being developed that run at much lower temperatures. One way of doing this is to use biocatalysts – the **enzymes** produced by microorganisms. Enzymes operate within a limited temperature range, above which they are **denatured** and no longer work. Each enzyme also works within a limited pH range. This limits the conditions that can be used for enzyme-catalysed processes.

A large heat exchanger works like a laboratory condenser. One liquid or gas flows through pipes surrounded by another liquid flowing in the opposite direction. The hotter liquid or gas heats the cooler fluid.

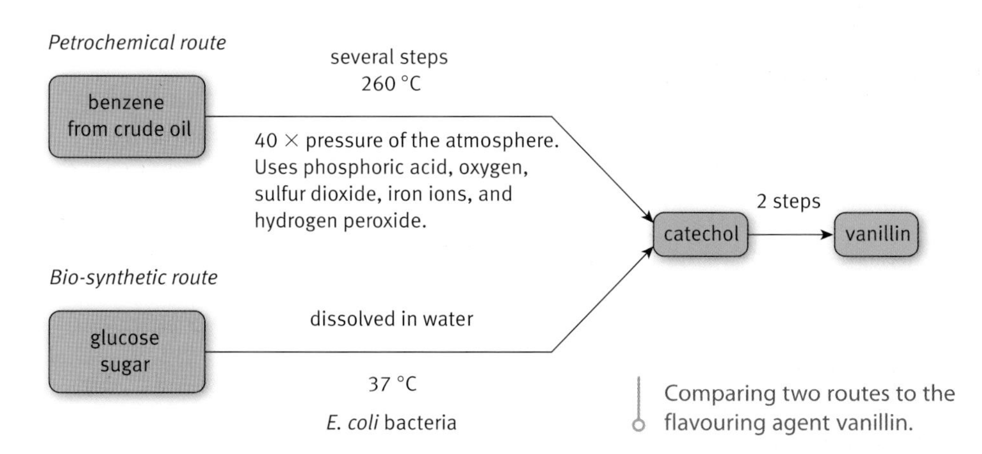

Comparing two routes to the flavouring agent vanillin.

Questions

5 The reaction involving cyanide in the older process for making the active ingredient for Roundup was exothermic. The replacement reaction in the newer process is endothermic. Suggest why this difference contributes to safety.

6 a Write a short paragraph to explain why the biosynthetic route to vanillin is 'greener' than the petrochemical route.

b Explain why it may not be a good idea to try and speed up the biosynthetic route by heating the reaction mixture.

A plant for recycling chemicals on a large scale.

Reducing waste and recycling

One of the principles of green chemistry is that it is better to prevent waste than to treat or clean up waste after it is formed.

One way of cutting down on waste is to develop processes with higher atom economies. Another way is to increase **recycling** at every possible stage of the life cycle of a chemical product. A third way is to find uses for by-products that were previously dumped as waste.

Recycling

Industries have always tried to recycle waste produced during manufacturing processes. Recycling is easier when the composition of the waste material is known.

A major manufacturer of polypropylene, Basell Polyolefins, used to burn unreacted propene. Much was burnt in an open flame so that not even the energy was recovered. This changed when the company installed a distillation unit to separate chemicals from the waste gases. The recovery unit cut the amount of waste. It collected over 3000 tonnes of propene per year.

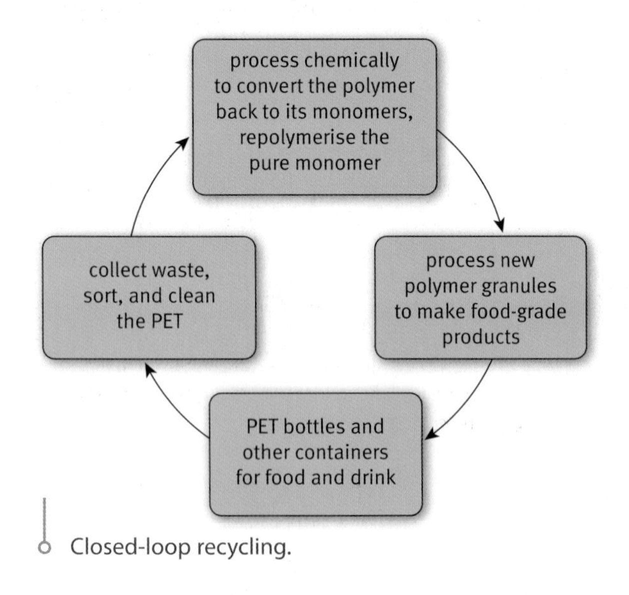

Closed-loop recycling.

Closed-loop recycling

Recycling is at its best when the waste material that is collected can be used to manufacture the same product with no loss in quality. With plastic waste this can be done by breaking down the waste into the monomers originally used to make the polymer. Several companies have developed processes for depolymerising the polyester (PET) in soft-drinks bottles. The result is fresh feedstock for making new polymer.

Open-loop recycling

In some cases, waste from one product is recovered and used in the manufacture of another, lower-quality product. This is open-loop recycling. It cuts down the amount of fresh feedstock needed, and the amount of waste going to landfill. But it is not as good value as closed-loop recycling.

Discarded PET soft-drinks bottles can be collected and fed through grinders that reduce them to flake form. The flake then proceeds through a separation and cleaning process that removes all foreign particles such as paper, metal, and other plastic materials.

The recovered PET is sold to manufacturers, who convert it into a variety of useful products such as carpet fibre, moulding compounds, and non-food containers. Carpet companies can often use 100% recycled polymer to make polyester carpets. PET is also spun to make fibre filling for pillows, quilts, and jackets.

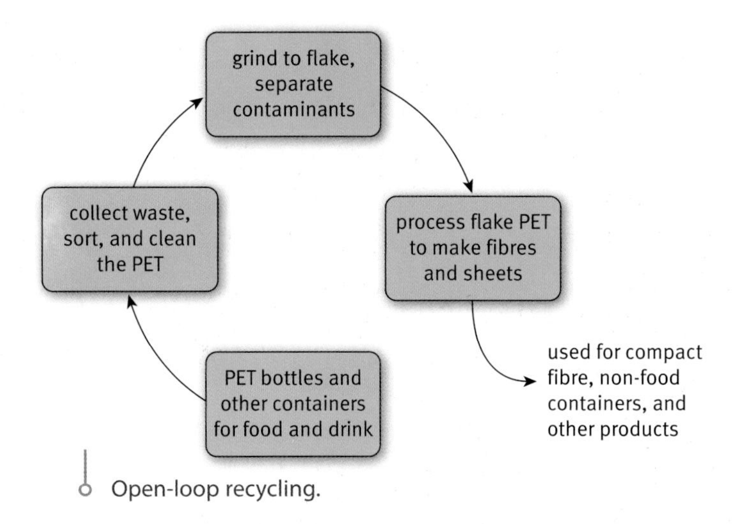

Open-loop recycling.

New uses for by-products

All chemical processes give a mixture of products: the one that the chemists want to make and others – the by-products. The process is more sustainable if the by-products can be used to make another product, so that less waste has to be dumped.

Titanium dioxide is an important pigment used in paint, plastics, paper, cosmetics, and toothpaste. By-products of the process for producing titanium dioxide include iron sulfate, which is sold to water companies to treat water, and gypsum, which is used to make plasterboard.

Titanium dioxide is the white pigment in the paint protecting a railway bridge in Newcastle-upon-Tyne.

Cutting pollution by waste

It is usually impossible to eliminate waste completely. This means that it is important to remove or destroy any harmful chemicals before waste is released into air, water, or landfills. On many sites, ground water must also be collected and processed, as it may contain traces of the chemicals made and used on the site.

Many manufacturing sites have a single processing plant for dealing with waste. A wide range of separation techniques may be used, including filtering, centrifuging, and distillation.

Waste may also be treated chemically to neutralise acids or alkalis, to precipitate toxic metal ions, or to convert chemicals to less harmful materials. Microorganisms or reed beds may be used to break down some chemicals.

Questions

7 a Explain in a short paragraph the difference between open-loop and closed-loop recycling.
  b Suggest one possible advantage and one possible disadvantage of each of these approaches to recycling.

8 Why is it important to carefully control waste from industrial processes?

Designing new catalysts

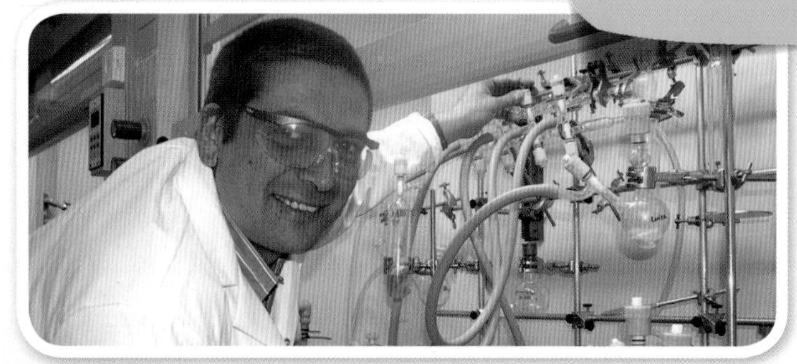
Matthew Davidson in his laboratory.

Matthew Davidson works at the University of Bath, where he designs new catalysts for the synthesis of polymers. His main strategy is to wrap an organic structure around a metal ion. He has helped a manufacturer to devise a catalyst for making a type of polyester by wrapping citrate ions around titanium ions. This catalyst produces a clear, rather than a yellowish, polymer and replaces a catalyst made of antimony, which is a toxic metal.

'The main thing about a catalyst is that it has to be highly reactive towards the starting materials but not to the products.'

Matthew enjoys making new complex molecules and then analysing their structure to see what makes them active. 'I have to think about two main things – firstly the size and shape of the catalyst molecules and secondly how they interact with the electrons of other molecules.'

The need for new catalysts

But why do we need new catalysts? Matthew says there are several reasons: 'First, many older catalysts were toxic or contained harmful metals. That is why we wanted to replace the antimony used to make polyesters.'

'Second, old catalysts may not be as efficient as possible. Our new titanium citrate catalyst is up to 15% more efficient than the traditional antimony catalyst. This allows more polymer to be made with less catalyst – good for both commercial and environmental reasons.'

'Third, there are many useful chemical transformations for which there are no suitable catalysts yet. New ones can help our lives by making new medicines or new polymers.'

Matthew Davidson uses physical models and computer models in his work.

Green chemistry and sustainable development

Many people think that green chemistry and sustainable development are the same thing, but there are differences. **Sustainable development** has a much wider brief.

Sustainable development is about how we organise our lives and work so that we don't destroy our planet, our most valued natural resource. It is about meeting the needs of the present without compromising the ability of future generations to meet their own needs.

Green chemistry is about the long-term sustainability of the planet and the short-term impacts of the chemical industry on our health and the environment. It is a way of thinking that can help chemists in research and production to make more eco-friendly and efficient products.

A green pain reliever

When Boots patented ibuprofen in the 1960s, there were six stages in the process of making the drug. The process had a low atom economy. It needed 75 atoms for each ibuprofen molecule formed. So, 42 of these atoms ended up as waste.

20 years later, the patent ran out. The Celanese Corporation used ideas about green chemistry to develop a new process with a more efficient atom economy, fewer harmful by-products, less waste, and catalysts that could be recovered and recycled.

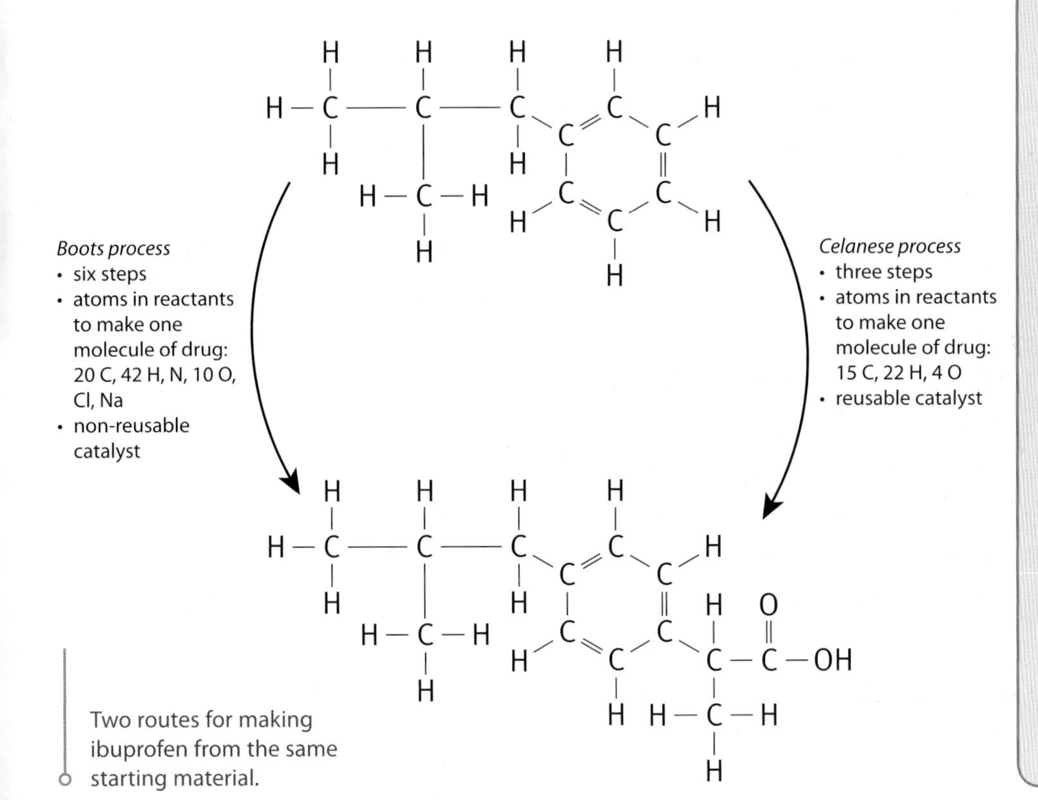

Two routes for making ibuprofen from the same starting material.

Key word

✔ **sustainable development**

Questions

9 Compare sustainable development and green chemistry. What is the difference?

10 What is the molecular formula for ibuprofen?

11 **a** What is the atom economy for making ibuprofen:
- by the Boots process?
- by the Celanese Corporation process?

b Calculate a new value of the atom economy for the Celanese process, if the ethanoic acid (CH_3COOH) formed as a by-product in one step is recycled and does not go to waste.

Topic 2: The chemistry of carbon compounds

Synthetic polymers make good nets to catch fish for food.

Spiders spin webs with a natural organic polymer to catch food.

There are more carbon compounds than there are compounds of all the other elements put together. The chemistry of carbon is so important that it forms a separate branch of the subject called **organic chemistry**.

Organic chemistry

The word 'organic' means 'living'. At first, organic chemistry was the study of compounds from plants and animals. Now that we know a wide variety of compounds can be made artificially, the definition of organic chemistry has been broadened. It now includes the study of synthetic compounds, such as polymers, drugs, and dyes.

Key word

- ✓ organic chemistry

Chains and rings

It helps to think of organic compounds being made up of a skeleton of carbon atoms supporting other atoms. Some of these other atoms may be reactive, while others are less so. In organic compounds, carbon is often linked to hydrogen, oxygen, nitrogen, and halogen atoms.

Carbon forms so many compounds because carbon atoms can join up in many ways, forming chains, branched chains, and rings. The chains can be very long, as in the polymer polythene. A typical polythene molecule may have 10 000 or more carbon atoms linked together. A polythene molecule is still very tiny, but much bigger than a methane molecule.

To make sense of the huge variety of carbon compounds, chemists think in terms of families, or series, of organic compounds.

Bonding in carbon compounds

The bonding in organic compounds is covalent. The structures are molecular. The structures can be worked out from knowing how many covalent bonds each type of atom can form. Carbon atoms form four bonds, hydrogen atoms form one bond, while oxygen atoms form two bonds.

Propane: a hydrocarbon with three carbon atoms in a chain.

Methylbutane: a hydrocarbon with a branched chain.

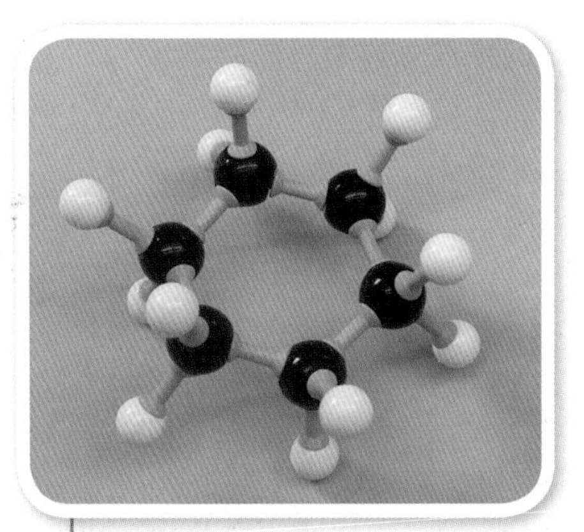

Cyclohexane: a hydrocarbon with a ring of carbon atoms.

Find out about

- the alkane series of hydrocarbons
- physical properties of alkanes
- chemical reactions of alkanes
- saturated and unsaturated hydrocarbons

CH_4 the molecular formula

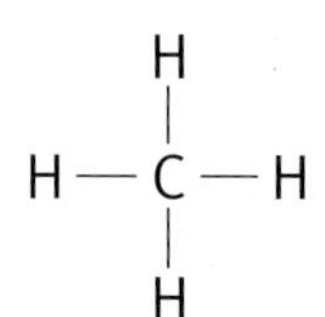

the structural formula showing the chemical bonds

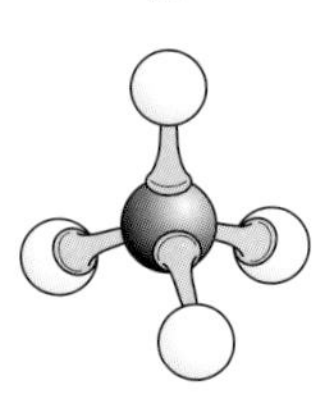

the ball and stick representation showing the tetrahedral shape of the molecule

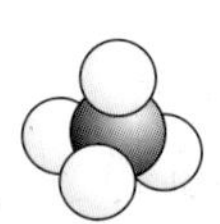

the space filled by the molecule

Ways of representing a molecule of methane.

Alkanes do not mix with water and are less dense than water. Here a dye colours the upper alkane layer.

The **alkanes** make up an important family of **hydrocarbons**. They are well known because they are the compounds in fuels such as natural gas, liquid petroleum gas (LPG), and petrol. The simplest alkane is methane. This is the main gas in natural gas.

The table below shows four alkanes.

Name	Molecular formula	Structural formula
methane	CH_4	H | H — C — H | H
ethane	C_2H_6	H H | | H — C — C — H | | H H
propane	C_3H_8	H H H | | | H — C — C — C — H | | | H H H
butane	C_4H_{10}	H H H H | | | | H — C — C — C — C — H | | | | H H H H

Physical properties of alkanes

The alkanes are oily. They do not dissolve in water or mix with it.

At room temperature, alkanes with small molecules (up to four carbon atoms) are gases, those with 4–17 carbon atoms are liquids, and those with more than 17 carbon atoms are solids.

Chemical properties of alkanes

Burning

All alkanes burn. Many common fuels consist mainly of alkanes. The hydrocarbons burn in air, forming carbon dioxide and water.

If the air is in short supply, the products may include particles of soot (carbon) and the toxic gas carbon monoxide.

Reactions with aqueous acids and alkalis

Alkanes do not react with common laboratory reagents such as acids or alkalis. The C—C and C—H bonds in the molecules are difficult to break and are therefore unreactive.

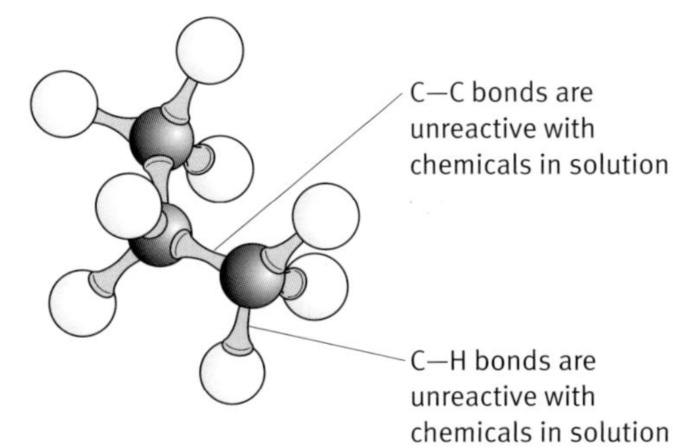

Alkanes are generally unreactive because the C—C and C—H bonds do not react with common aqueous reagents.

At room temperature the alkanes in candle wax are solid. The flame first melts them and then turns them to gas in the hot wick. The hot gases burn in air.

Saturated and unsaturated hydrocarbons

Alkenes are not very reactive. They are examples of **saturated** hydrocarbons because all the C—C bonds are single.

The **alkenes** are another important series of hydrocarbons. These are examples of **unsaturated** hydrocarbons. In unsaturated hydrocarbons there are C=C **double bonds** found between some carbon atoms. The presence of the double bond increases the reactivity of the hydrocarbon. Plastics such as polythene are made by polymerisation from the monomer **ethene**. It is the presence of the double bond that allows this to happen.

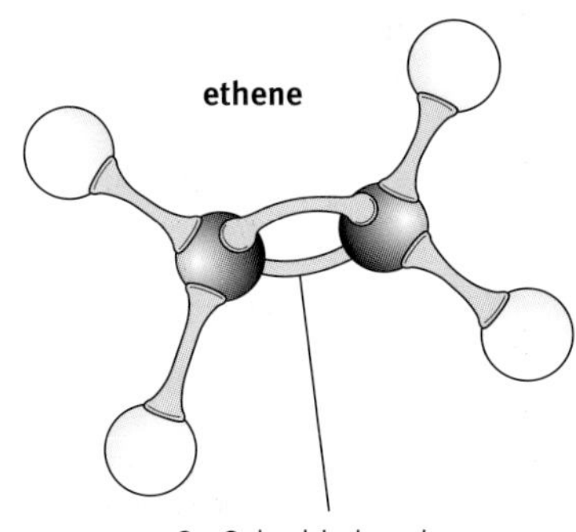

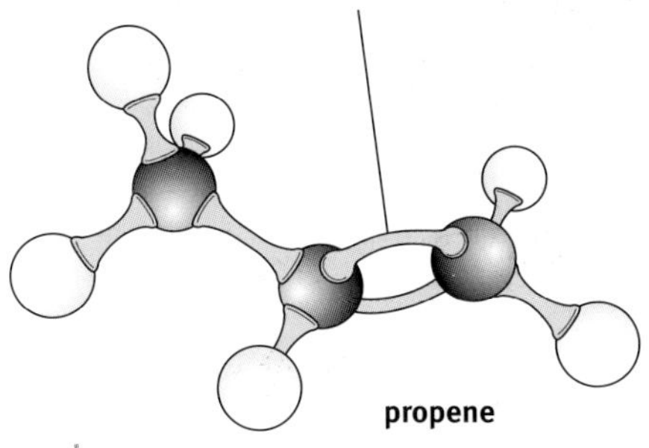

Ethene and propene are examples of alkenes. These are unsaturated hydrocarbons. The double bonds in unsaturated hydrocarbons are more reactive than single bonds.

Questions

1. a In which group of the periodic table does carbon belong?
   b How many electrons are there in the outer shell of a carbon atom?
   c Which groups in the periodic table include elements that form simple ions?
   d Is carbon likely to form simple ions?
2. Petrol contains octane molecules with the formula C_8H_{18}. Draw the structural formula of octane.
3. Write a word equation for propane burning in plenty of air. Write a balanced equation with state symbols for the same reaction.
4. Write a word equation for methane burning in a limited supply of air to form carbon monoxide and steam. Write a balanced equation with state symbols for the same reaction.
5. Explain why the ethene molecule is more reactive than the ethane molecule.

Key words

- ✓ **alkane**
- ✓ **hydrocarbon**
- ✓ **saturated**
- ✓ **unsaturated**
- ✓ **ethene**

Find out about

- physical properties of alcohols
- chemical reactions of alcohols

Ethanol (C_2H_5OH): a simple alcohol. Chemists name alcohols by changing the ending of the name of the corresponding alkane to '-ol'. Ethanol is the two-carbon alcohol related to ethane.

Propanol (C_3H_7OH): a three-carbon alcohol related to propane.

Uses of alcohols

Ethanol is the best-known member of the series of **alcohols**. It is the alcohol in beer, wine, and spirits. Ethanol is also a very useful solvent. It is a liquid that evaporates quickly, and for this reason it is used in cosmetic lotions and perfumes. Ethanol easily catches fire and burns with a clean flame, so it can be used as a fuel.

The simplest alcohol, methanol, can be made in two steps from methane (natural gas) and steam. This alcohol is important as a chemical feedstock. The chemical industry converts methanol to a wide range of chemicals needed to manufacture products such as adhesives, foams, solvents, and windscreen washer fluid.

Structures of alcohols

The first two members of the alcohol series are methanol and ethanol.

```
       H                   H    H
       |                   |    |
  H —  C — O — H      H —  C —  C — O — H
       |                   |    |
       H                   H    H
    methanol              ethanol
```

There are two ways of looking at alcohol molecules that can help us to understand their properties. On the one hand, an alcohol can be seen as an alkane with one of its hydrogen atoms replaced by an —OH group. On the other hand, the same molecule can be regarded as a water molecule with one of its hydrogen atoms replaced by a hydrocarbon chain.

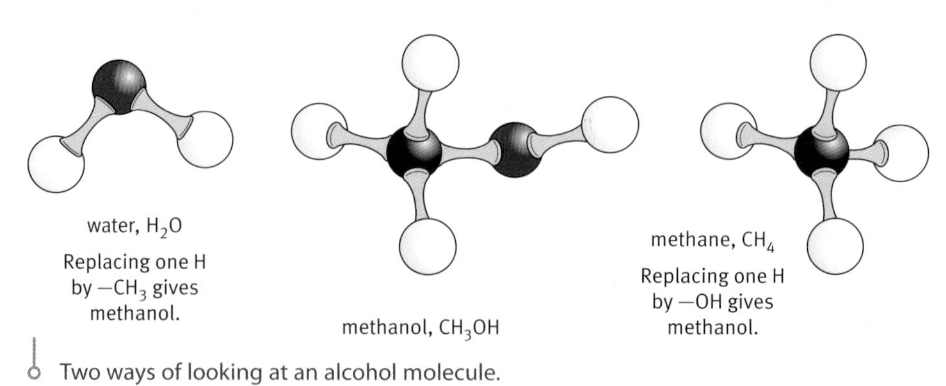

Two ways of looking at an alcohol molecule.

Physical properties

Methanol and ethanol are liquids at room temperature. Alkanes with comparable relative molecular masses are gases. This shows that the attractive forces between molecules of alcohols are stronger than they are in alkanes. The presence of an —OH group of atoms gives the molecules this greater tendency to cling together, like water.

Even so, the boiling point of ethanol (78°C) is below that of water (100°C). Ethanol molecules have a greater mass than water molecules, but the attractions between the hydrocarbon parts are very weak, as in alkanes.

Overall, ethanol molecules have less tendency to stick together than water molecules.

Similarly methanol and ethanol mix with water, unlike alkanes. This is because of the —OH group in the molecules. However, alcohols with longer hydrocarbon chains, such as hexanol ($C_6H_{13}OH$), do not mix with water because the oiliness of the hydrocarbon part of the molecules dominates.

Chemical properties

The —OH group is the reactive part of an alcohol molecule. Chemists call it the **functional group** of alcohols.

Burning

All alcohols burn. Methanol and ethanol are highly flammable and are used as fuels. These compounds can burn in air, to produce carbon dioxide and water, because of the hydrocarbon parts of their molecules.

Reaction with sodium

Alcohols react with sodium in a similar way to water. This is because both water molecules and alcohol molecules include the —OH group of atoms. With water, the products are sodium hydroxide and hydrogen. With ethanol, the products are sodium ethoxide and hydrogen.

$$2\,CH_3CH_2OH + 2Na \longrightarrow 2\,CH_3CH_2O^-Na^+ + H_2$$

sodium ethoxide

Only the hydrogen atom attached to the oxygen atom is involved in this reaction. The hydrogen atoms linked directly to carbon are inert, or unreactive.

The product has an ionic bond between the oxygen of the ethoxide ion and the sodium ion. Sodium ethoxide, like sodium hydroxide, is an ionic compound and a solid at room temperature.

Key words

- alcohols
- functional group

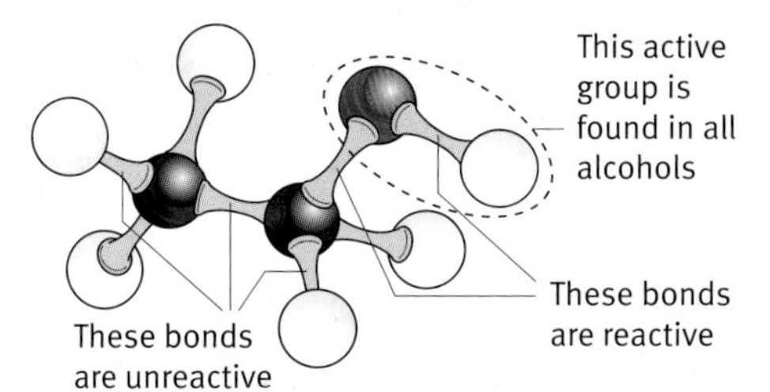

The number of carbon and hydrogen atoms does not have much effect on the chemistry of alcohols.

Some bonds in ethanol are more reactive than others. Alcohols are more reactive than alkanes because C—O and O—H bonds are more reactive than C—C and C—H bonds. The alcohols share similar chemical properties because they all have the —OH group in their molecules.

Questions

1. Produce a table for three alcohols similar to the table of alkanes in Section 2A.
2. a Use values of relative atomic masses from the periodic table in C4 Section A to show that propane and ethanol have a very similar relative mass.
   b Propane boils at −42 °C, but ethanol boils at 78 °C. Suggest an explanation for the difference.
3. Write a balanced equation for propanol burning.
4. Write balanced equations for the reactions of sodium with:
   a water
   b methanol.
   c ethane.

Find out about

- the production of ethanol
- the influence of green chemistry

Most ethanol produced in the world is used as a fuel. This is a rapidly growing market with the increasing popularity of biofuels as a source of renewable energy. Ethanol is also important for the chemical industry. It is used as a solvent in the manufacture of perfumes and pharmaceuticals, and as a feedstock in the production of acrylic polymers.

Ethanol can be produced by three different routes:

- fermentation
- biotechnology
- chemical synthesis.

The method used usually depends on what the ethanol is to be used for and the availability of feedstocks.

As the principles of green chemistry are continuously applied to each method, suitable modifications are made.

Bioethanol is the most widely used green car fuel in the world.

Fermentation

Most of the world's alcohol, 93%, is produced by the traditional method of **fermentation** of sugar with yeast. Ethanol produced by this method is mainly used as a fuel, with smaller amounts used in alcoholic beverages and the chemical industry.

Feedstocks

Common feedstocks are sugar cane, sugar beet, corn, rice, and maize. Large areas of land are needed to grow the crops, and only some parts of the plants can be fermented. The parts that cannot be fermented are used to make animal feeds and corn oil. Recent developments mean that more plant material can be fermented, and agricultural waste, paper mill sludge, and even household rubbish can be used for fermentation.

Sugar cane is a feedstock for the production of ethanol by fermentation.

The reaction

Cellulose polymers from the feedstock are heated with acid to break them down into simple sugars such as glucose. Glucose is then converted into ethanol and carbon dioxide. This reaction is catalysed by enzymes found in yeast:

$$\text{glucose} \longrightarrow \text{ethanol} + \text{carbon dioxide}$$

$$C_6H_{12}O_6 \longrightarrow 2C_2H_5OH + 2CO_2$$

The optimum temperature for the fermentation reaction with yeast is in the range 25–37 °C. At lower temperatures the rate of reaction is too slow, and at higher temperatures the enzymes are denatured. Enzymes are also affected by pH. This is because changes in pH can make and break bonds within and between the enzymes, changing their shape and therefore their effectiveness.

The concentration of ethanol solution produced by the fermentation process is limited to between 14 and 15% ethanol. If the ethanol concentration rises higher than 15%, it becomes toxic to the yeast, which is killed, and the fermentation stops.

If higher concentrations of ethanol are required, the mixture must be distilled. Spirits such as brandy and whisky contain about 40–50% ethanol, and are produced by **distillation**.

Distillation is a technique for separating a mixture of liquids by their boiling points. The boiling point of ethanol is 78.5 °C, while the boiling point of water is 100 °C. By heating the mixture to a temperature above 78.5 °C and below 100 °C the ethanol turns into a vapour; this is then condensed and collected, leaving much of the water behind.

A row of whisky stills. The whisky is distilled to increase the concentration of alcohol.

Energy balance

In order to produce bioethanol, energy needs to be used. This energy often comes from fossil fuels. The main stages that require energy are:

- producing fertilisers for the plants to grow
- transporting the feedstock to the factory
- processing the feedstock
- transporting the ethanol to its point of use.

When bioethanol is used as a fuel, energy is released. It is important that the amount of energy released from the fuel when it is burnt is greater than the amount of energy used in its production. This is the energy balance. The energy balance for fermentation is different for different feedstocks.

Feedstock	Energy balance	Comments
sugar cane	8.3–10.2	Sugar cane can only be grown in tropical and subtropical climates. It is the feedstock used for most bioethanol produced in Brazil.
sugar beet	1.4–2.1	Sugar beet is commonly used to produce bioethanol in the UK.
corn	1.3–1.6	Corn is commonly used to produce bioethanol in the US.

The energy balance of ethanol production by fermentation from different feedstocks.

The energy balance number tells you how many times more energy is released than is used in production. For example, with a corn feedstock, one unit of fossil fuel energy is required to create enough ethanol to release 1.3–1.6 units of energy when it is burnt. The energy balance for sugar cane from Brazil is much more favourable. The higher the energy balance, the more 'green' the process.

Questions

1. Explain why the concentration of ethanol solution made by fermentation will never reach 16%.
2. A new process for producing ethanol by fermentation uses a wheat feedstock. 42 MJ of energy is used to produce a gallon of ethanol, which releases 80 MJ of energy when it is burnt.
   a. What is the energy balance for this process?
   b. Is the energy balance better or worse than for corn?

Biotechnology can convert this waste biomass into ethanol.

Key words

- fermentation
- distillation
- biomass

Grangemouth petrochemical plant in Scotland is a major producer of synthetic ethanol.

Biotechnology

Yeast is good at converting glucose to ethanol; however, many plant feedstocks also contain other sugars that cannot be broken down by yeast. To make use of these feedstocks an alternative method is needed.

In 1987 Professor Lonnie Ingram, a microbiologist from the University of Florida, used biotechnology to genetically modify *E. coli* bacteria. He was looking for a way to make biofuels suitable for cars. He found that ethanol could be produced from a wide range of feedstocks. This method can be used to produce fuels and feedstocks for the chemical industry. At present only a very small proportion of bioethanol is produced in this way.

Further developments

In the continuing search for more efficient ways of producing ethanol, scientists have been researching other genetically modified microorganisms including bacteria, fungi, and yeast.

They have been developing new processes using:

- bacteria that can break down a wide range of sugars into ethanol
- fungi that can break down biomass into glucose for use in traditional fermentation
- yeast that can turn sugars other than glucose into ethanol
- yeast that can withstand higher concentrations of ethanol.

Feedstocks

A wide range of **biomass** waste can be used as a feedstock, including forestry and wood waste, rice hulls, and corn stalks.

The reaction

The genetically modified *E. coli* bacteria convert all plant sugars, not just glucose, into ethanol. The bacteria would normally produce ethanoic or lactic acid, but the modification means ethanol is produced instead:

$$\text{sugar} \longrightarrow \text{ethanol} + \text{carbon dioxide}$$

The optimum conditions for the reaction to occur once again lie within the temperature range 25–37 °C and the pH range 6–7.

Chemical synthesis

The UK is the world's largest producer of synthetic ethanol. It is produced by the petrochemical industry. Ethanol for use in the chemical industry is often made by the method shown in the next diagram.

Feedstocks

The main feedstock for producing synthetic ethanol is ethene. Ethene is produced by the cracking of ethane from natural gas. Ethene is also

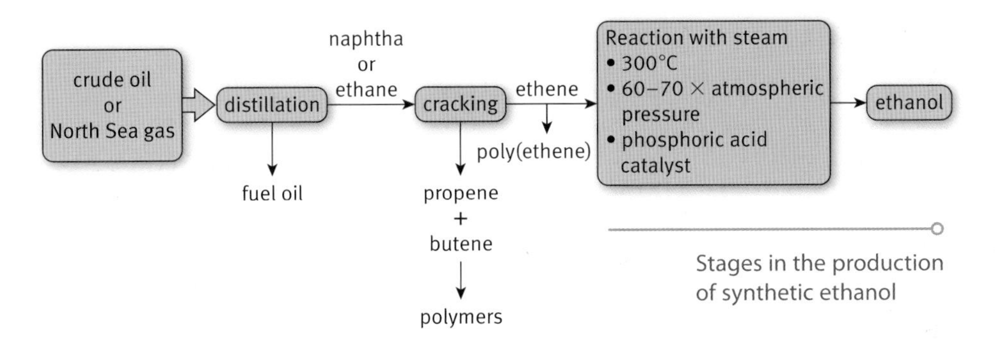

Stages in the production of synthetic ethanol

A cosmetics researcher smelling a prototype perfume. Perfume oils are usually diluted in a solvent. Synthetic ethanol can be used as a solvent in perfumes.

produced by cracking of naphtha from crude oil. The feedstock is not renewable. When oil and gas supplies eventually run out, an alternative feedstock will have to be found.

The reaction

Ethene reacts with steam in the presence of a phosphoric acid catalyst, at temperatures of about 300 °C and at 60–70 times atmospheric pressure:

$$\text{ethene} + \text{steam} \longrightarrow \text{ethanol}$$

$$C_2H_4 + H_2O \longrightarrow C_2H_5OH$$

The atom economy for the reaction is 100%, but some side reactions do occur, producing by-products such as polythene. Any unreacted molecules are recycled through the system again. The overall yield for the reaction is 95%.

Purifying the product

The end product is 96% ethanol, 4% water. It is really difficult to remove the last of the water and obtain 100% ethanol.

Older purification methods required a lot of energy to remove the water and sometimes introduced toxic chemicals into the system. New methods use special compounds called 'zeolites' that have tiny holes all over their surface. At room temperature, water molecules are absorbed onto the surface, leaving behind the larger ethanol molecules, which are too big to fit into the holes. The zeolites can then be dried and reused.

Questions

3 Suggest a reason why it is not safe to drink ethanol produced by the synthetic method.

4 Make a table listing the advantages and disadvantages of each method of producing ethanol.

5 Which method do you think has the largest impact on the environment? Give reasons for your answer.

6 Which method do you think is the 'greenest'? Use your answer to question 4 to explain your choice.

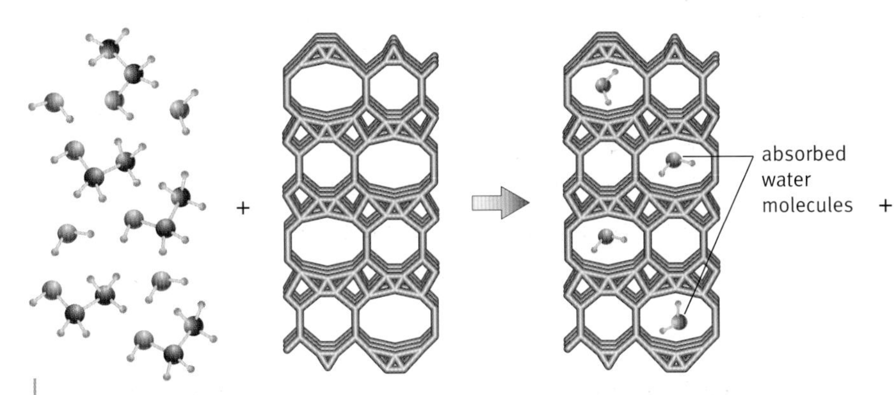

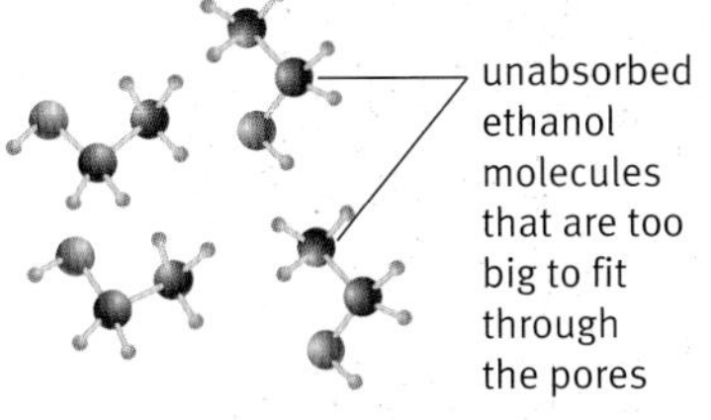

Zeolites can be used as a dehydrating agent as they only absorb the very small water molecules, leaving the ethanol molecules behind.

2D Carboxylic acids

Find out about

- structures and properties of organic acids
- acids in vinegar and other foods
- carboxylic acids as weak acids

The sting of a red ant contains methanoic acid. The traditional name for this acid is formic acid, from the Latin word for ant, *formica*.

Do you like a dash of dilute solution of ethanoic acid on your chips?

Acids from animals and plants

Many acids are part of life itself. These are the organic acids, many of which appear in lists of ingredients on food labels. Ethanoic acid, traditionally known as acetic acid, is the main acid in **vinegar**. Citric acid gives oranges and lemons their sharp taste.

Some of the acids with more carbon atoms have unpleasant smells. The horrible odours of rancid butter, vomit, and sweaty socks are caused by the breakdown of fats to produce organic acids, including butyric, hexanoic, and octanoic acids. Butyric acid is the traditional name for butanoic acid. The word 'butyric' comes from its origins in butter.

Structures and names of organic acids

The functional group in the molecules of organic acids is:

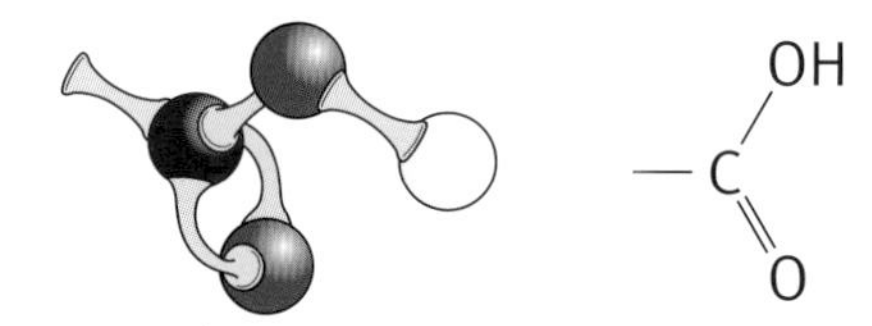

The series of compounds with this reactive group are the **carboxylic acids**. The chemical names of the compounds are related to the alkane with the same number of carbon atoms. The ending 'ane' becomes 'anoic acid'. So the systematic name for acetic acid, the two-carbon acid, is ethanoic acid.

Formation of vinegar

Oxidation of ethanol produces ethanoic acid. Vinegar is a dilute solution of ethanoic acid, and is manufactured by allowing solutions of alcohol to oxidise.

$$H-\underset{H}{\overset{H}{C}}-\underset{H}{\overset{H}{C}}-OH \xrightarrow{\text{oxidation}} H-\underset{H}{\overset{H}{C}}-C\begin{matrix}{}^{\nearrow O} \\ {}_{\searrow OH}\end{matrix}$$

Oxidation converts beer to malt vinegar. Cider oxidises to cider vinegar and wine to wine vinegar.

Acidity of carboxylic acids

Some acids ionise completely to give hydrogen ions when they dissolve in water. Chemists call them **strong acids**. Hydrochloric, sulfuric, and nitric acids are strong acids.

Carboxylic acids ionise to produce hydrogen ions when dissolved in water to a lesser extent than the strong acids. Only a small proportion of the molecules ionise so not all the hydrogens are released as ions into the solution. They are **weak acids**. This helps to explain why vinegar is pH3 but dilute hydrochloric acid is pH1.

In a molecule of ethanoic acid, there are four hydrogen atoms. Three are attached to a carbon atom and one to an oxygen atom. Only the hydrogen atom attached to oxygen is reactive. This is the hydrogen atom that ionises in aqueous solution.

$$CH_3COOH \xrightarrow{\text{water}} CH_3COO^- + H^+$$

ethanoic acid → ethanoate ion

Methanoic acid, HCOOH.

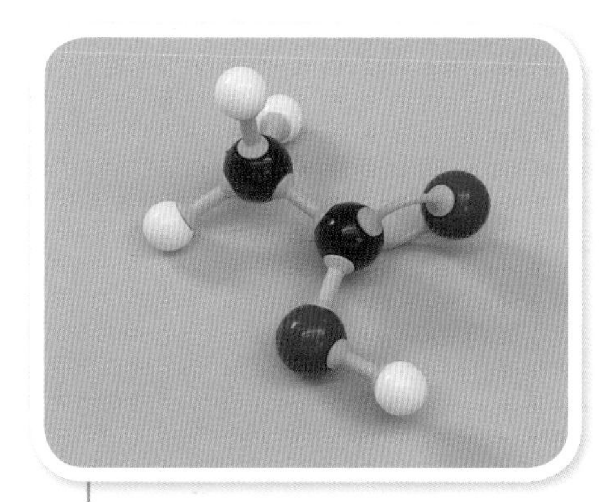
Ethanoic acid, CH_3COOH.

Ethanoic acid and the other carboxylic acids show the characteristic reactions of acids with metals, alkalis, and metal carbonates:

- acid + metal ⟶ salt + hydrogen
- acid + soluble hydroxide ⟶ salt + water
- acid + metal carbonate ⟶ salt + water + carbon dioxide

When ethanoic acid reacts with sodium hydroxide, the salt formed is sodium ethanoate.

There is an ionic bond between the sodium ion and the ethanoate ion:

$$CH_3COO^- \, Na^+$$

Key words

- vinegar
- carboxylic acid
- strong acid
- weak acid

Questions

1 Write these formulae (the table of alkane names in Section 2A will help you):
   a the structural formula of methanoic acid
   b the molecular formula of ethanoic acid
   c the structural formula of propanoic acid
   d the molecular formula of butanoic acid

2 Write word equations and balanced symbol equations for the reactions of methanoic acid with:
   a magnesium
   b potassium hydroxide solution
   c copper carbonate ($CuCO_3$).

3 A good way of removing the disgusting smell of butanoic acid from vomit on a carpet or inside a car is to sprinkle it with sodium hydrogencarbonate ($NaHCO_3$) powder. Write a word equation for the reaction that takes place. Can you explain why the smell might disappear after this reaction?

Find out about

- esters from acids and alcohols
- synthesis of an ester

Ethyl ethanoate (ethyl acetate) is a colourless liquid at room temperature. It has many uses. As well as flavouring food, it is a good solvent, used for decaffeinating tea and coffee and removing coatings such as nail varnish. It is an ingredient of printing inks and perfumes.

Fruity-smelling molecules

When you eat a banana, strawberry, or peach, you taste and smell the powerful odour of a mixture of **esters**. A ripe pineapple contains about 120 mg of the ester ethyl ethanoate in every kilogram of its flesh. There are also smaller quantities of other esters together with around 60 mg of ethanol.

Esters are very common. Many sweet-smelling compounds in perfumes and food flavourings are esters. Some drugs used in medicines are esters, including aspirin and paracetamol. The plasticisers used to make polymers such as PVC soft and flexible are esters. Esters are also used as solvents, such as nail-varnish remover.

Compounds with more than one ester link include fats and vegetable oils such as butter and sunflower oil. The synthetic fibres in many clothes are made of a polyester. The long chains in laminated plastics and surface finishes in kitchen equipment are also held together by ester links.

Ester formation

An alcohol can react with a carboxylic acid to make esters. The reaction happens on warming the alcohol and the acid in the presence of a small amount of a strong acid catalyst such as sulfuric acid.

$$CH_3COOH + CH_3OH \longrightarrow CH_3COOCH_3 + H_2O$$

ethanoic acid + methanol ⟶ methyl ethanoate + water

(In methyl ethanoate, the CH_3CO– part is labelled "ethanoate" and the –CH_3 part is labelled "methyl".)

Key words

- esters
- heat under reflux
- drying agent
- tap funnel

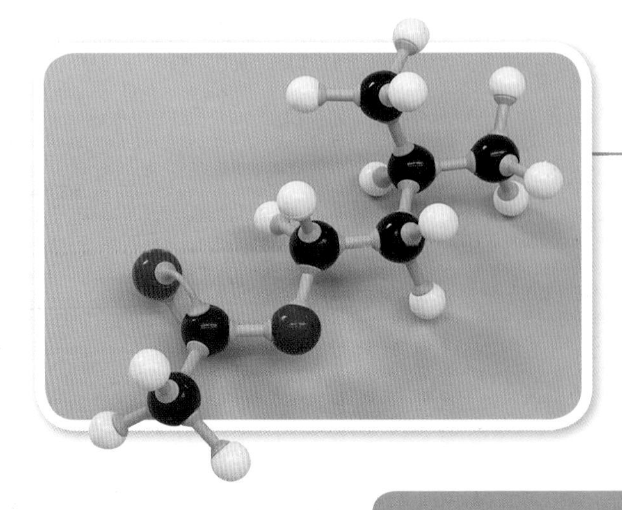

The ester with a very strong fruity smell, 3-methylbutyl ethanoate (3-methylbutyl acetate). It smells strongly of pear drops.

Making an ester

The synthesis of ethyl ethanoate on a laboratory scale illustrates techniques used for making a pure liquid product.

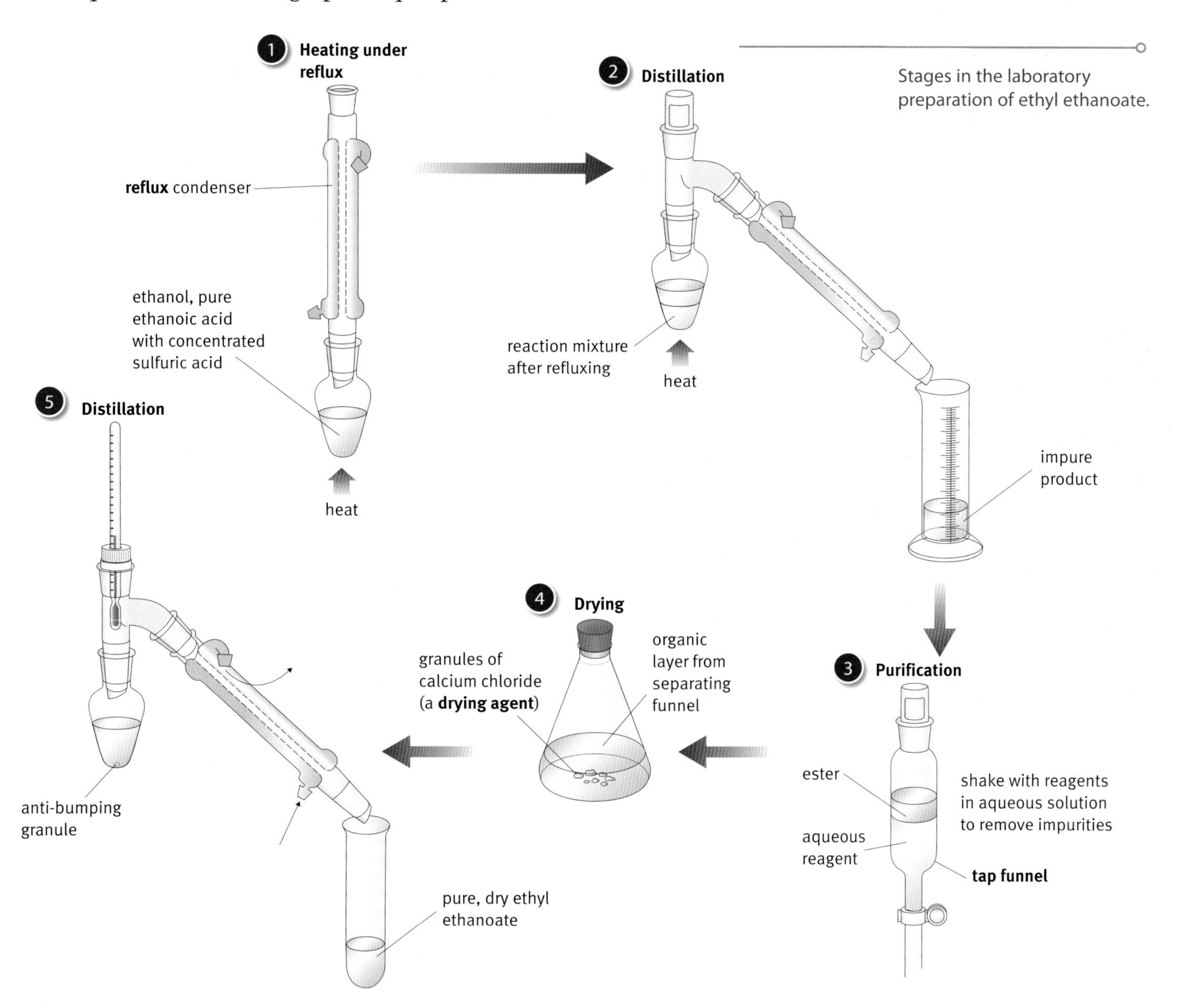

Stages in the laboratory preparation of ethyl ethanoate.

Questions

1 a To which atoms are the hydrogen atoms bonded in ethyl ethanoate?
  b Would you expect ethyl ethanoate to react with sodium?
  c Would you expect ethyl ethanoate to be an acid?

2 Explain what is happening at each step in the synthesis of ethyl ethanoate.

3 In step 1 of the synthesis of ethyl ethanoate:
  a what is the purpose of the condenser?
  b what is the purpose of the sulfuric acid?

4 Calculate the percentage yield if the yield of ethyl ethanoate is 50 g from a preparation starting from 42 g ethanol and 52 g ethanoic acid.

Chemistry in action

Looking and smelling good

Mimicking nature

Tony Moreton is a chemist working for The Body Shop. Fruit esters crop up all the time in his work: 'We use fruit esters in products that have a fruity smell. They have low molecular masses and low boiling points, which give them the volatility that makes them easy to smell. Their high volatility means they don't linger around for long, and they are referred to as "top notes" in a perfume.'

Fruit esters flavour these products.

'As with many organic chemicals, they are flammable, and with their high volatility as well, precautions have to be taken during manufacture.'

'The esters used in the industry are "nature identical", which means they are identical to materials found in nature but are made synthetically. Extracting the natural esters would cost about 100 times as much as making synthetic ones.'

Esters with a fruity smell

One of the suppliers of ingredients to The Body Shop is a Manchester-based company called Fragrance Oils. One of its products is a blackcurrant perfume concentrate that contains the ester ethyl butanoate. The company's perfumery director, Philip Harris, has been familiar with this chemical for a very long time: 'I remember buying strongly flavoured sweets called pineapple chunks, which tasted more or less exclusively of ethyl butanoate.'

'Ethyl butanoate has a strong pineapple aroma, but it's also reminiscent of all sorts of fruity aromas, so we use it in blackcurrant, strawberry, raspberry, apple, mango . . . everything fruity!'

'It's quite simply made from ethanol and butanoic acid. The only trouble is that the reaction can go in reverse. In the presence of alkali the ester can hydrolyse back to these starting products. As butanoic acid is extremely smelly, this isn't so good. We have to avoid using it in alkaline products.'

'I've always found it amazing that such an unpleasant smelling material could be used to produce such a delicious fruity smell.'

From fruit esters to soaps

Tony Moreton also uses esters with larger molecules found in various nuts. 'They are formed from glycerol and long-chain fatty acids. They are oily or waxy so are water resistant, soften the skin, and help retain moisture.'

'We can use these fruit esters to produce soap. Synthetic versions of these kinds of esters would be very expensive to make, so in this case we use the natural material.'

Philip Harris and Farzana Rujidawa working in front of the smelling booths at Fragrance Oils. Philip designed many of The Body Shop's fruit-based fragrances.

Find out about

- ✓ structures of fats and oils
- ✓ saturated and unsaturated fats

It is possible to have molecules with more than one ester link between alcohol and acid. Important examples are fats and oils. These compounds release more energy when oxidised than carbohydrates. This makes them important to plants and animals as an energy store.

The structures of fats and oils

The alcohol in fats and oils is **glycerol**. This is a compound with three —OH groups.

The carboxylic acids in fats and oils are often called **fatty acids**. These are compounds with a long hydrocarbon chain attached to a carboxylic acid group.

$$
\begin{array}{lllll}
 & H & & O & \\
 & | & & \| & \\
H- & C & -O- & C & -CH_2-CH_2-CH_2-CH_2-CH_2-CH_2-CH_2-CH_2-CH_2-CH_2-CH_2-CH_2-CH_3 \\
 & | & & O & \\
 & | & & \| & \\
H- & C & -O- & C & -CH_2-CH_2-CH_2-CH_2-CH_2-CH_2-CH_2-CH_2-CH_2-CH_2-CH_2-CH_2-CH_3 \\
 & | & & O & \\
 & | & & \| & \\
H- & C & -O- & C & -CH_2-CH_2-CH_2-CH_2-CH_2-CH_2-CH_2-CH_2-CH_2-CH_2-CH_2-CH_2-CH_3 \\
 & | & & & \\
 & H & & &
\end{array}
$$

The general structure of a compound in which glycerol has formed three ester links with three fatty acid molecules. In natural fats and oils the fatty acids may all be the same or they may be a mixture.

Glycerol, which is also called propan-1,2,3-triol.

Saturated and unsaturated fats

Animal **fats** are generally solids at room temperature. Butter and lard are examples. **Vegetable oils** are usually liquid, as illustrated by corn oil, sunflower oil, and olive oil.

Chemically the difference between fats and oils arises from the structure of the carboxylic acids. Stearic acid is typical of the acids combined in animal fats. All the bonds in its molecules are single bonds. Chemists use the term 'saturated' to describe molecules like this because the molecule has as much hydrogen as it can take. These saturated molecules are straight.

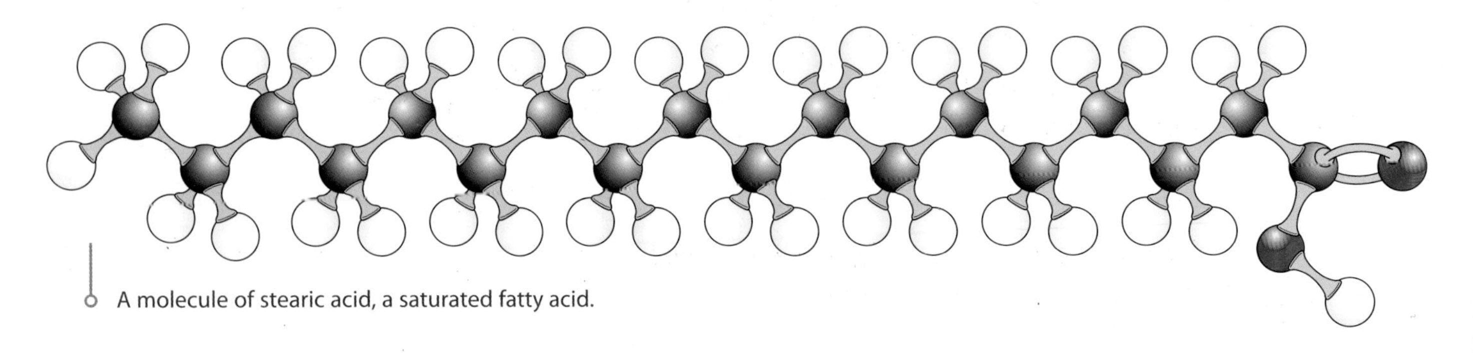

A molecule of stearic acid, a saturated fatty acid.

Esters made of glycerol and saturated fats have a regular shape. They pack together easily and are solid at room temperature.

Oleic acid is typical of the acids combined in vegetable oils. There is a double bond in each molecule of this acid. Oleic acid is unsaturated. The double bond means that the molecules are not straight. It is more difficult to pack together molecules made of glycerol and unsaturated fats. This means that they are liquid at room temperature.

Key words

- glycerol
- fatty acids
- fats
- vegetable oils

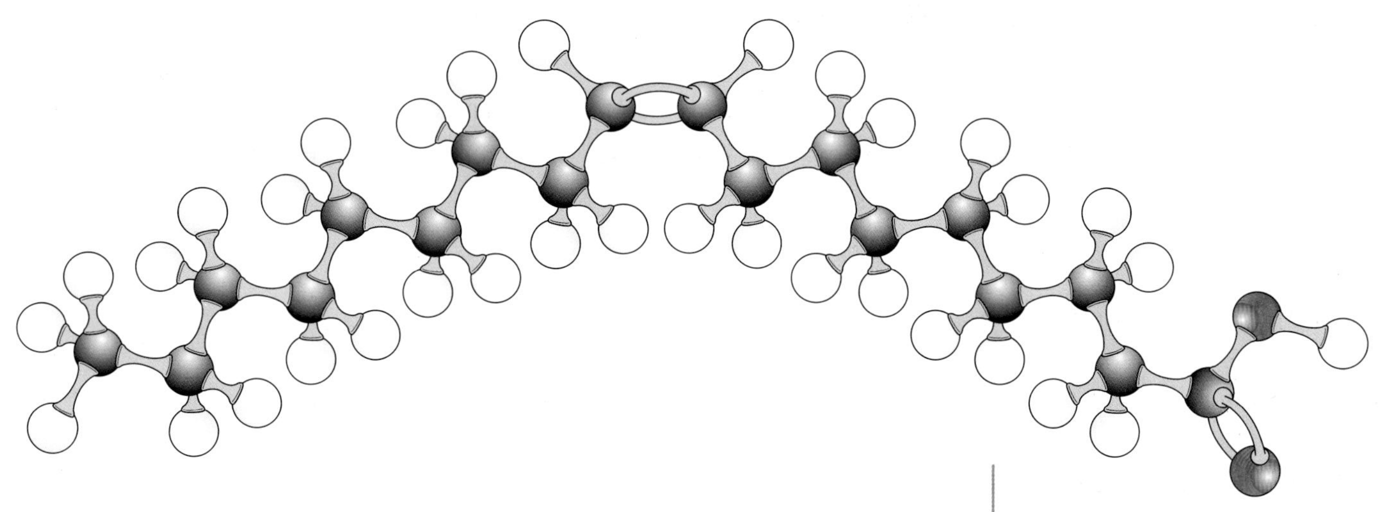

A molecule of oleic acid. The double bond means that there are carbon atoms that do not form four bonds with other atoms. Because there is not as much hydrogen as there would be with all single bonds, these molecules are 'unsaturated'.

Making soap from fats and oils

An ester splits up into an acid and an alcohol when it reacts with water. Chemists call this type of change hydrolysis. 'Hydro-lysis' is derived from two Greek words meaning 'water-splitting'.

$$\text{ester} + \text{water} \longrightarrow \text{acid} + \text{alcohol}$$

In the absence of a catalyst, this is a very slow change.

A strong alkali, such as sodium hydroxide, is a good catalyst for the hydrolysis of esters. Hydrolysis of fats and oils by heating with alkali produces soaps. Soaps are the sodium or potassium salts of fatty acids.

Questions

1. Chemists sometimes describe fats and oils as 'triglycerides'. Why is this an appropriate name for these compounds?
2. The molecular formula of an acid can be written C_xH_yCOOH. What are the values of x and y in:
   a. stearic acid?
   b. oleic acid?
3. Manufacturers state that some spreads are high in polyunsaturated fats. Suggest what the term 'polyunsaturated' means.

Fats, oils, and our health

Edible oils from plants.

Spreads made from vegetable oils.

The chemistry of fats and oils has triggered food scares. 'Saturated' fats and 'trans' fats are thought to be bad for people, while 'unsaturated' fats (especially polyunsaturated fats) and some 'omega' fatty acids are good. Many people use these terms with little idea of what they mean and limited understanding of the effects that these fats and fatty acids have on our health.

Hydrogenated vegetable oil

Margarine was originally a cheap substitute for butter. It can be made from vegetable oils that have been altered to 'harden' them so that they are solid at room temperature.

The first margarines were made by bubbling hydrogen through an oil with a nickel catalyst. The hydrogen hardened the oil by adding to the double bonds and turning it into a saturated fat.

This is a relatively cheap and easy process. But, research in the 1960s showed that these saturated fats could contribute to heart disease.

Trans and *cis* fats

Some margarine tubs, and bottles of vegetable oil, give information about *cis*- and *trans*-fatty acids. This refers to the arrangement of the atoms either side of the double bond in unsaturated fatty acids.

Trans-fatty acids can form during the hydrogenation process used to make some 'hard' margarines. In the 1990s an American scientist, Walter Willet, found evidence that too much *trans*-fat could aggravate heart disease. Not all scientists agreed with him, and the research continues.

If both parts of the hydrocarbon chain are on the same side of the double bond between carbon atoms, it is **cis**. Nearly all naturally occurring unsaturated fatty acids contain **cis** double bonds. If the two parts of the chain are on the opposite sides of the double bond, it is **trans**. The shape of a **trans**-unsaturated molecule is a bit like a saturated fatty acid. They are not as runny as **cis**-molecules.

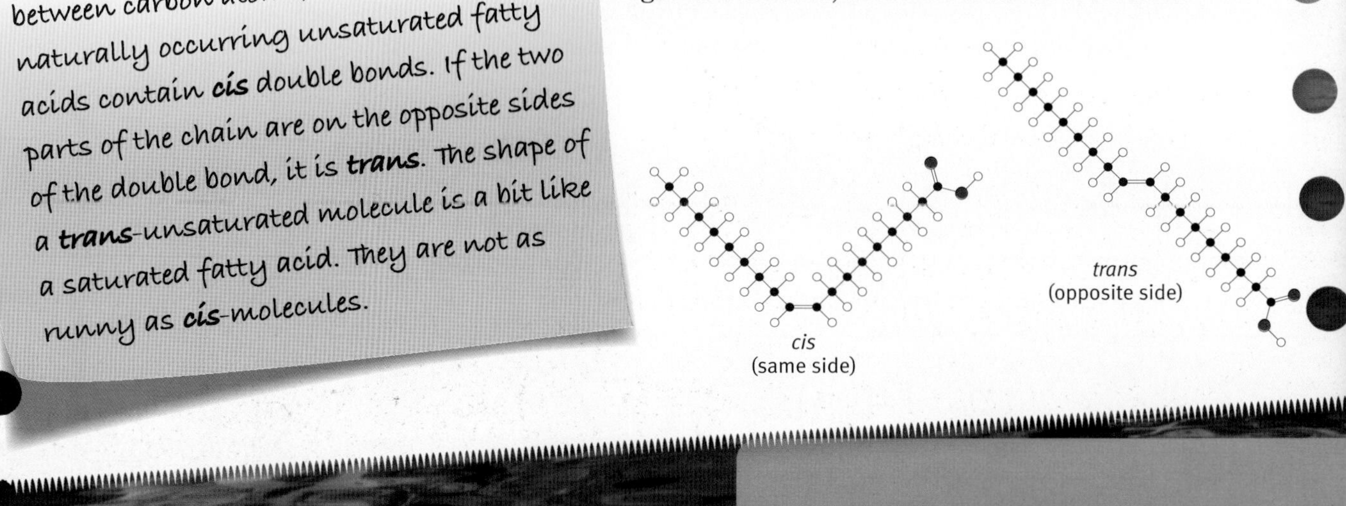

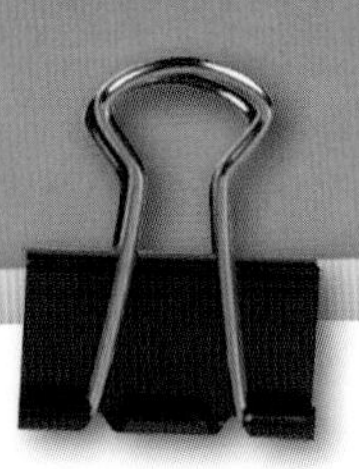

Hardening without hydrogen

Food scientists have found ways to turn vegetable oils into solid spreads without adding hydrogen. The fatty acids in vegetable oils are polyunsaturated. This means they contain many double bonds. The molecules of these fatty acids are not straight because of all the double bonds. So they do not pack together easily.

Imagine what would happen if you could swap the fatty acid chains around so that they all stack together more neatly and make a denser material. This is exactly what modern margarine makers do – using a catalyst, they make the fatty acids rearrange themselves. David Allen works for one of the suppliers to a big supermarket chain: 'It's a bit like musical chairs in that they all change places. Instead of music you have a catalyst, or an enzyme, and the right conditions.'

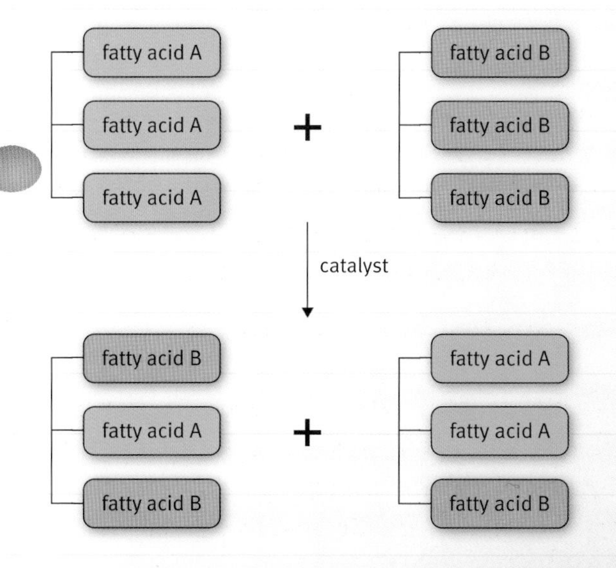

In the presence of a catalyst the fatty acids can swap places between molecules of fats and oils. This results in a mixture. The diagram shows just two of the possible products. This can raise the melting point by several degrees.

Omega-3 and omega-6 oils

Omega (ω) is the last letter in the Greek alphabet. The letter is used in a naming system that counts from the last carbon atom in the chain of a fatty acid molecule – the carbon atom furthest from the —COOH group. If you count in this way, omega-3 fatty acids have the first double bond between carbon atoms 3 and 4. Omega-6 fatty acids have the first double bond between carbon atoms 6 and 7.

Some of these omega compounds are essential fatty acids because our bodies need them but cannot make them. People need to take them in through their diet. According to Professor John Harwood of Cardiff University: 'You need a ratio of about three times omega-6 to one omega-3. But most people in Western countries take in far too much omega-6 and not enough omega-3 – more like a ratio of 15 to 1.'

'Omega-3 acids are found in oily fish and can reduce pain in joints. They are also important in the developing brains of the very young and in the brains of the very old, and have implications for heart disease. Omega-6 acids are found in margarines.'

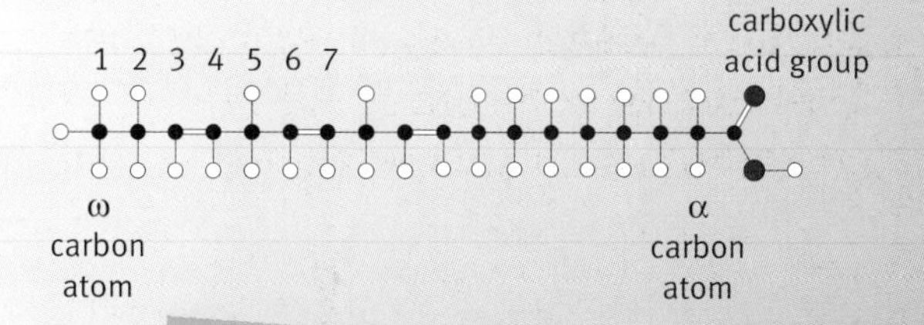

Linolenic acid has three double bonds. They are all cis. This form of linolenic acid is an omega-3 fatty acid. Another form of linolenic acid is an omega-6 fatty acid. Omega-6 acid has double bonds between carbon atoms numbered 6 and 7, 9 and 10, and 12 and 13.

Topic 3: Energy changes in chemistry

The thermite reaction is highly exothermic. The energy released causes the iron produced in the reaction to be molten. It can be used to weld rail tracks together.

A fireball from detonating gunpowder. Explosions are very fast reactions. They need to be carefully controlled to make sure they are safe.

When a reaction takes place, atoms are rearranged and bonds are broken and made. When this happens, energy is taken in and given out. It is important for chemists to understand the energy changes that occur during a reaction so that they can control how fast the reaction happens.

Energy in and out

Some chemical reactions take in energy, while others release energy. Exothermic reactions give out energy and are put to use all the time. An exothermic reaction between petrol and air causes energy to be released, making a car engine work. Scientists need to understand these energy changes in order to put them to good use in a controlled way.

How fast?

It is important for anyone trying to control changes to understand how fast a reaction will happen. Reactions that are too slow tie up equipment and people's time for too long. This costs money. But reactions that are too fast can be hazardous.

Catalysts can be used to speed up reactions that are too slow. Understanding how catalysts work is enabling scientists to make the chemical industry more sustainable.

Molecular theories

Chemists explain their observations with the help of theories about the behaviour of atoms and molecules. Atoms and molecules are too small to see, so chemists use models to help develop their theories.

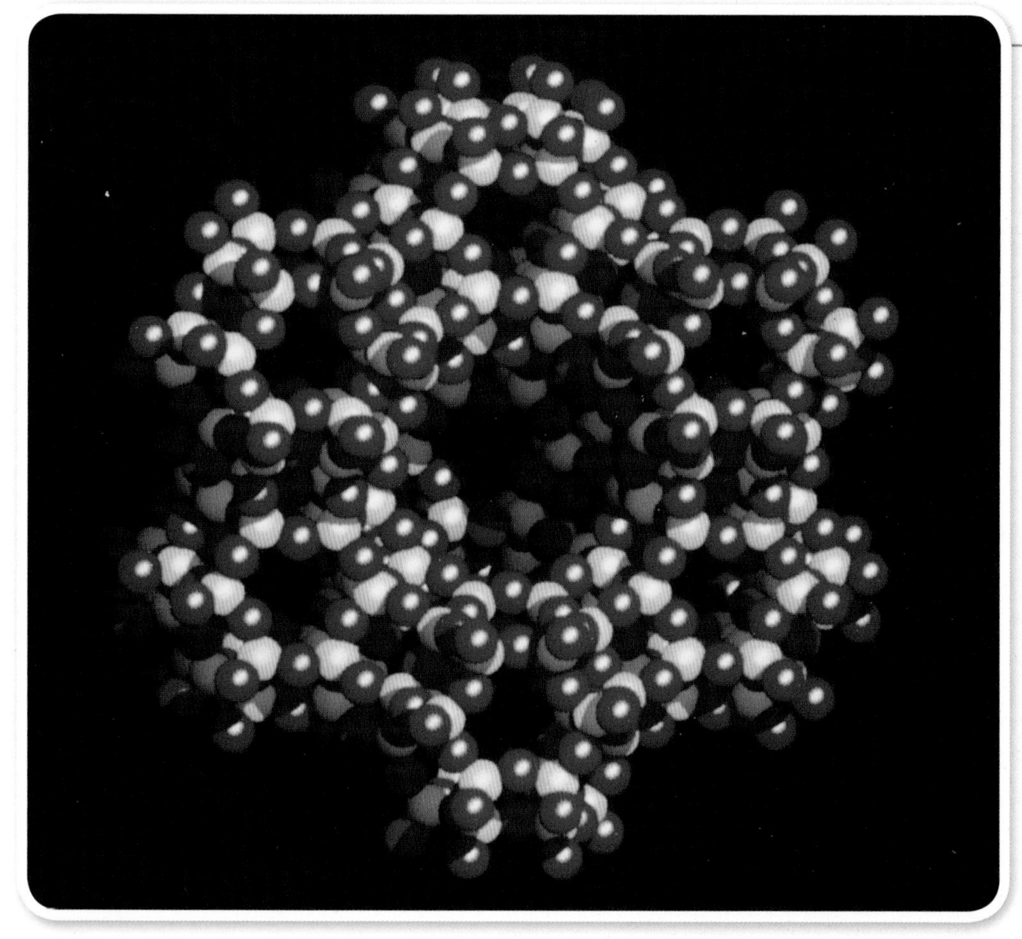

Computer graphic of a model of a zeolite crystal. The yellow atoms are either silicon or aluminium atoms. The red atoms are oxygen. Zeolites are catalysts used to control reactions in the petrochemical industry. Chemists can make synthetic zeolites with crystal structures designed to catalyse particular reactions.

Find out about

- exothermic and endothermic changes
- energy-level diagrams
- bond breaking and bond forming
- calculating energy changes

All chemical changes give out or take in energy. The study of energy changes is central to the science of explaining the extent and direction of a wide variety of changes. Understanding these energy changes also helps chemists to control reactions.

In C6 Section E, you were introduced to **exothermic** and **endothermic** reactions and used **energy-level diagrams** to try and explain the energy changes.

Exothermic reactions

Many reactions, such as burning and respiration, are exothermic. They give out energy to the surroundings.

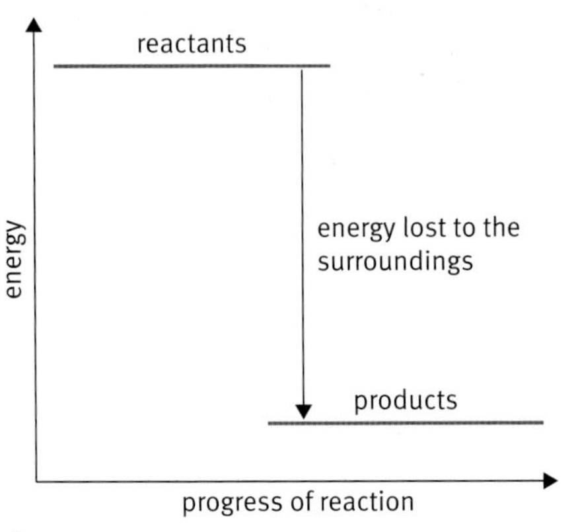

Energy-level diagram for an exothermic reaction.

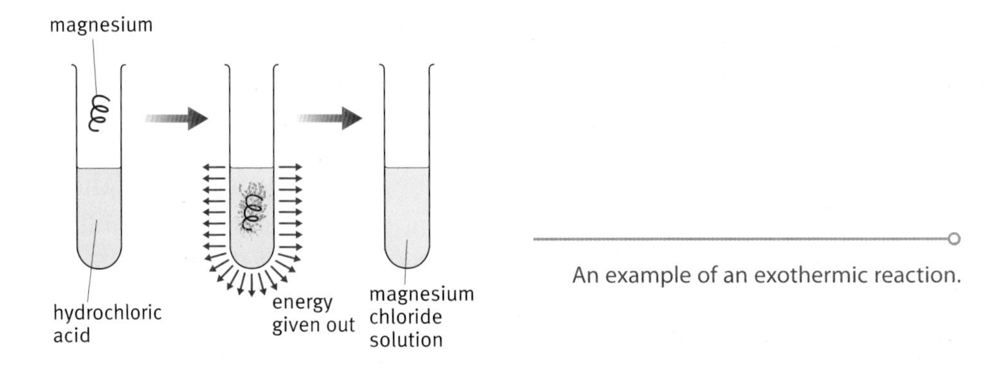

An example of an exothermic reaction.

In an exothermic reaction, the energy of the products is less than the energy of the reactants.

Endothermic reactions

Endothermic reactions, such as photosynthesis, take in energy from the surroundings.

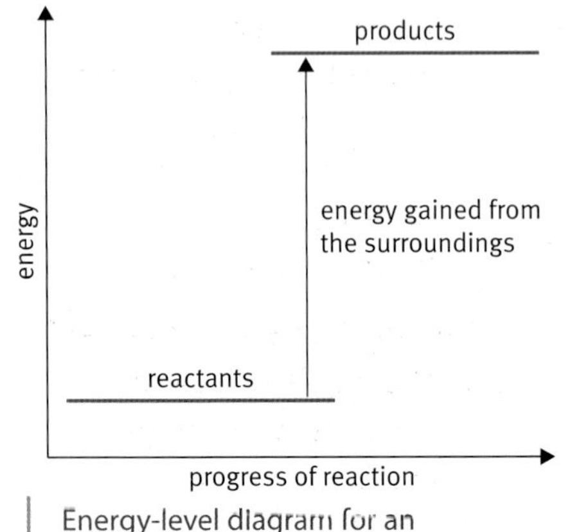

Energy-level diagram for an endothermic reaction.

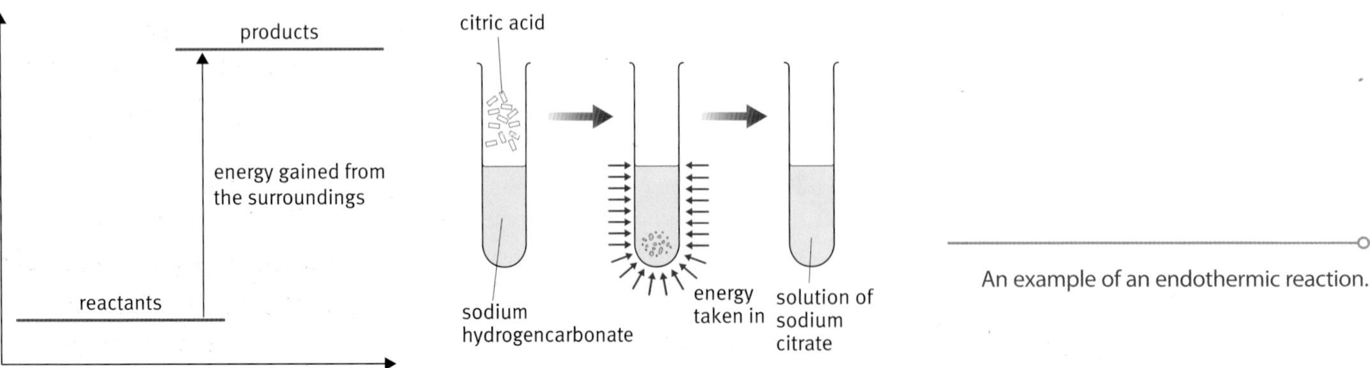

An example of an endothermic reaction.

In an endothermic reaction, the energy of the products is greater than the energy of the reactants.

How much energy?

There is growing interest in hydrogen as a fuel because water is the only product of its reaction with oxygen. This is a highly exothermic reaction, but at room temperature the two gases do not react. It takes a hot flame or an electric spark to heat up the mixture enough for the reaction to start.

This is because, in all reactions, regardless of whether they are exothermic or endothermic, some of the chemical bonds in the reactants have to be broken before new chemical bonds in the products can be formed.

Think of chemical bonds as tiny springs. In order to get hydrogen to react with oxygen, the tiny springs joining the atoms in the molecules have to be stretched and broken. This takes energy, as you will know if you have ever tried to stretch and break an elastic band.

The product is water. This is created as new bonds form between oxygen atoms and hydrogen atoms. Bond formation releases energy – just like relaxing a spring. So what decides whether a chemical reaction is exothermic or endothermic? It is the difference between the energy taken in to break bonds and the energy given out as new bonds form.

The strength of the chemical bonds that break and form during a reaction determines the size of the overall energy change, and whether it is exothermic or endothermic.

Key words

- exothermic
- endothermic
- energy-level diagram

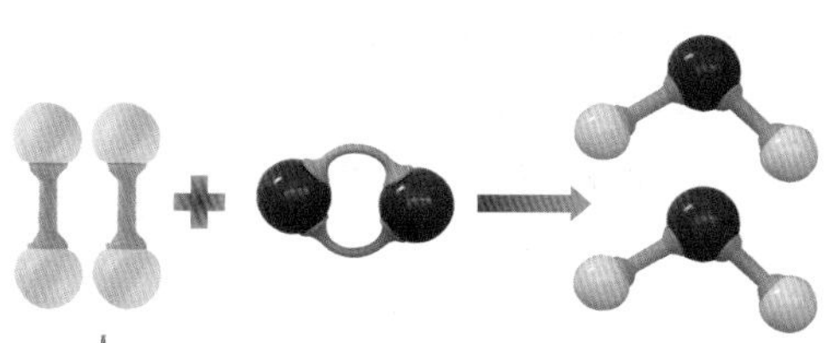

Two H—H bonds and one O=O bond break when hydrogen reacts with oxygen. The atoms recombine to make water as four new O—H bonds form.

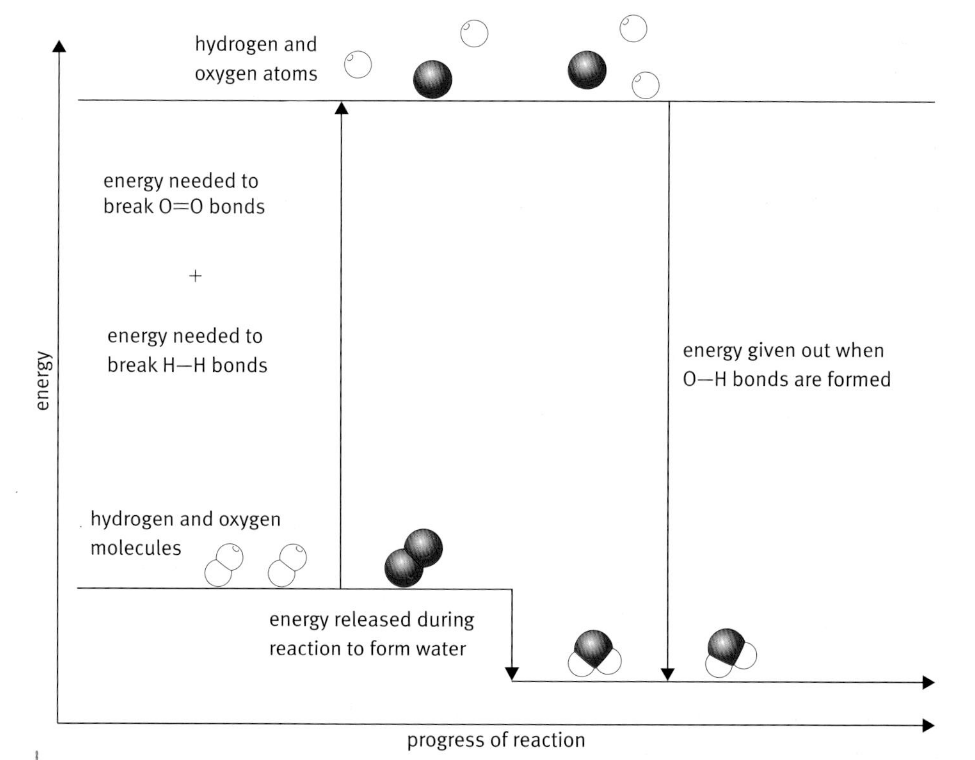

Two H—H bonds and one O=O bond break when hydrogen reacts with oxygen. The atoms recombine to make water as four new O—H bonds form.

Questions

1 Classify these changes as exothermic or endothermic:
  - a petrol burning
  - b water turning to steam
  - c water freezing
  - d sodium hydroxide neutralising hydrochloric acid.

2 Turning 18 g ice into water at 0 °C requires 6.0 kJ of energy. Use an energy-level diagram to show this change.

Key word

- bond strength

More energy is given out when the bonds form than is taken in when the bonds break, and so the reaction overall is exothermic. The energy given out keeps the mixture hot enough for the reaction to continue.

Energy change calculations

The overall energy change that takes place during a chemical reaction can be calculated if the strength of all the chemical bonds in the reactants and the products are known. So, to calculate the energy change that takes place during the formation of steam from hydrogen and oxygen, the following data is needed. The units of **bond strength** are kilojoules (kJ).

$H_2(g) + O_2(g) \longrightarrow 2H_2O(g)$

Bond	Energy change for the formula masses (kJ)
H—H	434
O=O	498
O—H	464

From the energy-level diagram, we can see that during the reaction, two H—H and one O=O bonds are broken, and four O—H bonds are formed.

So the energy needed to break the bonds = 2 × (H—H) + 1 × (O=O)

$= (2 \times 434) + 498 = 1366\,\text{kJ}$

The energy given out as the new bonds are formed = 4 × (O—H)

$= 4 \times 464 = 1856\,\text{kJ}$

Overall energy change

= energy needed to break bonds – energy given out as new bonds are formed

$= 1366 - 1856 = -490\,\text{kJ}$

The negative sign shows that this reaction is exothermic. Remember that in an exothermic reaction, the products have less energy than the reactants.

Question

3 Hydrogen burns in chlorine.

$H_2(g) + Cl_2(g) \longrightarrow 2HCl(g)$

a Which bonds are broken during the reaction?

b Which bonds are made during the reaction?

c Use the data in the table to calculate the overall energy change for the reacting masses shown in the equation.

Bond	Energy change for the formula masses (kJ)
H—H	434
Cl—Cl	242
H—Cl	431

d Is this reaction exothermic or endothermic?

e Draw an energy-level diagram for the reaction.

Molecular collisions

In a mixture of hydrogen gas and oxygen gas the molecules are constantly colliding. Millions upon millions of collisions happen every second. If every collision led to a reaction, there would immediately be an explosive reaction.

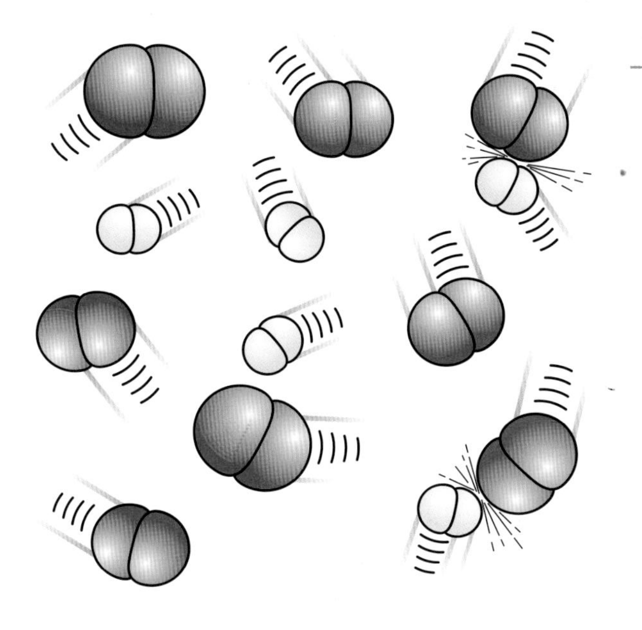

A mixture of hydrogen and oxygen molecules. The molecules that are colliding may react to form new molecules, but only if they have enough energy to start breaking bonds.

Activation energies

It is not enough for the hydrogen and oxygen atoms to collide. Bonds between atoms must break before new molecules can form. This needs energy. For every reaction, there is a certain minimum energy needed before the process can happen. This minimum energy is called the **activation energy**. It is like an energy hill that the reactants have to climb before a reaction will start. The higher the hill, the more difficult it is to get the reaction started.

The collisions between molecules have a range of energies. Head-on collisions between fast-moving particles are the most energetic. If the colliding molecules have enough energy, the collision is 'successful', and a reaction occurs.

Fast and slow reactions

The course of a reaction is like a high-jump competition. The bar is set at a height such that only a few competitors with enough energy can jump it and land safely the other side. The chemical equivalent is shown in the diagram on the next page.

Find out about

- collisions between molecules
- activation energies

Question

1 a Concentration is one of four factors that affect the rates of chemical reactions. Identify the three other factors that affect rates.

b Give examples of measurements that chemists make to collect data to show how changes in concentration affect the rate of a reaction.

c What is the theory that chemists use to explain how changes in concentration affect reaction rates?

d Use this example to explain the difference between data and an explanation.

The activation energy for a reaction. The size of the activation energy is usually less than the energy needed to break all the bonds in the reactant molecules because new bonds start forming while old bonds are breaking, returning energy to the reaction mixture.

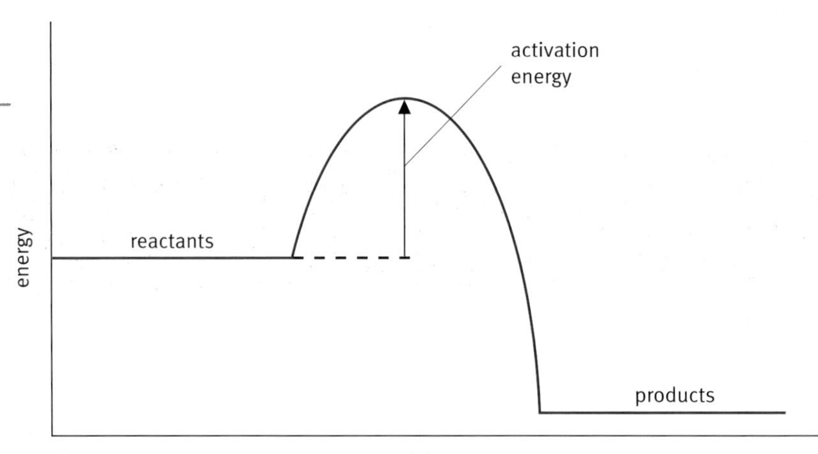

Key word

- activation energy

In this diagram for an exothermic reaction, the activation energy can be thought of as the height of the high-jump bar, and the products as the landing area.

If the high-jump bar is low, many competitors are successful. If it is high, the success rate is much less. In chemical reactions, if the activation energy is low, a high proportion of collisions have enough energy to break bonds, and the reaction is fast even at low temperatures.

Reactions in which the activation energy is high are very slow at room temperature, because only a small fraction of collisions have enough energy to cross the activation energy barrier. Heating the mixture to raise the temperature gives the molecules more energy. In the hot mixture, more molecules have enough energy to react when they collide.

Using a catalyst provides an alternative route for a chemical reaction with a lower activation energy. The energy of the reactions and the energy of the products does not change, but the activation energy is smaller. This means a higher proportion of collisions have enough energy to cross the activation energy barrier.

Questions

2 **a** Why is a spark or flame needed to light a Bunsen burner?
b Why does the gas keep burning once it has been lit?

3 **a** Adding a catalyst to a reaction mixture means that the activation energy for the change is lower. Explain why this speeds up the reaction even if the temperature does not change.
b Copy the energy-level diagram on this page and add in a new line to show the effect of adding a catalyst.

Chemistry in action

The explosives expert

All chemists secretly love controlled explosions, even if they do not admit it openly. The word 'control' is important. To be useful, explosions have to be controlled. Designing and understanding that control is part of Jackie Akhavan's job.

Jackie Akhavan works for Cranfield University as an explosives chemist. Her work has applications in quarrying and mining, bomb disposal, demolition, and fireworks.

Two types of explosive

'There are two types of high explosive,' explains Jackie. 'Primary explosives have low activation energies, while secondary explosives have higher activation energies.'

Therefore it takes less energy to initiate primary explosives. They are more sensitive to an external stimulus such as friction or impact. This makes them more dangerous to handle than secondary explosives. Secondary explosives are more difficult to initiate.'

New explosives and detonators

Jackie is a polymer chemist. She has worked on making polymer-bonded secondary explosives. 'A polymer explosive can be manufactured into sheets that resemble plasticine. The explosive can be wrapped around an old bomb or a pipe, and a detonator pushed into the sheet. On initiation, the explosive cuts the metal into two pieces.'

Jackie has also been involved in new ways of providing the activation energy to make detonators start explosions. This used to be done using an electrical current, but these types of detonator are vulnerable to initiation by unwanted electromagnetic radiation from overhead pylons or thunderstorms. 'We've developed a new detonator that is secondary initiated with a laser pulsed through a fibre-optic cable. It is safe to handle and can't be set off by unwanted electromagnetic radiation.'

Safety

Safety is always important in chemistry and is essential in Jackie's work: 'I am not allowed to detonate an explosion myself as I don't have the specific training – even though I understand the chemistry.' It is illegal as well as dangerous to carry out unauthorised experiments with explosive chemicals.

Topic 4: Reversible reactions and equilibria

Changing the temperature of tubes containing a mixture of nitrogen compounds determines which way and how far the reaction goes. When the tube is warmed, more of the brown NO_2 gas forms and the gas in the tube becomes darker. When the tube is cooled, more of the colourless N_2O_4 gas forms and the gas in the tube becomes paler.

Water and ice are at equilibrium at the melting point of the ice and the freezing point of the water. Molecules of ice are escaping into the water (melting) and molecules of water are being captured onto the surface of the ice (freezing).

There are many questions that scientists try to answer when explaining changes to chemicals and materials. This topic tackles the challenging questions 'Which way?' and 'How far?'

Which way?

'Which way?' is an important question for chemists to ask if they are going to control reactions to get the right results.

Cooks also have to understand how to control conditions to get the direction right, to make sure they produce a good meal. Otherwise they might serve up runny jelly or frozen smoked salmon. Not what the customer ordered!

How far?

'How far?' is an important question for anyone trying to get the maximum yield from a chemical reaction.

The study of carboxylic acids in Topic 2D shows that these are weak acids. They do ionise, but only to a slight extent.

Answering the question 'How far?' helps to explain why some acids, like organic acids, are weak, while other acids, such as the mineral acids, are strong.

Chemists have to be able to answer the questions 'Which way?' and 'How far?' in order to design pH-balanced shampoos and other toiletries. Controlling the pH of shampoos helps to protect the skin and eyes.

Reversible changes

Find out about

- reactions that go both ways
- factors affecting the direction of change

Questions

1. a Write a symbol equation to show water turning into steam.
   b Write another equation to show steam condensing to water.
   c Write a third equation to show the changes in parts a and b as a single, reversible change.
2. The pioneering French chemist Lavoisier heated mercury in air and obtained the red solid mercury oxide. He also heated mercury oxide to form mercury and oxygen.
   a How can both of these statements be true?
   b Why should you not try to repeat the experiment?
3. a Write an equation to show the reversible reaction of carbon monoxide gas with steam to form carbon dioxide and hydrogen.
   b In your equation, what happens in the forward reaction?
   c In your equation, what happens in the backward reaction?

Some changes go only in one direction. For example, the reactions that happen to a raw egg in boiling water cannot be reversed by cooling the egg. To produce a soft-boiled egg, a cook has to check that it stays in the water for just the right amount of time. Other processes in the kitchen are easily reversed. A table jelly sets as it cools but becomes liquid again on warming. Chemists, like cooks, have to understand how to control conditions to get reactions to go far enough and in the right direction.

Burning methane in air is an example of an **irreversible change**. The gas burns to form carbon dioxide and water. It is then pretty well impossible to turn the products back into methane and oxygen.

Reversible changes of state

In contrast, melting and evaporating are familiar **reversible processes**. Heating turns water into steam, but water re-forms as steam condenses on cooling.

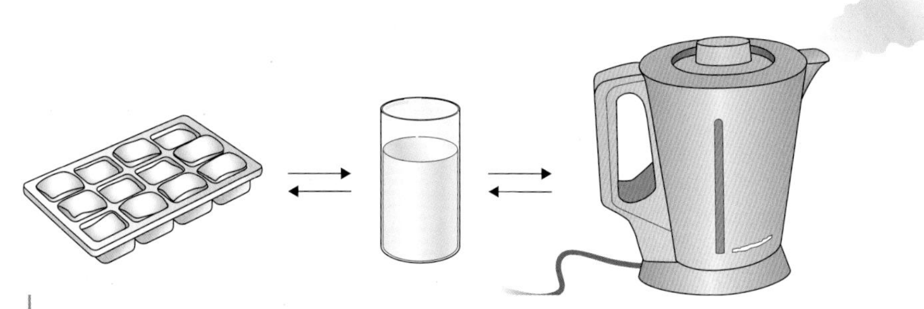

Two familiar reversible processes.

Heating turns ice into water:

$$H_2O(s) \longrightarrow H_2O(l)$$

Ice reforms if water cools to 0 °C or below:

$$H_2O(l) \longrightarrow H_2O(s)$$

Combining these two equations gives:

$$H_2O(s) \rightleftharpoons H_2O(l)$$

Many chemical reactions are also reversible. A reversible reaction can go forwards or backwards depending on the conditions. The direction of change may vary with the temperature, pressure, or concentration of the chemicals.

Temperature and the direction of change

Heating decomposes blue copper sulfate crystals to give water and anhydrous copper sulfate, which is white:

$$CuSO_4.5H_2O(s) \longrightarrow CuSO_4(s) + 5H_2O(l)$$

Add water to the white powder after cooling, and it changes back into the hydrated form. As it does so it turns blue again and gets very hot:

$$CuSO_4(s) + 5H_2O(l) \longrightarrow CuSO_4.5H_2O(s)$$

Temperature also affects the direction of change in the formation of ammonium chloride. At room temperature ammonia gas and hydrogen chloride gas react to form a white solid, ammonium chloride:

$$NH_3(g) + HCl(g) \longrightarrow NH_4Cl(s)$$

Gentle heating decomposes ammonium chloride back into ammonia and hydrogen chloride:

$$NH_4Cl(s) \longrightarrow NH_3(g) + HCl(g)$$

Concentration and the direction of change

This equation describes the reaction between iron and steam:

$$3Fe(s) + 4H_2O(g) \longrightarrow Fe_3O_4(s) + 4H_2(g)$$

The change from left to right (from reactants to products) is the forward reaction. The change from right to left (from products to reactants) is the backward reaction.

The forward reaction is favoured if the concentration of steam is high and the concentration of hydrogen is low.

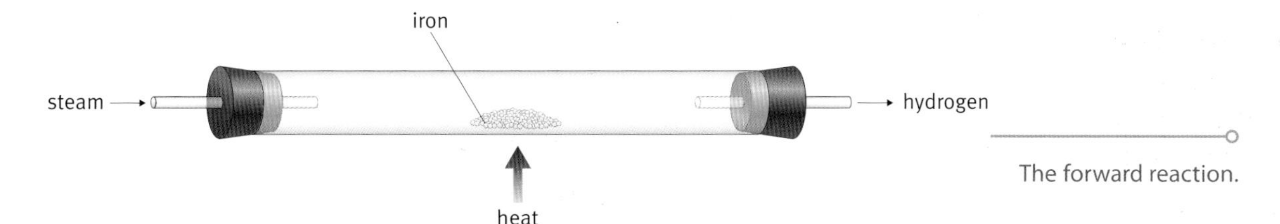

The forward reaction.

The backward reaction is favoured if the concentration of hydrogen is high and the concentration of steam is low:

$$Fe_3O_4(s) + 4H_2(g) \longrightarrow 3Fe(s) + 4H_2O(g)$$

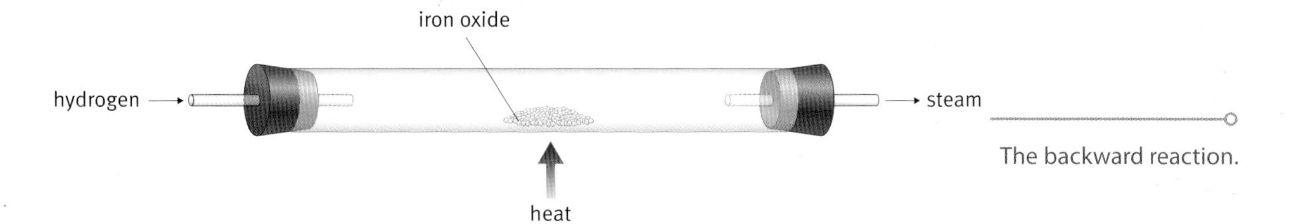

The backward reaction.

Key words

- irreversible change
- reversible process

Clouds of solid ammonium chloride form where ammonia and hydrogen chloride gases meet above concentrated solutions of the two compounds on glass stoppers.

Find out about

- chemical equilibrium
- dynamic equilibrium
- strong and weak acids

Equilibrium

Reversible changes often reach a state of balance, or **equilibrium**. A solution of litmus in water at pH 7 is purple because it contains a mixture of the red and blue forms of the indicator. Similarly, melting ice and water are at equilibrium at 0 °C. At this temperature, the two states of water coexist with no tendency for all the ice to melt or all the water to freeze.

When reversible reactions are at equilibrium, neither the forward nor the backward reaction is complete. Reactants and products are present together and the reaction appears to have stopped. Reactions like this are at equilibrium. Chemists use a special symbol in equations for reactions at equilibrium: $\rightleftharpoons$

So at 0 °C,

$$H_2O(s) \rightleftharpoons H_2O(l)$$

The question 'How far?' asks where the equilibrium point is in a reaction. At equilibrium the reaction may be well to the right (mainly products), well to the left (mainly reactants), or at any point between these extremes.

Reaching an equilibrium state

A mixture of two solutions of iodine helps to explain what happens when a reversible process reaches a state of equilibrium.

Iodine is slightly soluble in water but much more soluble in a potassium iodide solution in water. The solution with aqueous potassium iodide is yellow–brown. Iodine is also soluble in organic solvents (such as hexane, a liquid alkane), in which it forms a violet solution. Aqueous potassium iodide and the organic solvent do not mix.

Approaching the equilibrium state starting with all the iodine in the liquid alkane.

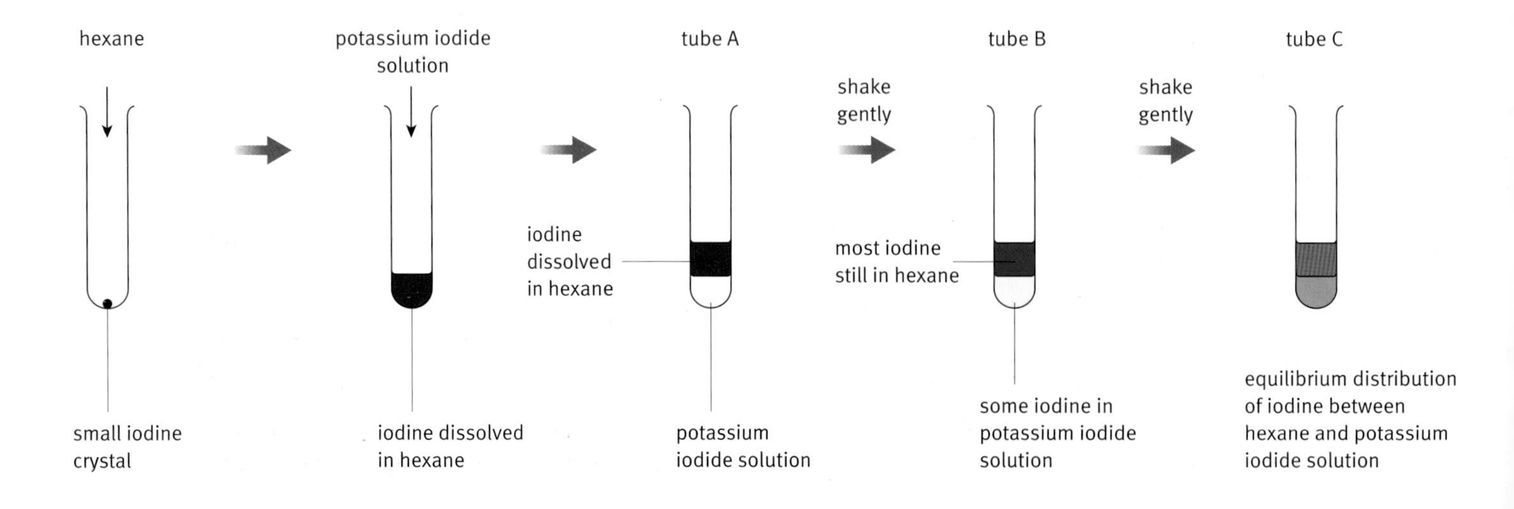

Graph 1 shows how the iodine concentrations in the two layers change with shaking. In tube C, the iodine is distributed between the organic and aqueous layers and there is no more change. In this tube there is an equilibrium:

$$I_2(\text{organic}) \rightleftharpoons I_2(\text{aq})$$

Key words

- equilibrium
- dynamic equilibrium

Graph 1

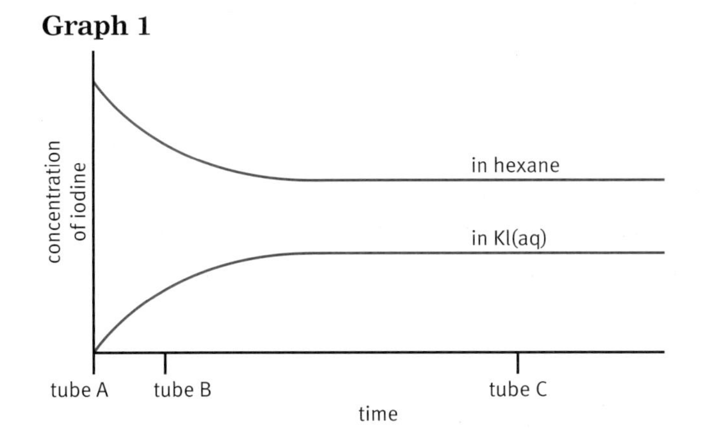

The change of concentration of iodine with time in the mixture, starting with all the iodine in the liquid alkane.

The same equilibrium can be reached starting with all the iodine dissolved in potassium iodide solution rather than hexane.

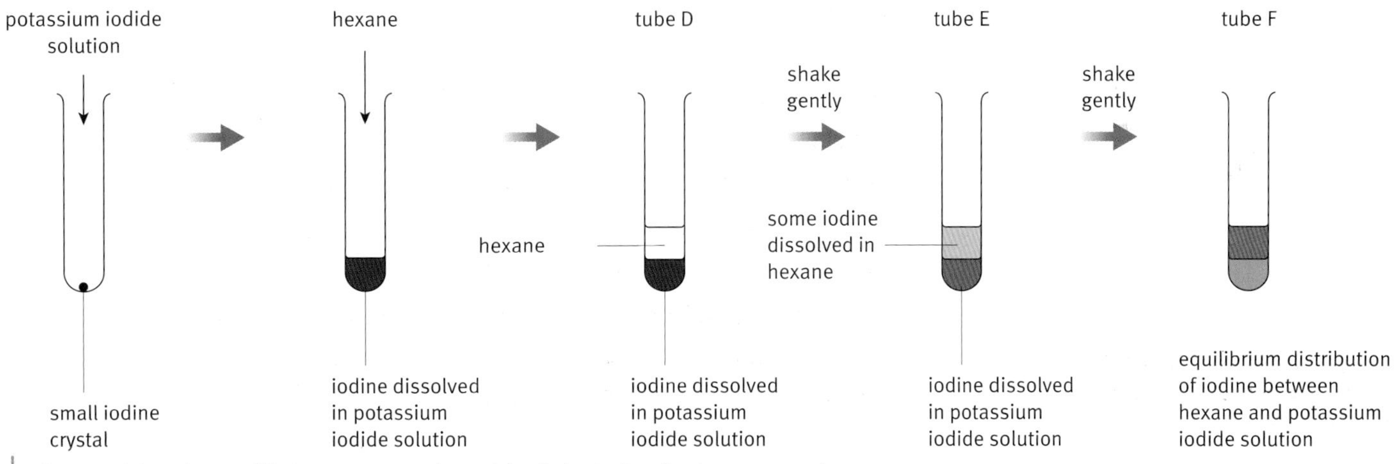

Approaching the equilibrium state starting with all the iodine in the aqueous layer.

Graph 2 shows how the iodine concentration in the two layers change with shaking.

Tube F looks just like tube C. Tube F is also at equilibrium: equilibrium mixtures in the two tubes are the same. This illustrates two important features of equilibrium processes:

- At equilibrium, the concentrations of reactants and products do not change.
- An equilibrium state can be approached from either the 'reactant side' or the 'product side' of a reaction.

Graph 2

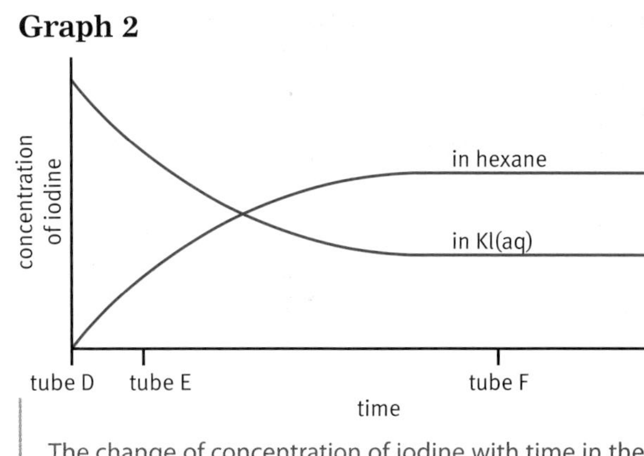

The change of concentration of iodine with time in the mixture, starting with all the iodine in the aqueous layer.

Dynamic equilibrium

The diagram below gives a picture of what happens to the iodine molecules if you shake a solution of iodine in an organic solvent with aqueous potassium iodide (see tube A on page 242).

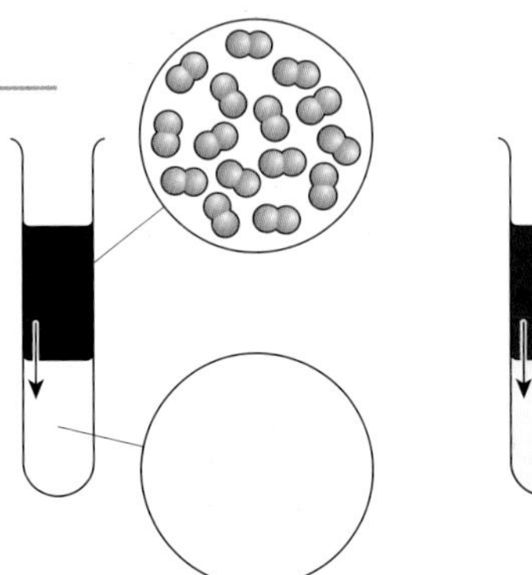

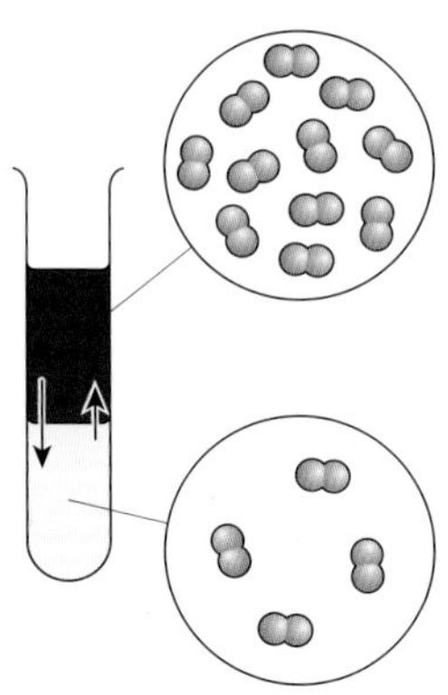

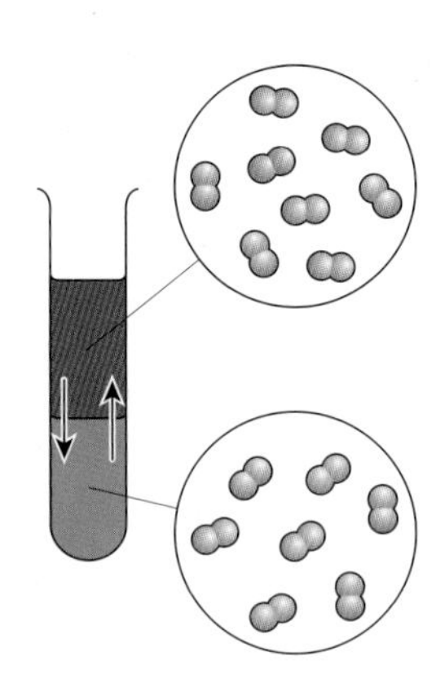

Iodine molecules reaching dynamic equilibrium between two solvents. The solvent molecules are far more numerous. They are not shown.

All the iodine starts in the upper, organic layer. At first, when the solution is shaken, movement is in one direction (the forward reaction) as some molecules move into the aqueous layer. There is nothing to stop some of these molecules moving back into the organic layer. This backward reaction starts slowly because the concentration in the aqueous layer is low. So to begin with, the overall effect is that iodine moves from the organic to the aqueous layer. This is because the forward reaction is faster than the backward reaction.

As the concentration in the organic layer falls, the rate of the forward reaction goes down. As the iodine concentration in the aqueous layer rises, the rate of the backward reaction goes up. There comes a point at which the two rates are equal. At this point both forward and backward reactions continue, but there is no overall change because each layer is gaining and losing iodine at the same rate. This is **dynamic equilibrium**.

Questions

1 Under what conditions are these in equilibrium:
   - a water and ice?
   - b water and steam?
   - c salt crystals and a solution of salt in water?

2 a Why do iron and steam not reach an equilibrium state when they react as shown in the forward reaction diagram on page 241?
   - b Suggest conditions in which a mixture of iron and steam would react to reach an equilibrium state.
   - c Which chemicals would be present in an equilibrium mixture formed from iron and steam?

3 Explain in your own words what is meant by the term 'dynamic equilibrium'.

Ammonia and green chemistry

Ammonia

Ammonia, NH_3, is an extremely important and interesting bulk chemical. It is able to both sustain life through increased food production from the use of fertilisers, and to destroy life through its exploitation as an explosive.

Demand for explosives as well as fertilisers has led to the development of chemical processes to 'fix' nitrogen, to produce nitrogen compounds such as ammonia, nitric acids, and nitrates.

Nitrogen gas is all around us in the air but for a long time chemists did not know it was there because it is very unreactive. The low reactivity of nitrogen means that fixing it into nitrogen compounds is not easy.

Producing ammonia

At first, ammonia was obtained from the distillation of coal and the by-products of other industrial processes. However, in the early 20th century, German scientists Fritz Haber and Carl Bosch discovered and developed a new process for making ammonia. They reacted nitrogen from the air with hydrogen. They had achieved **nitrogen fixation**.

During the First World War, the **Haber process** became very important in Germany because ammonia was needed to make explosives. Before the war, explosives were made from nitrates imported from Chile. But the British navy blockade stopped imports and so a different production method was needed.

Today, ammonia is produced in more than 80 countries around the world. About 130 million tonnes of ammonia is produced annually. Over 50% is produced in developing countries such as China and India.

Fertilisers: benefits and costs

The production of ammonia has increased significantly over the past 60 years. More fertilisers are being produced to achieve the increase in food production that is needed to support the world's growing population.

The ability to fix nitrogen has affected society, and has also had an impact on the environment. The increased availability of fertilisers has led to changes in land use. Less land is needed to provide food for more people and so larger towns and cities can be supported. The fixing of nitrogen also affects the natural nitrogen cycle. For example, the overuse of fertilisers such as ammonium nitrate can lead to excess concentrations of nitrogen compounds being washed into the rivers by rain. This can lead to increased growth of algae, which upsets ecosystems.

Find out about

- ammonia and the Haber process
- nitrogen fixation
- Le Chatelier's principle
- the influence of green chemistry on the ammonia industry

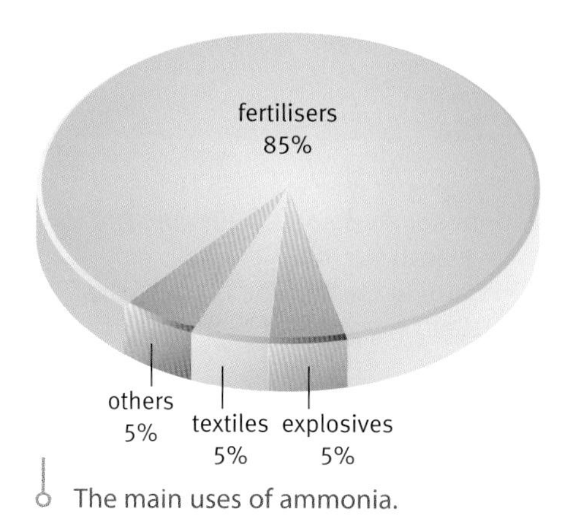

The main uses of ammonia.

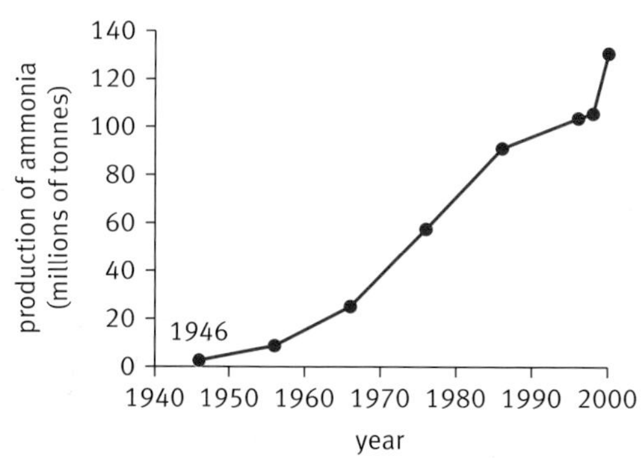

Worldwide production of ammonia. Since the 1960s, production of ammonia has grown at a rapid rate.

Questions

1. What is the main use of ammonia?
2. What impact do you think the Haber process had on the First World War?

Fritz Haber won the 1918 Nobel Prize for his process. A number of French scientists refused their awards at the time in protest against Haber's war work for Germany.

The Haber process

The basis of the Haber process is a reversible reaction between nitrogen and hydrogen gas. Nitrogen is obtained from the air, and the main feedstock for hydrogen is natural gas or methane.

$$\text{nitrogen} + \text{hydrogen} \rightleftharpoons \text{ammonia}$$

$$N_2(g) + 3H_2(g) \rightleftharpoons 2NH_3(g)$$

The reactant gases are compressed to about 200 times atmospheric pressure, heated to about 450 °C, and passed over an iron catalyst. Haber and Bosch systematically tested about 20 000 catalysts, before finding the right one. Finally they found an iron ore containing traces of alkali metals that worked.

The atom economy for the Haber process is 100%, since there are no by-products. All the starting atoms end up in the ammonia molecules. But the yield is only 15%. The reaction yield is increased by recycling the unreacted nitrogen and hydrogen gas.

In this reversible reaction, there has to be a compromise between a high yield and a high rate of reaction. The process must be economically viable without using too much energy.

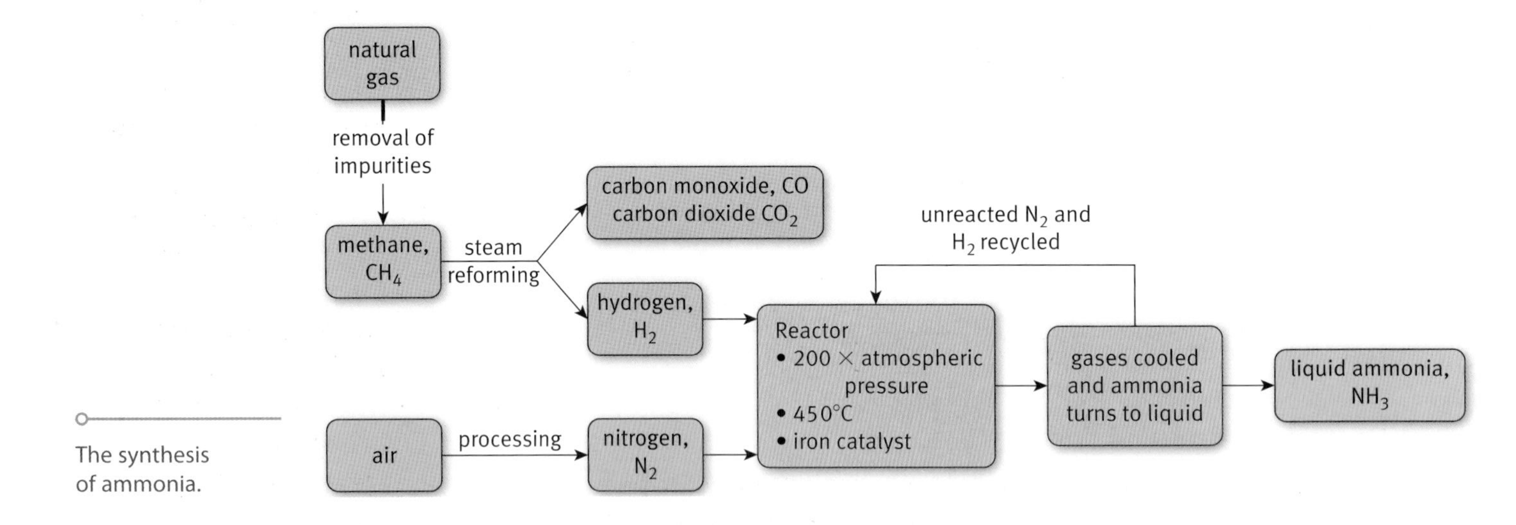

The synthesis of ammonia.

Getting the conditions right

Reversible reactions can reach a state of equilibrium. Changing the conditions of temperature and pressure can alter the proportions of reactants and products in a mixture of chemicals at equilibrium.

The effect on an equilibrium mixture of changing conditions can be predicted using **Le Chatelier's principle**. It states that:

'When the conditions change, an equilibrium mixture of chemicals responds in a way that tends to counteract the change.'

Effect of pressure

In the reaction to make ammonia there are four molecules of gas on the left-hand side of the equation but only two molecules of gas on the right. An equilibrium mixture of the gases responds to an increase in pressure by changing to make more ammonia and less nitrogen and hydrogen. This is what Le Chatelier's principle predicts because reducing the number of molecules tends to lower the pressure.

Effect of temperature

The reaction of nitrogen with hydrogen is exothermic. The reverse reaction, the decomposition of ammonia, is endothermic. Le Chatelier's principle predicts that if the temperature of an equilibrium mixture of the gases rises, then the equilibrium changes in a way that takes in energy because this tends to lower the temperature. So at a higher temperature there is less ammonia and more nitrogen and hydrogen in the equilibrium mixture.

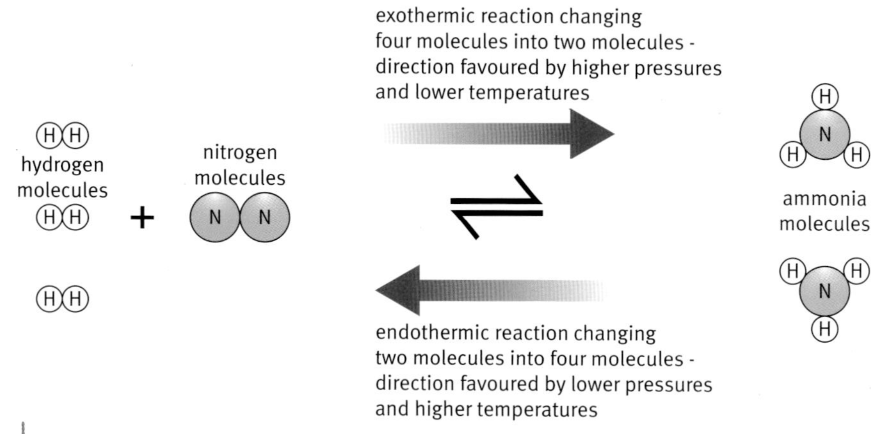

The position of the equilibrium is affected by temperature and pressure. The conditions used for the Haber process are a compromise to achieve a good yield at a reasonable cost.

Conditions in industry

In the Haber process the mixture of gases flows continuously through the reactor. The gases are only in contact with the catalyst for a short time. This means that the mixture never gets all the way to equilibrium.

The conditions chosen for the Haber process in industry are a compromise that balances chemical efficiency with cost and safety. The higher the pressure, the higher the yield of ammonia as the gas mixture approaches an equilibrium state. But high-pressure plants are expensive to build and run. They can also be more hazardous for plant operators. The lower the temperature, the higher the posssible yield of ammonia, but the reaction becomes too slow to be economic.

Key words

- nitrogen fixation
- Haber process
- Le Chatelier's principle

Questions

3 Suggest possible consequences for the environment of the large-scale manufacture of ammonia.

4 In the Haber process there is a continuous flow of reactants through the reactant chamber, rather than batches of reactants. Explain:
  a why this has an advantage regarding the amount of reactants used
  b how the ammonia is separated from the flow of reactants.

5 Suggest reasons why the Haber process can become uneconomic if the operators try to increase the yield of ammonia by:
  a making the pressure even higher
  b lowering the temperature.

6 At a pressure of 200 times atmospheric pressure and at 450 °C, an equilibrium mixture of nitrogen, hydrogen, and ammonia contains about 40% ammonia. In an industrial plant working under these conditions the mixture of gases leaving the reactor is only about 15% ammonia. Explain why.

Key words

- limiting factor
- triple bond
- nitrogen cycle
- nitrogenase

Nitrogen fixation occurs in the nodules of legumes such as clover.

Nitrogen fixation: a really important process

Nitrogen fixation is the process by which nitrogen is taken from its unreactive molecular form (N_2) in the air and converted into nitrogen compounds such as ammonia, nitrate, and nitrogen dioxide.

The growth of all organisms depends on the availability of mineral nutrients, especially nitrogen. Nitrogen is often the **limiting factor** for growth in environments where there is suitable climate and availability of water to support life.

Most organisms are unable to use nitrogen directly from the atmosphere because its **triple bond** makes it so stable. Some organisms are able to fix nitrogen directly from the air. These include different types of bacteria found in the roots of some plants.

Nitrogen is also fixed during chemical reactions that occur in the air during lightning flashes, and by industrial methods such as the Haber process. A significant proportion of all nitrogen fixation is as a result of non-biological processes.

Type of fixation		N_2 fixed (10^{12} g per year, or 10^6 metric tons per year)
non-biological	industrial	about 50
	combustion	about 20
	lightning	about 10
	TOTAL	*about 80*
biological	agricultural land	about 90
	forest and non-agricultural land	about 50
	sea	about 35
	TOTAL	*about 175*

Data from various sources, compiled by DF Bezdicek & AC Kennedy, in *Microorganisms in Action* (eds JM Lynch & JE Hobbie). Blackwell Scientific Publications, 1998.

Fixing nitrogen by organisms

Nitrogen from the air is 'fixed' at normal temperatures and pressures by some organisms. The enzyme **nitrogenase** acts as a catalyst to allow the reaction to take place. Nitrogenase contains clusters of iron, molybdenum, and sulfur (Fe/Mo/S). Nitrogenase converts nitrogen to ammonia:

$$N_2 + 6H^+ + 6e^- \longrightarrow 2NH_3$$

The best known examples of nitrogen fixing in plants are the root nodules of legumes such as peas, beans, and clovers.

The nitrogen cycle

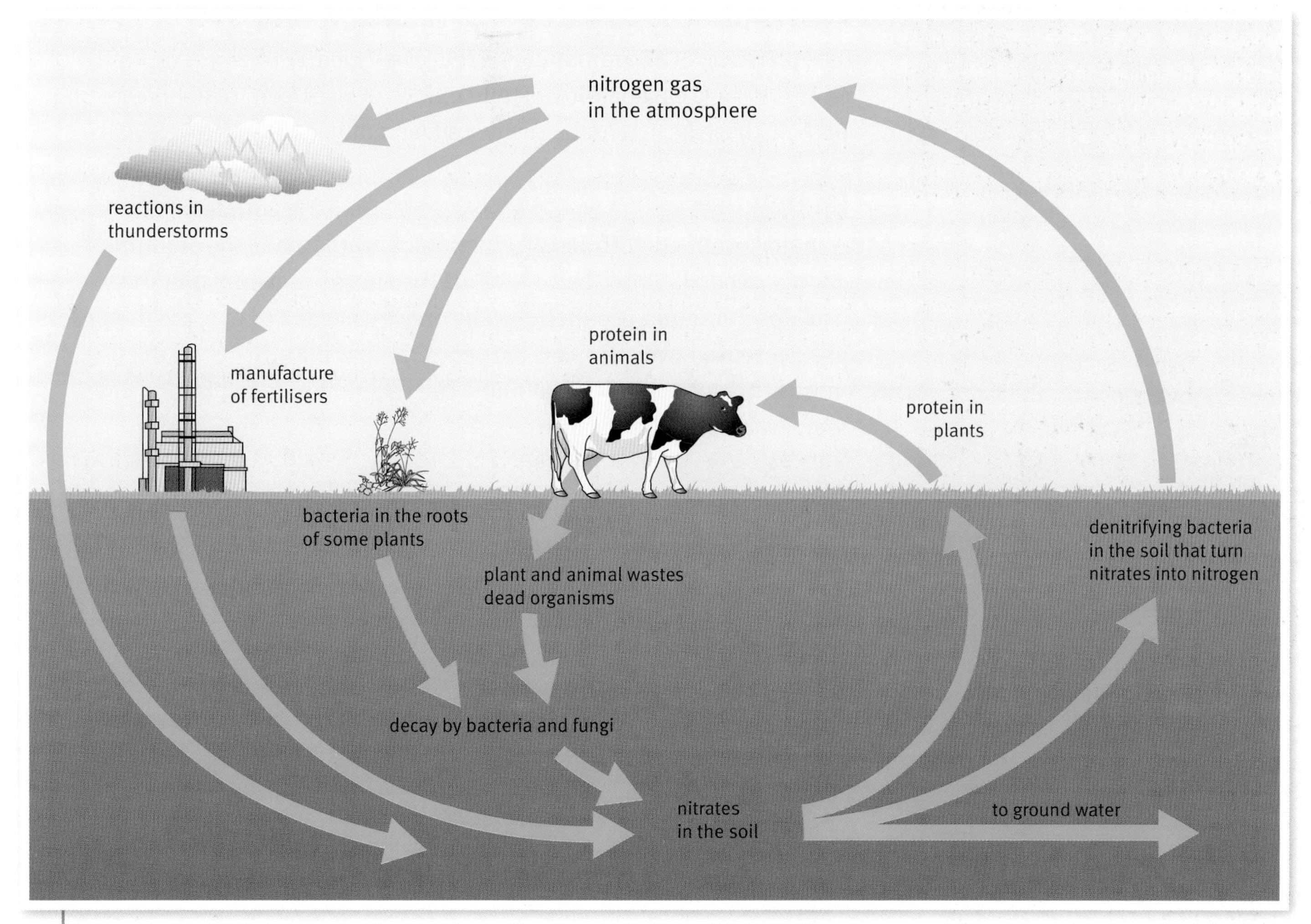

Natural and human activities contribute to the nitrogen cycle in the environment.

The **nitrogen cycle** consists of nitrogen fixation processes, which remove nitrogen from the atmosphere, and denitrification processes, which return nitrogen to the atmosphere.

Industrial nitrogen fixation has significantly increased the amount of fixed nitrogen, disrupting the natural nitrogen cycle. This has led to problems for the environment and human health. High levels of nitrates in rivers and lakes can cause the rapid growth of algae, which can damage ecosystems. Nitrates can also get into drinking water and be harmful to human health.

Restoring the balance in the nitrogen cycle is an important challenge for sustainable development.

Questions

7 Explain why nitrogen fixation is important.

8 Explain why plants can be short of nitrogen when there is so much nitrogen in the air.

9 The total amount of nitrogen fixed per year is about 255×10^{12} g. What percentage of this total is:
 a non-biological?
 b industrial?

10 Which process in the nitrogen cycle would need to increase in order to balance non-biological nitrogen fixation?

Can nitrogen fixation become any greener?

Environmental impact of the Haber process

The production of ammonia is a relatively clean process. The only emissions are carbon dioxide and oxides of nitrogen. In a modern plant, both of these gases can be recovered or reduced to very low levels.

Energy use

A major problem with the production of ammonia is the amount of energy needed. More than 1% of all the energy consumed in the world is used for ammonia production. The energy needed to operate the process has decreased over the past 100 years. This is mainly due to the use of more efficient catalysts that allow the reaction to take place at lower temperatures and pressures.

When the Haber process was introduced, the energy used in the production of ammonia dramatically decreased. Application of the principles of green chemistry in more recent years has ensured there is still a downward trend.

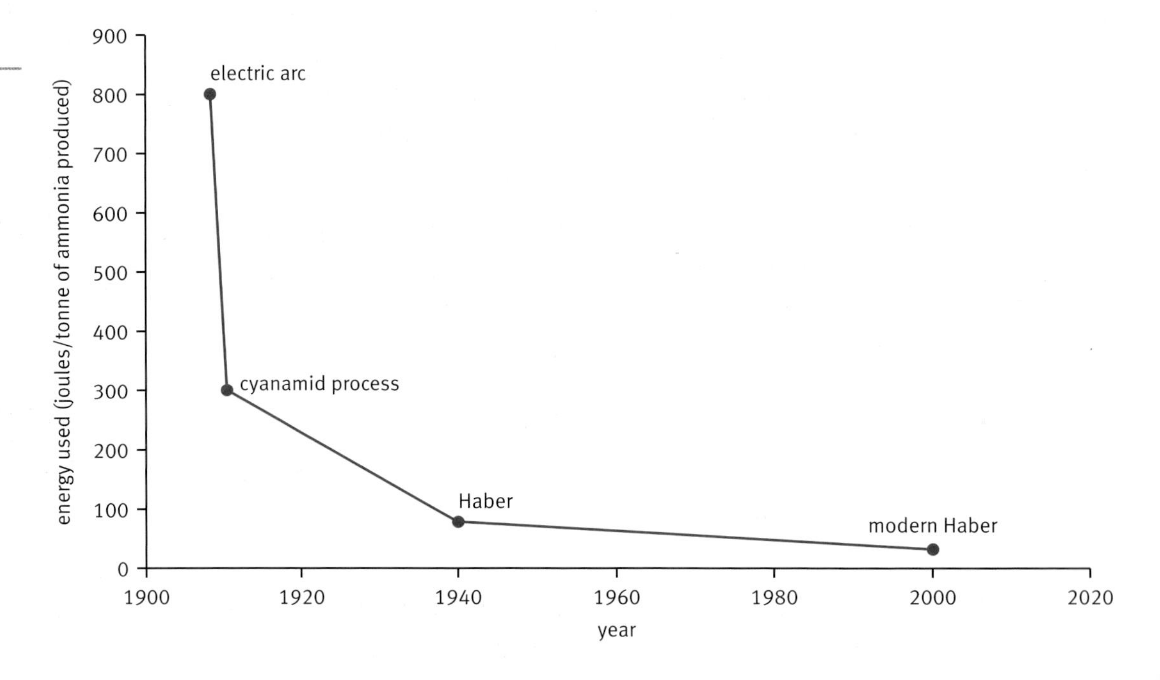

Future feedstocks

Currently, hydrogen is extracted from methane by steam reforming. If fossil fuel sources of methane run low, alternative ways of producing hydrogen will be needed. The electrolysis of water could become a major source of hydrogen. This feedstock would depend on a cheap and renewable electricity supply, such as hydroelectricity or solar power.

Using renewable energy sources for ammonia production would also reduce the amount of greenhouse gases entering the atmosphere.

New catalysts

The search for new catalysts is an important area of current research and development. The higher the catalytic activity of a catalyst, the more efficient it is at synthesising ammonia.

The Kellogg Advanced Ammonia Process (KAAP)

In 1992, M. W. Kellogg and the Ocelot Ammonia Company started ammonia production using a new ruthenium catalyst deposited on an active carbon support. With this catalyst a pressure of 40 times atmospheric pressure can be used for ammonia production, instead of 200 times atmospheric pressure.

The ruthenium catalyst is more active than the original iron-based catalyst and yields of about 20% ammonia can be achieved.

This new technology can be fitted to existing ammonia plants, saving money and energy. The newer catalyst is more expensive, but this is outweighed by other cost savings.

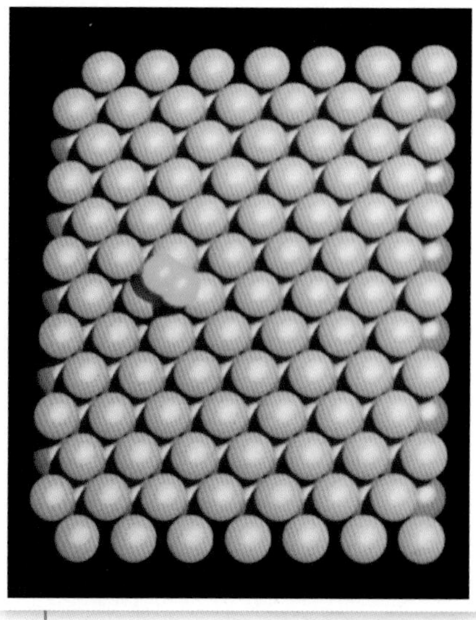

Model of a nitrogen molecule held on a layer of iron atoms at the surface of the usual iron catalyst for making ammonia. Nitrogen and hydrogen molecules react when brought together on the catalyst surface. Replacing iron atoms by ruthenium atoms gives a more effective catalyst.

The Haldor–Topsøe catalyst

In 2000, scientists at a Danish company announced the discovery of a new commercially viable catalyst for the Haber reaction. The new compounds contain iron, molybdenum, nitrogen, nickel, and cobalt. They appear to be two or three times more efficient than the current commercial, iron-based catalysts at the same operating conditions.

The new catalysts are cheaper than the ruthenium-based catalysts of the KAAP. The same Haldor–Topsøe team have also produced new ruthenium catalysts that are 2.5 times more active than current ruthenium catalysts.

Learning from nature

Chemists are keen to learn about nitrogen fixation from nature. Studies of the enzyme nitrogenase have shown that it contains clusters of iron, molybdenum, and sulfur.

Chemists have been successful in making similar artificial clusters that show catalytic activity. By producing and using new catalysts that mimic natural enzymes, it may be possible in the future to produce ammonia at room temperature and pressure. This, of course, would lead to even lower energy use during production.

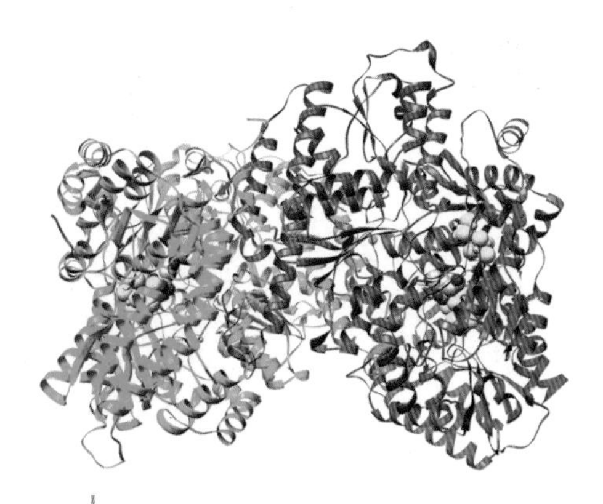

A computer-generated image representing the enzyme nitrogenase.

Questions

11 How are the principles of green chemistry influencing the development of industrial nitrogen fixation? (You may need to look back to page 200 to remind yourself about green chemistry.)

12 The enzyme nitrogenase fixes nitrogen at a pressure of 1 atmosphere, the KAAP ruthenium catalyst fixes nitrogen at 40 times atmospheric pressure, and the iron catalyst of the Haber process fixes nitrogen at 200 times atmospheric pressure.

Explain why processes that give a good yield at a lower pressure are more sustainable.

Topic 5: Chemical analysis

The LGC in Teddington is the UK's largest independent analytical laboratory. Its work covers food and agriculture, oil and chemicals, the environment, health care, life sciences, and law enforcement. It organises proficiency tests to check the performance of analytical laboratories.

The business of analysis

Analytical measurement is important. It is essential to ensure that the things we use in our everyday lives do us good and not harm. Over £7 billion is spent each year on chemical analysis in the UK.

Health and safety

The responsibility for health and safety in a laboratory is shared by everybody. Observing these regulations helps to keep accidents down to a minimum and the risk of injury low. Laboratories have their own health and safety regulations and codes of practice. Many have health and safety officers.

Looking after equipment

Equipment must be kept in good working condition. It has to be serviced regularly. Measuring instruments are checked at regular intervals.

Equipment should be cleaned properly after use and stored correctly. This is particularly important for fragile pieces of equipment such as glassware.

Accreditation

Analytical laboratories must show that they can do the job. Like the things tested in them, all laboratories must meet standards. Their standards are checked by the United Kingdom Accreditation Service (UKAS).

Analysts use proficiency tests to assess their work. Each laboratory receives identical samples to analyse. They send their results back to the organisers, who evaluate them. The laboratories are not named in the report, but results are coded so that a laboratory can recognise its results and see how well it has done.

International standards

There are international standards too. The International Olympic Committee (IOC), for example, accredits 27 laboratories to test blood and urine samples from athletes. The laboratories are all over the world and analyse 100 000 samples each year.

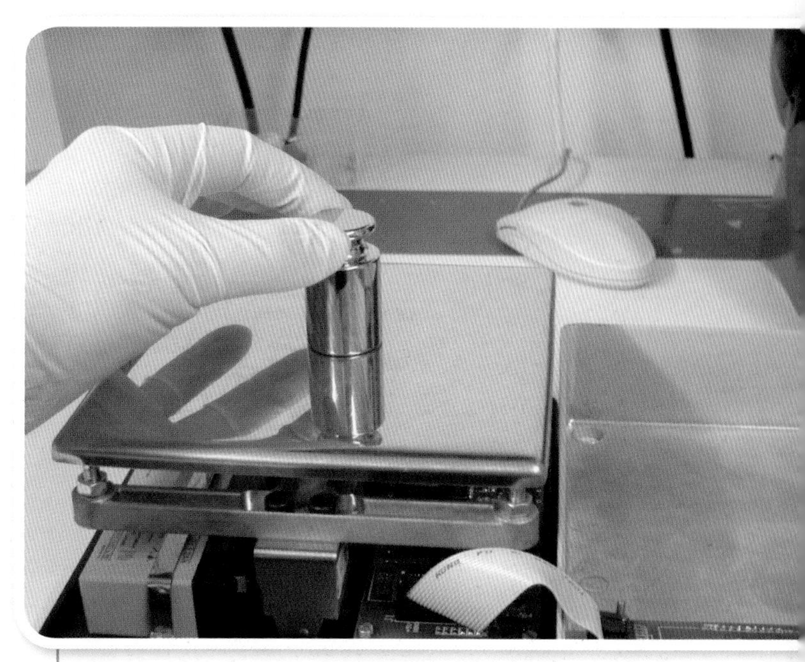

This electronic balance, like all equipment, must be maintained correctly and tested.

An analyst testing for stimulants in one of the laboratories accredited by the IOC.

Stages in an analysis

Find out about

- qualitative and quantitative methods
- steps in analysis

Choosing an analytical method

The first step is to pick a suitable method of analysis. A **qualitative** method can be used if the aim is simply to find out the chemical composition of the specimen. However, if the aim is to find out how much of each component is present, then a **quantitative** approach is essential. Each step of the analysis must be carried out as set down in the agreed standard procedures.

Analysts study samples of blood and urine to look for the presence of banned substances in athletes and other sports people. This can include qualitative analysis to identify drugs, and quantitative analysis to find out how much of a banned substance there is in a sample.

Taking a sample

The analysis must be carried out on a **sample** that represents the bulk of the material under test. This can be hard to achieve with an uneven mixture of solids, such as soil, but much easier when the chemicals are evenly mixed in solution, as in urine.

Measuring out laboratory specimens for analysis

Analysts take a sample and measure out accurately known masses or volumes of the material for analysis. It is common to carry out the analysis with two or more samples in parallel to check on the reliability of the final result. These are **replicate samples**.

Dissolving the sample

Many analytical methods are based on a solution of the specimens. With the analysis of acids or alkalis that are soluble in water, this is not a problem. It can be much more difficult to prepare a solution when analysts are working with minerals, biological specimens, or polymers.

Questions

1. What is the difference between a quantitative and qualitative analytical method? Give an example of each method.
2. Suggest reasons why it is important for analysts to follow standard procedures.
3. Samples are often dissolved in a solvent. Why is this?

Measuring a property of the sample in solution

When determining quantities, analysts look for a property to measure that is proportional to the amount of chemical in the sample. With an acid, for example, the approach is to find the volume of alkali needed to neutralise it. The more acid present, the greater the volume of alkali needed to neutralise it.

Calculating a value from the measurements

An understanding of chemical theory allows analysts to convert their measurements to chemical quantities. Given the equation for the reaction, and the concentration of the alkali, it is possible to calculate the concentration of the acid from the volume of alkali needed to neutralise it.

Key words

- qualitative
- quantitative
- sample
- replicate sample

Analysts preparing blood and urine samples for analysis. They are using a type of spectrometer to measure the concentrations of iron and zinc in the samples.

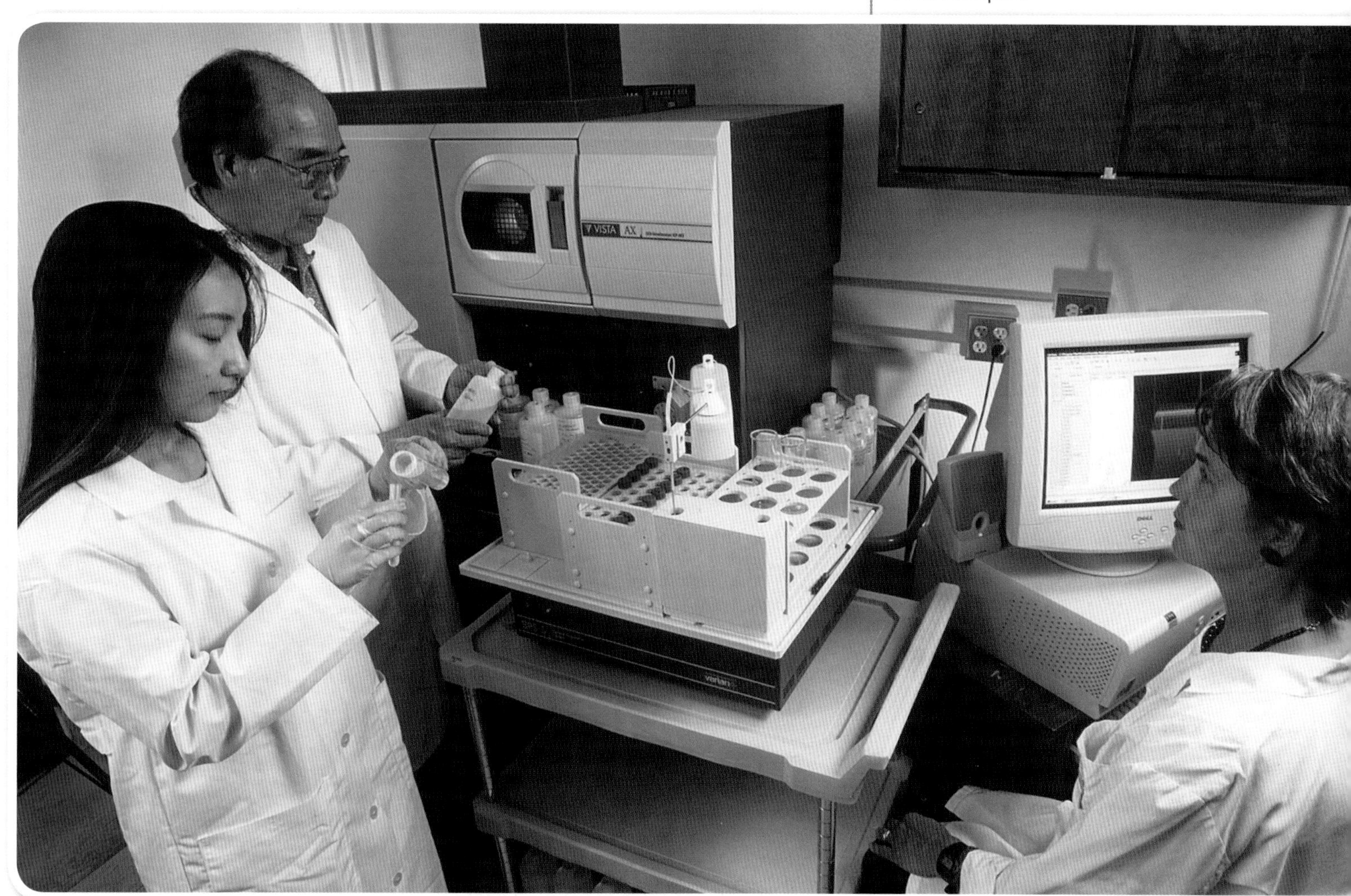

Estimating the reliability of the results

Analysts have to state how much confidence they have in the accuracy of their results. Comparing the values obtained from two or three replicate samples helps.

5B Sampling

Find out about

- ✓ collecting, storing and analysing samples

Analysts work with samples of materials. Rarely do they analyse the whole thing. How big the samples need to be depends on the analyses to be carried out.

Representative samples

The samples the analyst chooses must be **representative**. In other words, the samples should give an accurate picture of the material as a whole.

The composition of a homogeneous material is uniform throughout, like a milk chocolate bar.

The composition of a heterogeneous material varies throughout it, like a chocolate bar made in layers.

Scientists have to decide:

- how many samples, and how much of each, must be collected to ensure they are representative of the material
- how many times an analysis should be repeated on a sample to ensure results are reliable
- where, when, and how to collect the samples of the material
- how to store samples and transport them to the laboratory to prevent the samples from 'going off', becoming contaminated, or being tampered with.

Analysing water

Think about analysing water from two different sources. One is bottled water bought at a local supermarket. The other is from a local stream.

Key word
- representative sample

The bottled water is clear. There are no solids in suspension. It is likely to be tested for dissolved metal salts. The water is homogeneous, so only a single sample is needed. However, to check a batch of bottles, the analyst would take samples from a number of bottles. How much is needed depends on the test. There are no storage or transport problems. The bottle can be opened in the laboratory. This is a straightforward sampling problem.

Water from the stream may be cloudy. It may contain small creatures. It may be tested for a range of things, from the concentrations of dissolved chemicals to the number and variety of living organisms. Samples may vary from one part of the stream to another. They are likely to be heterogeneous. The time of year when samples are collected will have an effect on the water's composition. Also, samples need to be stored and taken back to the laboratory for analysis. This is a complex sampling problem that needs careful planning.

Water from a stream is likely to be heterogeneous. Samples may vary from one part of stream to another.

Bottled water is homogeneous. To test the water in the bottle, only a single sample is needed.

Questions

1. Why must the sample be representative of the bulk material under test?
2. Why is it good practice to take two or more samples?
3. Suggest how an analyst should go about sampling when faced with the following problems. In each case, identify the difficulties of taking representative samples. Suggest ways of overcoming the difficulties.
   - a Measuring the concentration of chlorine disinfectant in a swimming pool.
   - b Checking the purity of citric acid supplied to a food processor.
   - c Detecting banned drugs in the urine of athletes.
   - d Monitoring the quality of aspirin tablets made by a pharmaceutical company.
   - e Determining the level of nitrates in a farmer's field.
4. Why is the way in which samples are stored important?

Collecting samples

A life of grime

All local authorities have Environmental Health Officers. Most of their work involves checking restaurants, cafés, and food shops, but they do many other things as well, including checking pollution levels. The photograph shows Ralph Haynes of Camden Council in London taking soil samples.

Ralph says: 'There is concern that chromium salts from an old metal plating factory nearby may have contaminated the soil. I am taking 1 kg samples and putting them in plastic containers. I am taking care to label it properly. There is a British Standard on labelling, you know: it's called BS5969.'

'I'm taking care to take a sample from where I can actually see the change in the soil. I'm also going to take a sample from where I can't see it. Then I'm going to take samples from anywhere I think people could be at risk – such as gardens where children play.'

'Soil samples will keep for a while, but I'm also going to take water samples, and these must get to the lab within a couple of days.'

Ralph Haynes taking a soil sample in Camden.

Sporting samples

Sports men and women are often asked to provide urine samples to check if they have been taking drugs. Sometimes these cases hit the headlines, and allegations have been made in the past that urine samples have been tampered with.

The scientists who carry out the analysis at Kings College, London have to be sure that they have the right sample and that it has been correctly stored and labelled.

- First, the athlete has to produce the sample in front of a testing officer, who has to actually see the urine leaving the athlete's body and ensure there has been no cheating.
- With the testing officer watching, the athlete is allowed to pour the sample into two bottles. They seal the bottles themselves so that they feel assured no one else has tampered with them.
- The bottles are labelled with a unique code rather than the athlete's name, so the lab does not know the identity of the athlete. The bottles are sent to the lab by courier in secure polystyrene packaging.
- At the lab, one bottle is analysed immediately and the other stored in the freezer in case there is a query at a later date.

'We send the results to the Sports Council,' says Richard Caldwell, one of the analysts. 'It's someone at the Council who tallies up which bottle was collected from which person. It's quite interesting when we have a positive result and we find out from the press a few months later who it was!'

Labelled urine samples from athletes ready for testing.

Find out about

- principles of chromatography
- paper chromatography
- thin-layer chromatography

There are several types of **chromatography**. At the cheap-and-simple end is paper chromatography, which can be done with some blotting paper and a solvent. At the expensive end is gas chromatography, which involves high-precision instruments. All types of chromatography work on similar principles.

Chromatography can be used to:

- separate and identify the chemicals in a mixture
- check the purity of a chemical
- purify small samples of a chemical.

Principles of chromatography

Chromatography depends on the movement of a **mobile phase** through a fixed medium called the **stationary phase**. The analyst adds a small sample of the mixture to the stationary phase. As the mobile phase moves through the stationary phase, the chemicals in the sample move between the mobile and stationary phases.

For each chemical in the mixture there is a dynamic equilibrium as the molecules distribute themselves between the stationary phase and mobile phase. If a chemical in the mixture is attracted more to the mobile phase it moves faster. If a chemical is attracted more to the stationary phase it moves more slowly. Since each chemical in the mixture is attracted differently they move at different speeds and are separated.

- The chemical moves quickly if the position of equilibrium favours the mobile phase.
- The chemical moves slowly if the position of equilibrium favours the stationary phase.

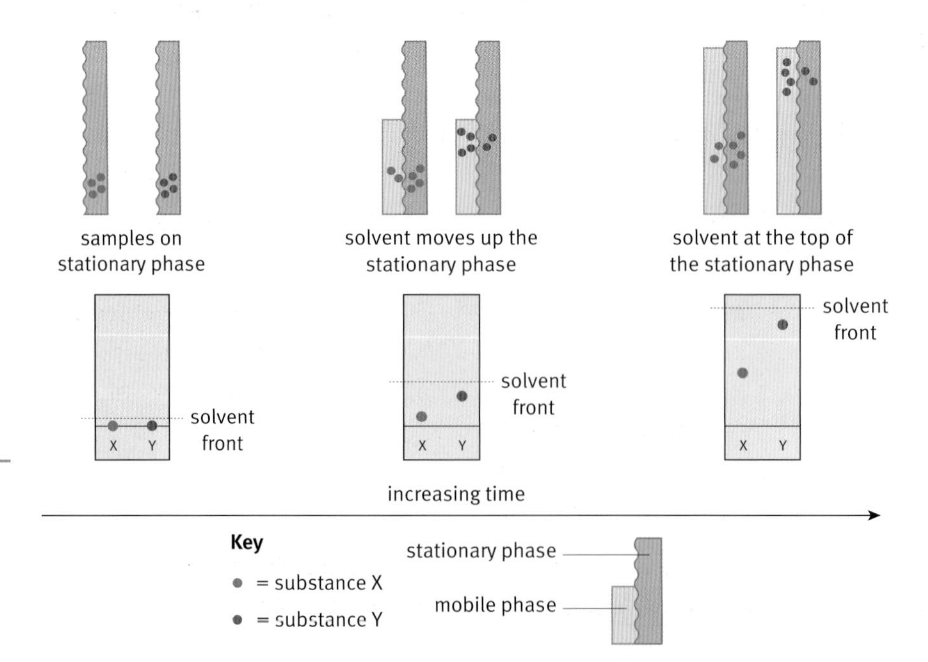

For each chemical there is a dynamic equilibrium for the molecules as they distribute themselves between the two phases. How quickly substances move through the stationary phase depends on the position of equilibrium.

Paper and thin-layer chromatography

Paper chromatography and thin-layer chromatography (TLC) are used to separate and identify substances in mixtures. The two techniques are very similar.

These techniques do not require expensive instrumentation, but are limited in their use. Paper chromatography is very rarely used. TLC is 'low technology', but it can be useful before moving on to more complex techniques. TLC is quick, cheap, and only requires small volumes of solution. A large number of samples can be run at once.

Key words

- chromatography
- mobile phase
- stationary phase
- aqueous
- non-aqueous

Organic reactions are monitored to find out at what point the reaction is complete, and purification can begin. TLC is simple and quick, so it is often used to monitor the progress of organic reactions and to check the purity of products.

Forensic laboratories may use TLC to analyse dyes extracted from fibres and when testing for controlled drugs, cannabis in particular.

Stationary and mobile phases

In paper chromatography the stationary phase is the paper, which does not move. In TLC the stationary phase is an absorbent solid supported on a glass plate or stiff plastic sheet.

In both paper chromatography and TLC, the mobile phase is a solvent, which may be one liquid or a mixture of liquids. Substances are more soluble in some solvents than others. For example, some substances dissolve well in water, while others are more soluble in petrol-like, hydrocarbon solvents. With the right choice of solvent, it becomes possible to separate complex mixtures.

Chemists call solutions in water **aqueous** solutions. The term **non-aqueous** describes solvents with no water in them.

Questions

1 Describe in your own words how chromatography works.

2 Name two substances that dissolve better in water than in hydrocarbon (or other non-aqueous) solvents.

Key words

- reference materials
- chromatogram
- solvent front
- locating agent
- retardation factor (R_f)

Chromatography plates must be spotted carefully. Small, concentrated spots are needed. Their starting position should be marked.

Preparing the paper or plate

The sample is dissolved in a solvent. This solvent is not usually the same as the mobile phase.

A small drop of the solution is put on the paper, or TLC plate, and allowed to dry, leaving a small 'spot' of the mixture.

If the solution is dilute, further drops are put in the same place. Each is left to dry before the next is added. This produces a small spot with enough material to analyse. The separation is likely to be poor if the spot spreads too much.

One way of identifying the chemicals in the sample is to add separate spots of solutions of substances suspected of being present in the unknown mixture. These are called **reference materials**.

Running the chromatogram

The analyst adds the chosen solvent (the mobile phase) to a chromatography tank and covers it with a lid. After the tank has stood for a while, the atmosphere inside becomes saturated with solvent vapour.

The next step is to place the prepared paper or TLC plate in the tank, checking that the spots are above the level of the solvent.

The solvent immediately starts to rise up the paper or plate. As the solvent rises, it carries the dissolved substances through the stationary phase. Covering the tank ensures that the solvent does not evaporate.

The chromatography paper, or TLC plate, is taken from the tank when the solvent gets near the top. The analyst then marks the position of the **solvent front**.

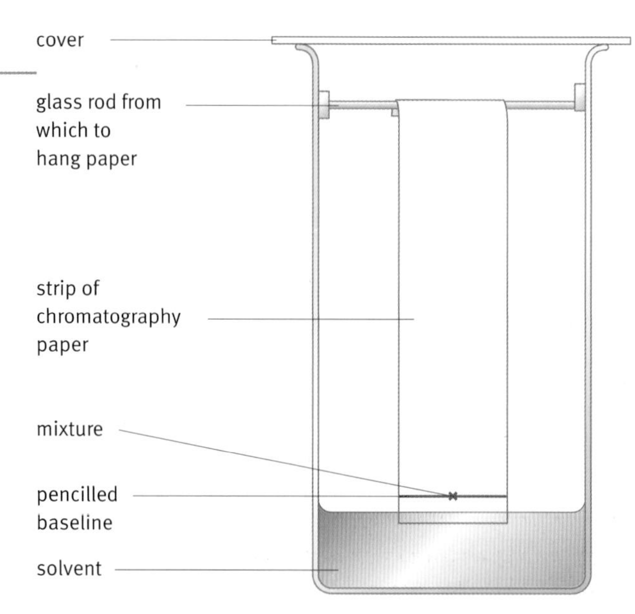

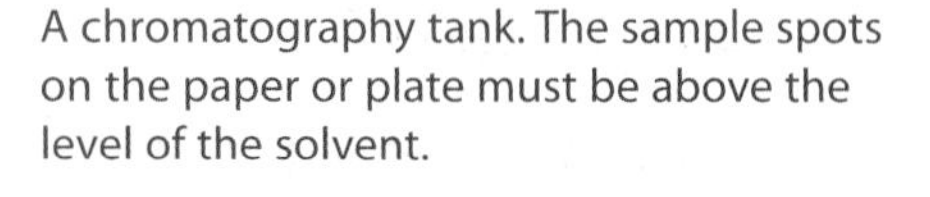

A chromatography tank. The sample spots on the paper or plate must be above the level of the solvent.

Locating substances

There is no difficulty marking the positions of coloured substances. All the analyst has to do is outline the spots in pencil and mark their centres before the colour fades.

There are two ways to locate colourless substances:

- Develop the chromatogram by spraying it with **a locating agent** that reacts with the substances to form coloured compounds.
- Use an ultraviolet lamp with TLC plates that contain fluorescers, so that the spots appear violet in UV light.

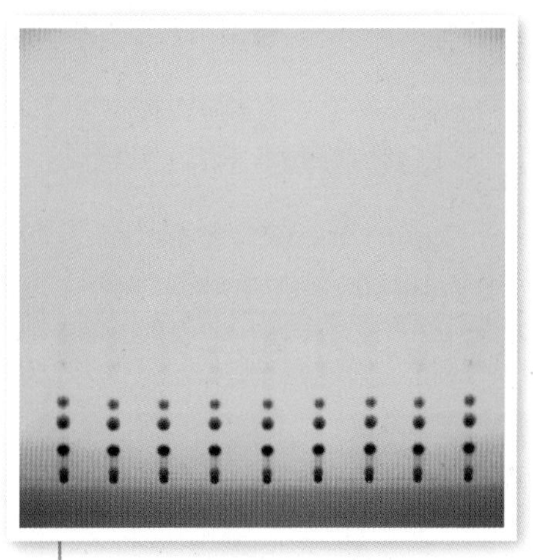

'Invisible' spots can often be seen under a UV lamp.

Interpreting chromatograms

Chemicals may be identified by comparing spots with those from standard reference materials.

A chemical may also be identified by its **retardation factor** (R_f). This does not change, provided the same conditions are used. It is calculated, using the following formula, by measuring the distance travelled by the substance:

$$R_f = \frac{\text{distance moved by chemical}}{\text{distance moved by solvent}} = \frac{y}{x}$$

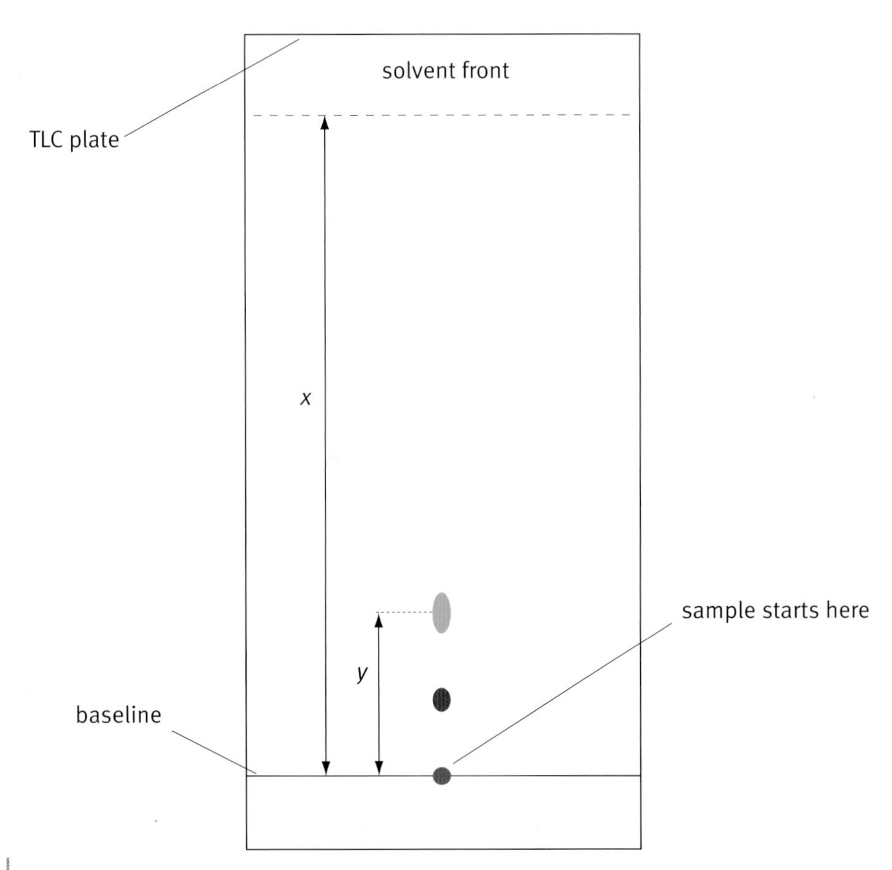

Retardation factors (R_f) can be calculated by measuring the distances travelled by chemicals in the sample and by the solvent.

Questions

3 Why is it sometimes necessary to 'develop' a chromatogram? How can this be done?

4 Why is it sometimes useful to use thin-layer chromatography plates that have been impregnated with fluorescers?

5 What are reference materials used for?

6 Paper chromatography is used to separate a mixture of a red and a blue chemical. The blue compound is more soluble in water while the red chemical is more soluble in the non-aqueous chromatography solvent. Sketch a diagram to show the chromatogram you would expect to form.

Gas chromatography

Find out about

- gas chromatography
- retention times

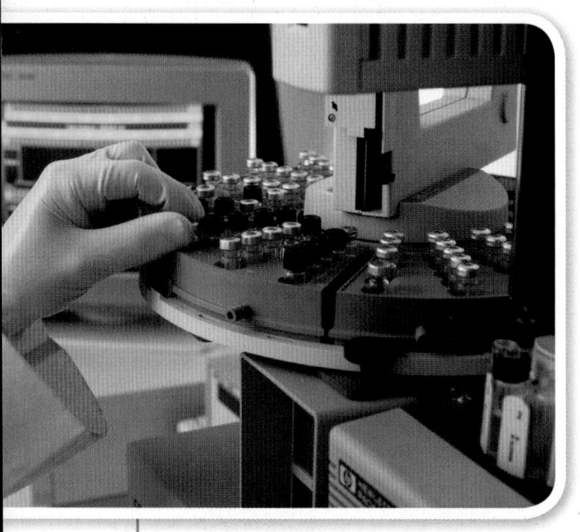

This scientist uses GC to check the quality of water from a river.

Key word

- retention time

Gas chromatography (GC) is used to separate complex mixtures. The technique separates mixtures much better than paper or thin-layer chromatography (TLC).

This technique is also more sensitive than paper chromatography or TLC, which means it can detect small quantities of compounds. That is why it is usually preferred to paper chromatography or TLC. The technique not only identifies the chemicals in a mixture but also can measure how much of each is present.

Understanding the limits of detection for a technique can be very important, otherwise an analyst can report that a contaminant is absent when it is in fact present, but at too low a concentration to be detected. Careful research has been necessary to find out the detection limits for such chemicals as pesticide residues in food.

Stationary and mobile phases

The principles of GC are the same as for paper chromatography and TLC. A mobile phase carries a mixture of compounds through a stationary phase. Some compounds are carried through more slowly than others. This is because they have different boiling points or a greater attraction for the stationary phase. Because they travel at different speeds, the compounds can be separated and identified.

The mobile phase is a gas such as helium. This is the **carrier gas**.

The stationary phase is a thin film of a liquid on the surface of a powdered solid. The stationary phase is packed into a sealed tube,

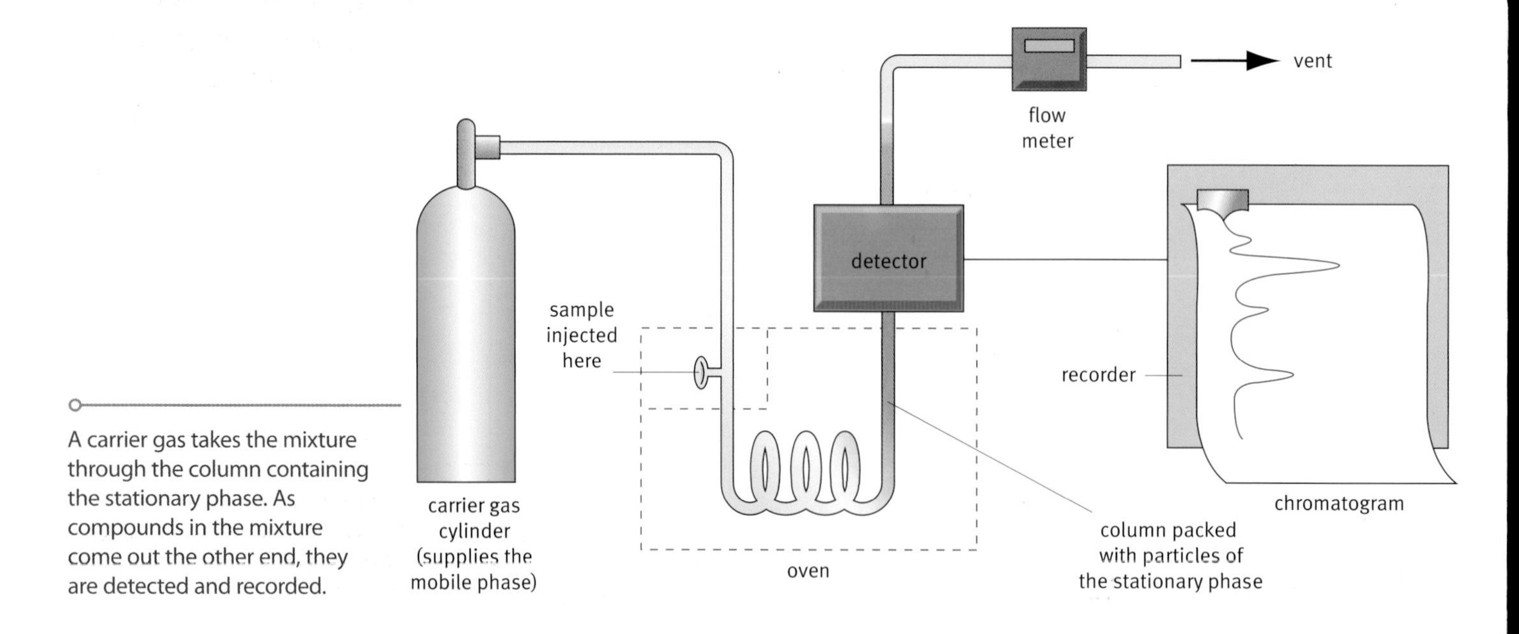

A carrier gas takes the mixture through the column containing the stationary phase. As compounds in the mixture come out the other end, they are detected and recorded.

which is the column. The column is long and thin. Some columns are 25 m long but only 0.25 mm in diameter.

Only very small samples are needed. The analyst uses a syringe to inject a tiny quantity of the sample into the column. Samples are generally gases or liquids.

The column is coiled inside an oven, which controls the temperature of the column. This means that it is possible to analyse solids if they can be injected in solution and then turn to a vapour at the temperature of the column.

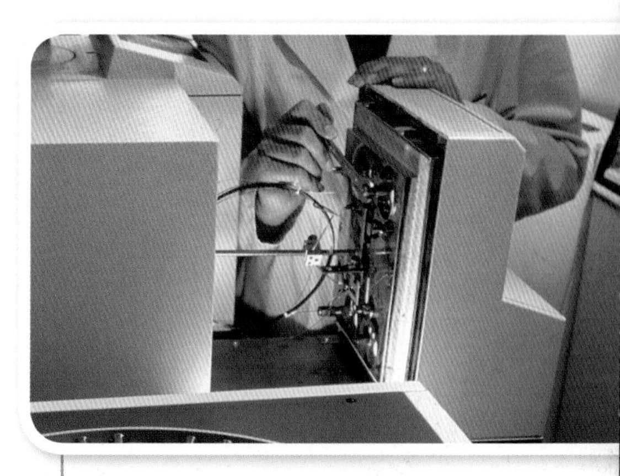

The coiled column inside the oven of a GC instrument. The detector is connected to a computer and the chromatogram appears on the screen.

Separation and detection

Once the column is at the right temperature, the carrier gas is turned on. Its pressure is adjusted to get the correct flow rate through the column. The analyst injects the sample at the start of the column where it enters the oven. The chemicals in the sample turn to gases and mix with the carrier gas. The gases pass through the column.

In time, the chemicals from the sample emerge from the column and pass into a detector. The chemicals can be identified using mixtures of known composition.

Interpreting chromatograms

The detector sends a signal to a recorder or computer when a compound appears. A series of peaks, one for each compound in the mixture, make up the chromatogram. The position of a peak is a record of how long the compound took to pass through the column. This is its **retention time**. The height of the peak indicates how much of the compound is present.

Questions

1 Look at the chromatogram on this page.
   a How many components have been separated?
   b Estimate the retention time of each component.
   c Which component was present in the largest quantity? Which one was present in the smallest quantity?

2 Why is it important to understand the detection limits of a technique?

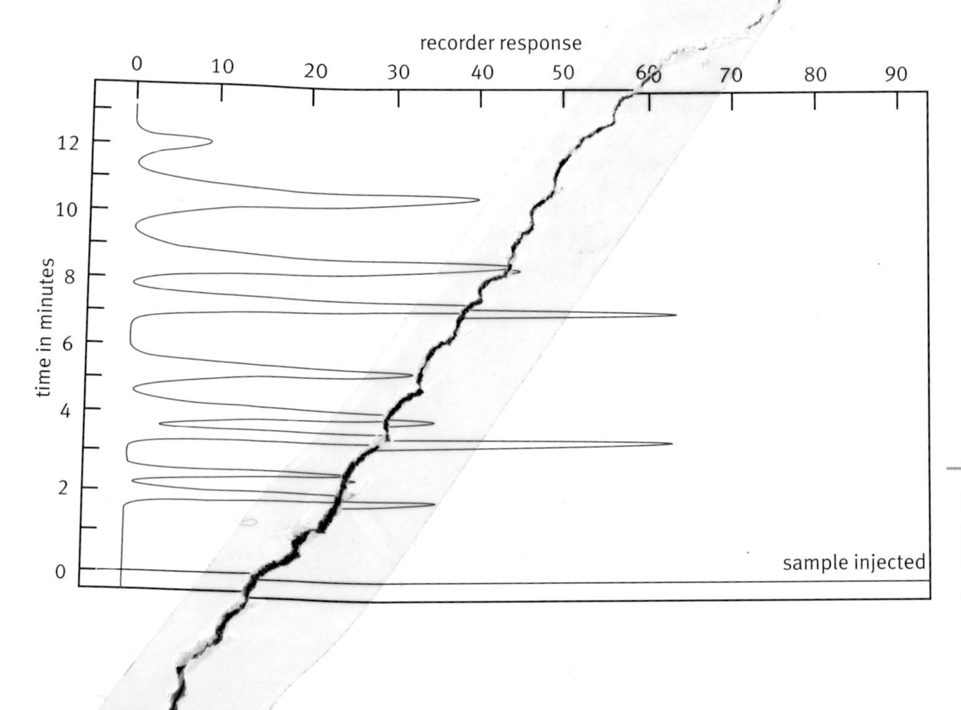

Each compound in the mixture appears as a peak in the chromatogram. The time it takes to get through helps the scientist to identify it. The height of the peak enables the scientists to say how much there is.

Chemical archaeology

Richard Evershed is a professor at Bristol University. He has used gas chromatography all his career: 'During my PhD, I used gas chromatography to study the chemical messages insects use to communicate. After my PhD, I moved to the University of Bristol, where I used the same technique to look at the organic chemicals preserved in ancient rocks originating from organisms that lived many millions of years ago.'

Analytical chemist Anna Mukherjee carrying out a GC analysis at the University of Bristol.

Ancient traces

Richard recognised that gas chromatography could also be used to identify the remains of fats, waxes, and resins preserved at archaeological sites.

He has used this technique to analyse organic residues trapped in the walls of very old cooking pots to find out what people ate in the past. He uses gas chromatography to separate the mixture of fats and waxes in the residues. By gradually increasing the temperature of the gas chromatography column, he is able to separate compounds with different boiling points. When the temperature reaches the boiling point of a chemical, it turns into gas and is carried by the carrier gas to a flame, where it burns to produce an electrical signal. The separated compounds appear as a series of peaks on the chromatogram.

Richard says: 'We made a real breakthrough when we found that we could identify traces of butter in 6000-year-old pottery from prehistoric Britain. This showed us that milking animals is a very ancient practice.'

Analysis of the chemicals absorbed into old pots gives clues to the food that our ancestors cooked.

A whiff of cabbage

Traces of cooked cabbage can survive for a long time. Richard found this to be true when investigating a set of pots dating from late Saxon times. The pots are over a thousand years old. 'I've found traces of cabbage preserved in the pot wall. It's the natural wax you can see on the surface of the cabbage, which is released during boiling. We can extract the same waxes from modern supermarket cabbages, and the gas chromatography traces look pretty much identical.'

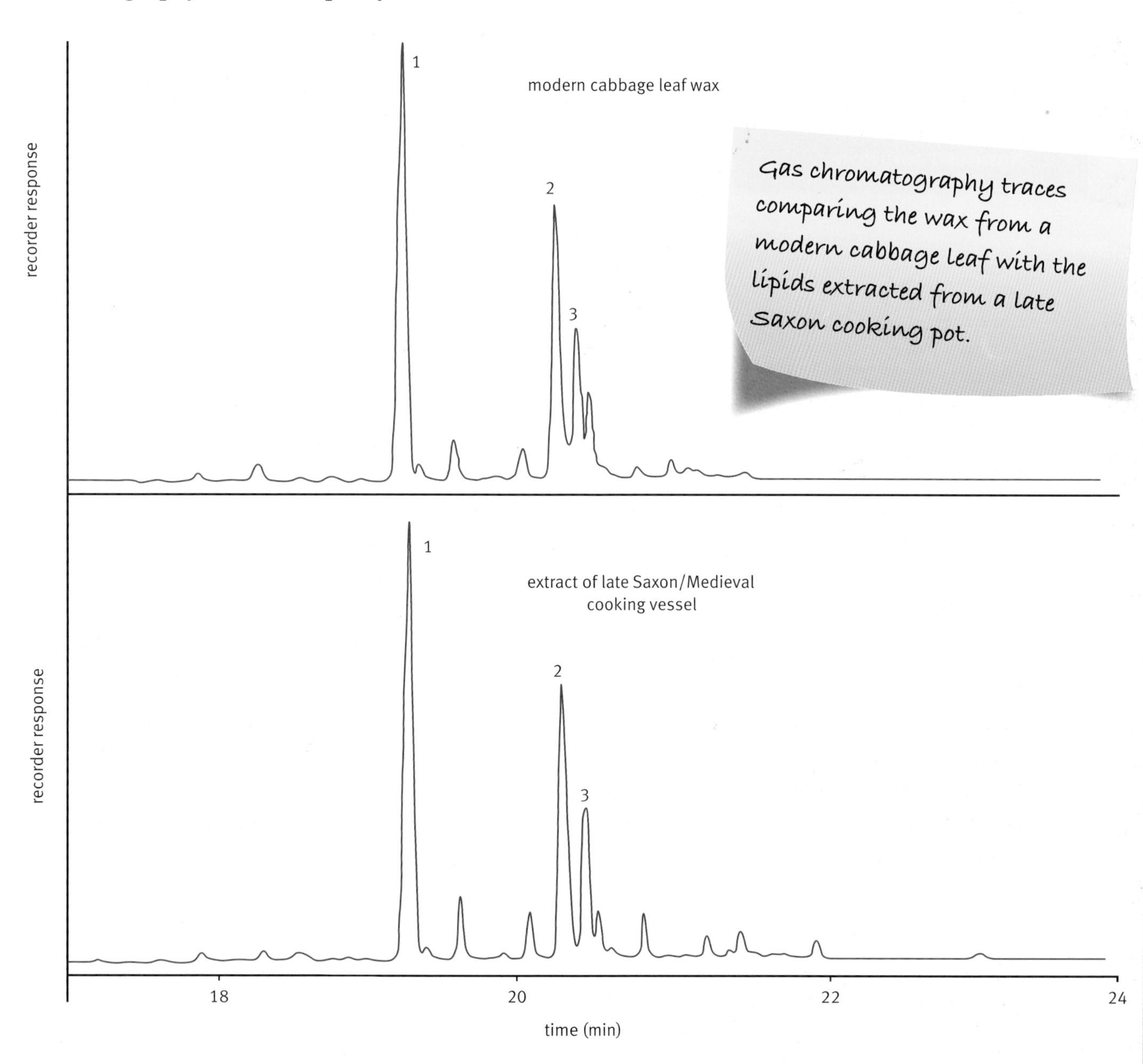

Gas chromatography traces comparing the wax from a modern cabbage leaf with the lipids extracted from a late Saxon cooking pot.

Titrations

Find out about

- acid–alkali titrations
- standard solutions

Key words

- titration
- pipette
- end point
- burette
- standard solution

A **titration** is a quantitative technique based on measuring the volumes of solutions that react with each other. Chemists use titrations to measure concentrations and to investigate the quantities of chemicals involved in reactions. Titrations are widely used because they are quick, convenient, accurate, and easy to automate.

Titration procedure

In a typical titration, an analyst uses a **pipette** (or a burette) to transfer a fixed volume of liquid to a flask. In an acid–base titration to find the concentration of an acid, this might be 20 cm^3 of the solution of acid.

Next, the analyst adds one or two drops of a coloured indicator. The indicator is chosen to change colour sharply when exactly the right amount of alkali has been added to react with all the acid. The indicator works because there is a very sharp change of pH at this point, which is called the **end point.**

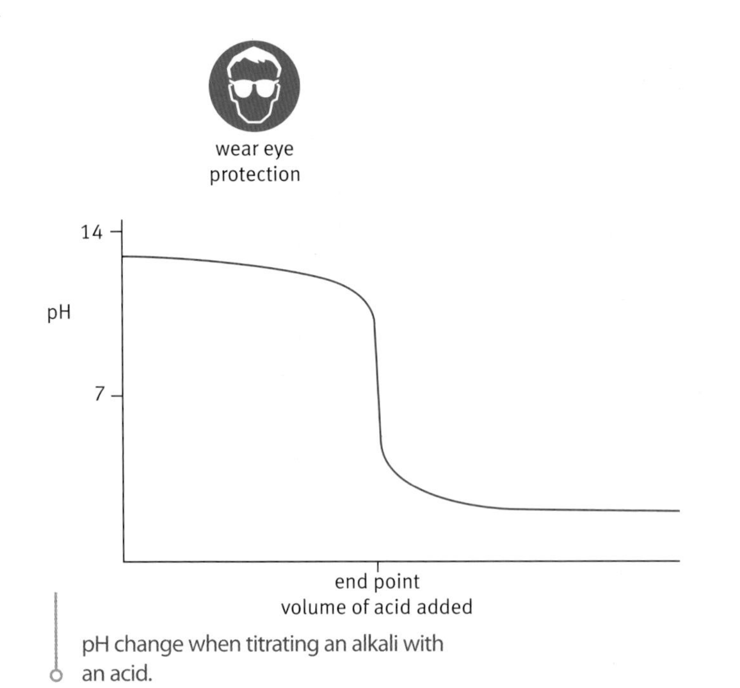

pH change when titrating an alkali with an acid.

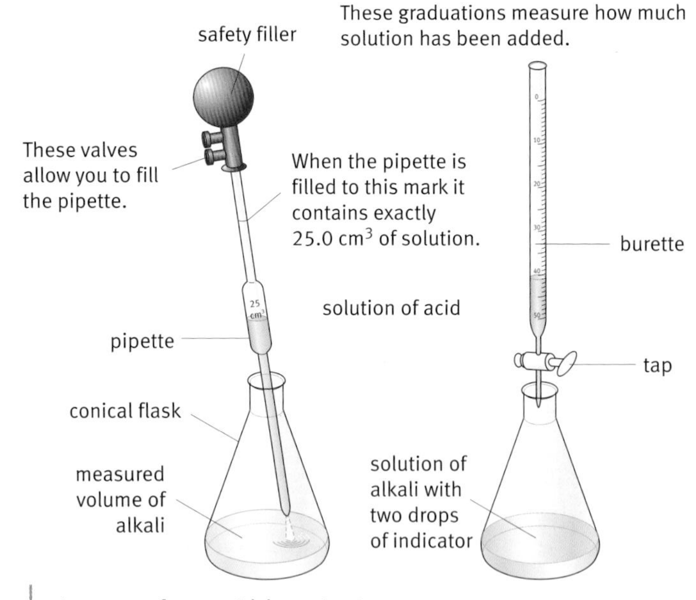

Apparatus for an acid–base titration.

The analyst has a **burette** ready containing a solution of acid with a concentration that is known accurately. Then the analyst runs the acid from a burette into the alkali a little at a time until the indicator changes colour. Reading the burette scale before and after the titration shows the volume of alkali added.

It is common to do a rough titration first to get an idea where the end point lies, and then to repeat the titration more carefully, adding the acid drop by drop near the end point. The analyst repeats the titration two or three times as necessary to achieve consistent values for the volume of alkali added.

Remember

1 litre = 1 dm^3 = 1000 cm^3

$$\text{concentration (g/dm}^3\text{)} = \frac{\text{mass (g)}}{\text{volume (cm}^3\text{)}}$$

Preparing accurate solutions

The accuracy of a titration can be no better than the accuracy of the solutions used to make the measurements. If chemists know the concentration of a solution accurately, they call it a **standard solution**, because it can be used in analysis to measure the concentrations of other solutions.

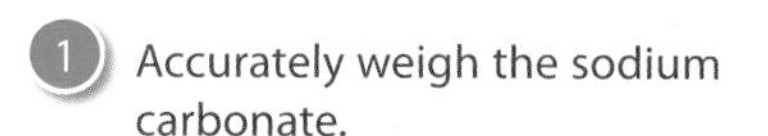

1 Accurately weigh the sodium carbonate.

2 Dissolve the solute in a small amount of solvent, warming it if necessary.

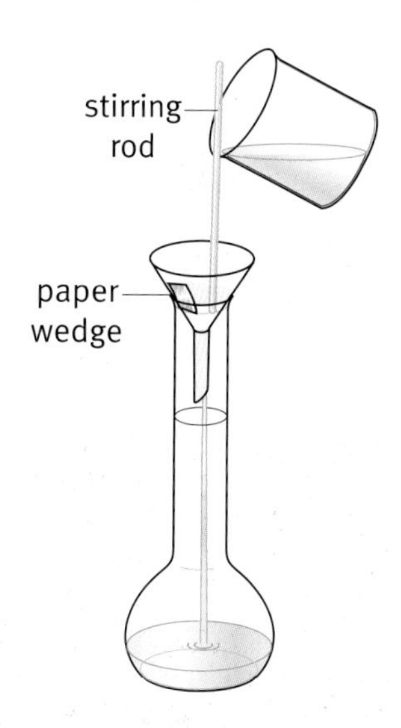

3 Transfer the sodium carbonate solution to a graduated flask.

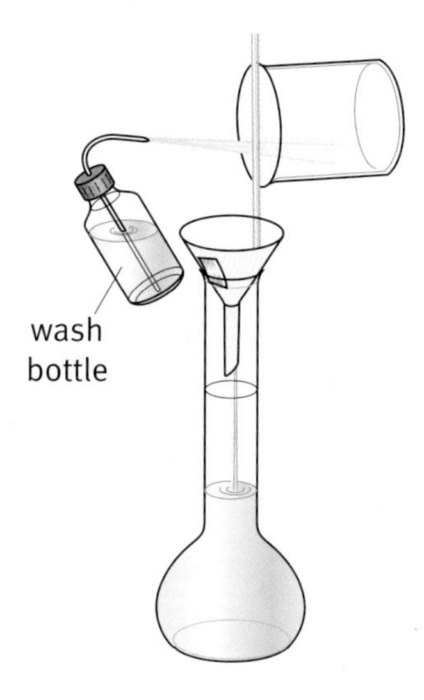

4 Rinse all the solution into the flask with more solvent.

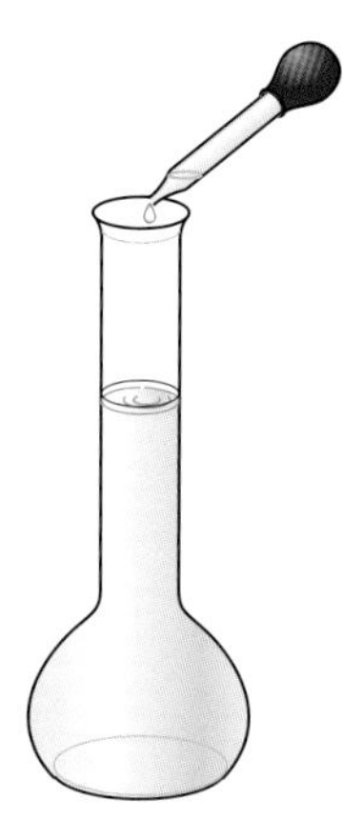

5 Add solvent drop by drop to make up the volume to the mark on the flask

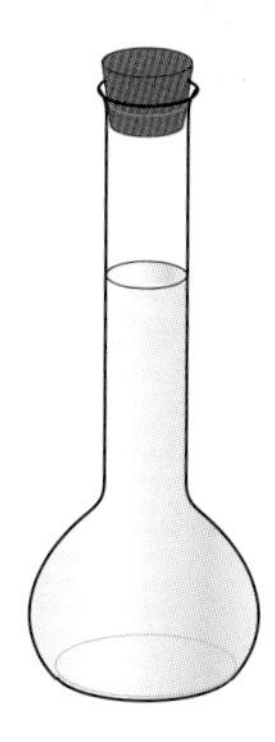

6 Stopper and shake the flask.

The procedure for making up a standard solution with an accurately known concentration.

Questions

1 What is the concentration of these solutions in grams per litre (g/dm^3):

- a a solution of sodium carbonate made by dissolving 4.0 g of the solid in water and making the volume up to 500 cm^3 in a graduated flask?
- b a solution of citric acid made by dissolving 2.25 g of the solid in water and making the volume up to 250 cm^3 in a graduated flask?

2 What is the mass of solute in these samples of solutions?

- a A 10 cm^3 sample of a solution of silver nitrate with a concentration of 2.55 g/dm^3.
- b A 25 cm^3 sample of a solution of sodium hydroxide with a concentration of 4.40 g/dm^3.

Titrating acids in food and drink

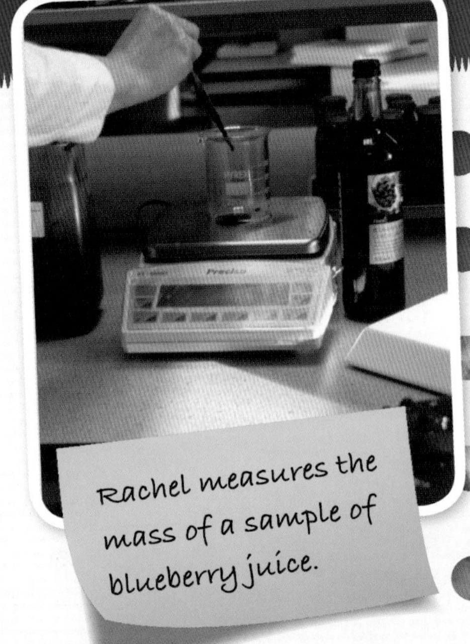

Rachel measures the mass of a sample of blueberry juice.

Rachel Woods is the quality control manager for Danisco, a company that manufactures ingredients for food and soft drinks. Acids are very important in her work. Some acids occur naturally in the fresh ingredients; others are added as preservatives or to improve flavour.

The quantity of acid in any food or drink is important. Think of a soft drink – too much acid and it tastes sour; too little and it might be insipid. With just the right amount it tastes refreshing and fruity,and just the right amount of acid makes a drink seem more thirst-quenching and satisfying.

Automated titrations

One of the most important acids in Rachel's work is citric acid. She regularly has to test ingredients and finished products to check their citric acid content. She uses a titration machine. Here she is testing a sample of blueberry juice. First, using a dropping pipette, she takes a sample of the juice from the container into a beaker. It is the same beaker in which she will carry out the titration, which makes things so much quicker and simpler. She has weighed exactly 4.30 grams.

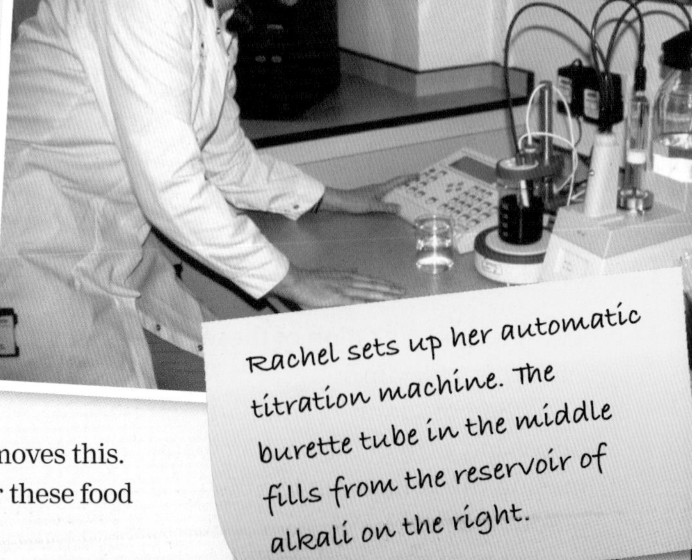

Rachel sets up her automatic titration machine. The burette tube in the middle fills from the reservoir of alkali on the right.

She adds boiled water to the juice to bring it up to the 300 cm^3 mark. She does not use water straight from the tap as it has dissolved calcium hydrogencarbonate in it. Boiling the water removes this. For many titrations distilled water is necessary, but for these food samples ordinary boiled water is fine.

All Rachel has to do is put the pH probe and tube from the burette into the beaker and put it on the stand.

It has a magnetic stirrer, so she does not even have to swirl the flask. She presses a button, and the burette tube starts to fill from the reservoir bottle. The concentration of her sodium hydroxide, NaOH, solution is 8.0 g/dm^3.

When the burette is full, it slowly pumps the NaOH solution into the beaker. Because blueberry juice contains a natural indicator, you can see a colour change from purply–red to dark-blue–grey as it reaches the end point.

The exact point is not easy to see, but that does not matter, as the machine works by measuring the pH of the solution. It measures the quantity of NaOH required to bring the blueberry juice up to a pH of 8.3. Not that Rachel has to measure that volume – far from it: all she has to do is look at the readout on the screen. It shows a little graph and calculates the percentage of citric acid in the juice. This sample is 6.08% citric acid, which is exactly what it should be, and using the machine, Rachel can test as many samples as she needs very quickly and simply.

The beaker with the blueberry juice in the titration machine. A magnetic stirrer mixes the juice with the alkali added by the burette. The probe dipping into the solution measures the pH.

The hands-on method

Not all of Rachel's titrations can be done on the machine. Here she is using the traditional method to test the concentration of a sample of butanoic acid.

Butanoic acid (C_3H_7COOH) is an important part of butter and cheese flavours, so it is used in things like cheese-and-onion crisps, and some margarines. Used in this way, butanoic acid is great, but unfortunately in large quantities it smells terrible.

Rachel fills the burette with the acid solution. She keeps her eye level with the zero mark on the burette. She wants the meniscus to sit exactly on the line.

It is not Rachel's favourite titration, but she is so used to it she manages to keep smiling. Butanoic acid is associated with fats, so it is not soluble in water. Instead it is dissolved in ethanol.

The next job is to pipette exactly 20.0 cm^3 of warm potassium hydroxide, KOH, solution into a clean flask. The concentration of the alkali is accurately known. Again Rachel keeps her eye level with the mark on the pipette as she uses the valve on the filler to adjust the level. A tiny drop of solution is always left in the end of the pipette, but Rachel resists the temptation to blow it out. As she adds a few drops of the indicator, phenolphthalein, the alkaline solution turns a stunning shade of shocking pink.

Rachel uses a pipette to run a measured volume of the standard potassium hydroxide into a titration flask.

Now the titration begins. With her right hand she swirls the flask, and with her left hand she gently releases the tap on the burette to let 1 cm^3 of the acid into the flask at a time. She keeps her eye on the flask all the time. Because she does this titration so regularly, she knows when the end point is coming up. It happens suddenly. First the solution in the centre of the flask goes colourless, then the pink colour disappears altogether. At that precise point Rachel closes the tap on the burette and takes a note of the reading: 40.2 cm^3. She repeats the titration at least once more.

Interpreting the results

Relative formula masses: C_3H_7COOH = 88 and KOH = 56

The equation for the reaction in the titration flask is

$$\underset{88\,g}{C_3H_7COOH} + \underset{56\,g}{KOH} \longrightarrow C_3H_7COOK + H_2O$$

Concentration of the potassium hydroxide solution = 11.2 g/dm^3

In 20.0 cm^3 of the KOH solution there is

$$\frac{20}{1000} \times 11.2\,g = 0.224\,g$$

If 56 g KOH reacts with 88 g of the acid, then 0.224 g reacts with

$$\frac{0.224}{56} \times 88\,g = 0.352\text{ g butanoic acid}$$

This amount of acid was present in 40.0 cm^3 = 0.040 dm^3 of butanoic acid solution. So the concentration of the butanoic acid solution is

$$\frac{0.352\,g}{0.040\,dm^3} = 8.80\,g/dm^3$$

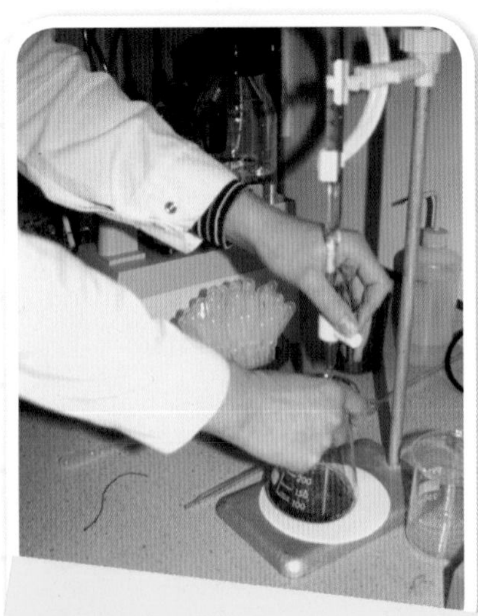

Rachel runs the butanoic acid from the burette into the flask during the titration.

Evaluating results

Scientists need to be able to make sense of analyses and tests. This means that they have to be able to interpret their significance and say what they show. Scientists must also judge how confident they are about the accuracy of results.

Find out about

- systematic and random sources of uncertainty
- accuracy and precision

Measurement uncertainty

All measurements have an uncertainty. This means that scientists usually give results within a range. For example, they may analyse the purity of a drug and give the answer as 99.1 ± 0.2%. This means that the average value obtained from analyses of several samples was 99.1%. The precise value is uncertain. The scientists are confident that the true value lies between 98.9% and 99.3%. To show this, they quote the results as 99.1 (the mean) ± 0.2%.

Errors of measurement are not mistakes. Mistakes are failures by the operator and include such things as forgetting to fill a burette tip with the solution, or taking readings from a sensitive balance in a draught. Mistakes of this kind lead to outliers in results, and should be avoided by people doing practical work.

Types of uncertainty

There are two general sources of **measurement uncertainty**: systematic errors and random errors.

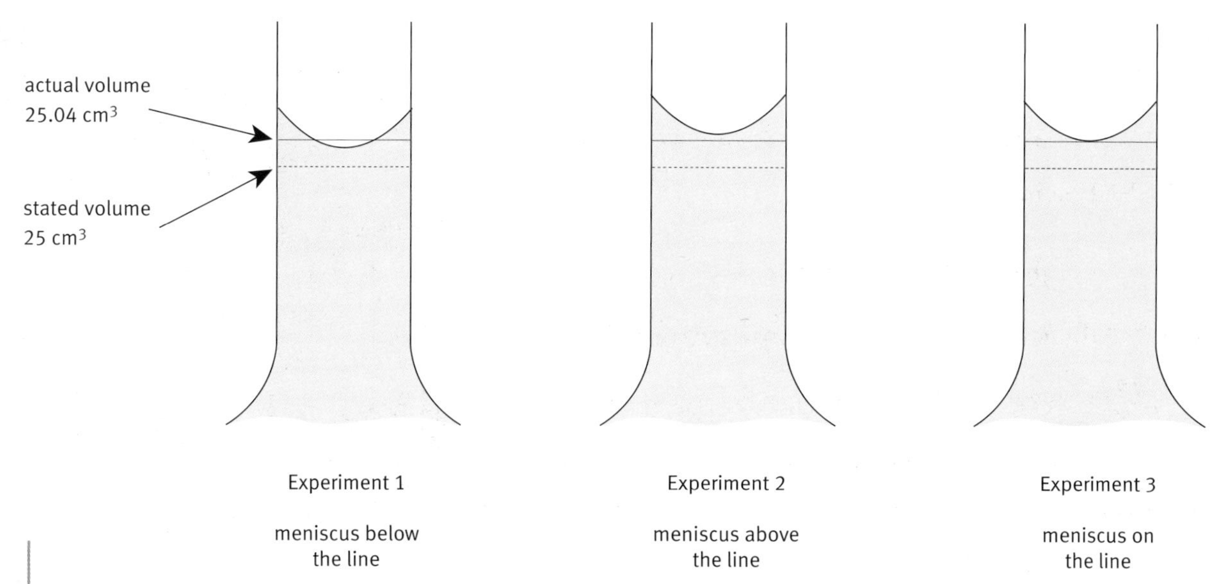

Systematic and random errors in the use of a pipette. The manufacturing tolerance for a 25-cm^3 grade B pipette is ±0.06 cm^3. This can give rise to a systematic error. Every time an analyst uses the pipette, the meniscus is aligned slightly differently with the graduation mark. This gives rise to random error.

Random error means that the same measurement repeated several times gives different values. This can happen, for example, when making judgements about the colour change at an end point or when estimating the reading from a burette scale.

Systematic error means that the same measurement repeated several times gives values that are consistently higher than the true value, or consistently lower. This can result from incorrectly calibrated measuring instruments, or from making measurements, at a consistent, but wrong, temperature.

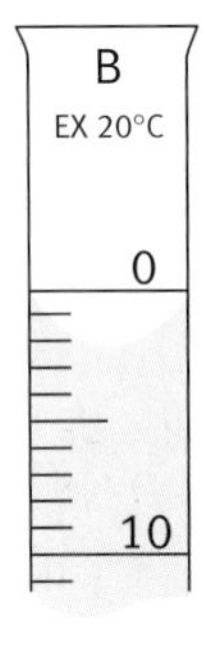

It is difficult to determine accurately the volume of liquid in a burette if the meniscus lies between two graduation marks.

The material used to prepare a standard solution may not be 100% pure.

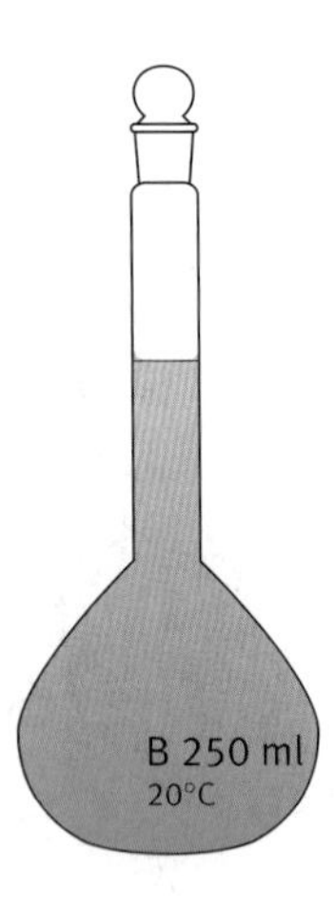

A 250 cm^3 volumetric flask may actually contain 250.3 cm^3 when filled to the calibration mark owing to permitted variation in the manufacture of the flask.

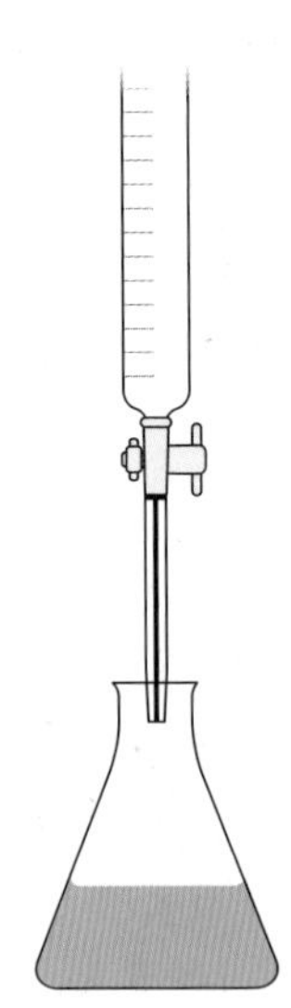

It is difficult to make an exact judgement of the end point of a titration (the exact point at which the colour of the indicator changes).

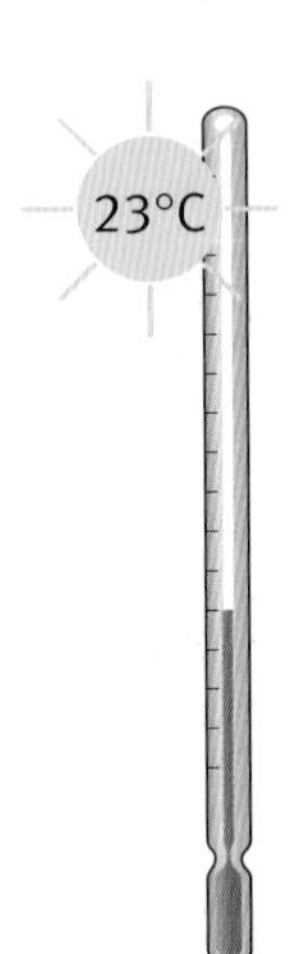

The burette is calibrated by the manufacturer for use at 20 °C. When it is used in the laboratory the temperature is 23 °C. This difference in temperature will cause a small difference in the actual volume of liquid in the burette when it is filled to a calibration mark.

The display on a laboratory balance will only show the mass to a certain number of decimal places.

Sources of uncertainty in analysis by titration.

An analysis or test is often repeated to give a number of measured values, which are then averaged to produce the result.

- **Accuracy** describes how close this result is to the true or 'actual' value.
- **Precision** is a measure of the spread of measured values. A big spread indicates a greater uncertainty than a small spread.

Key words

- ✓ measurement uncertainty
- ✓ accuracy
- ✓ precision

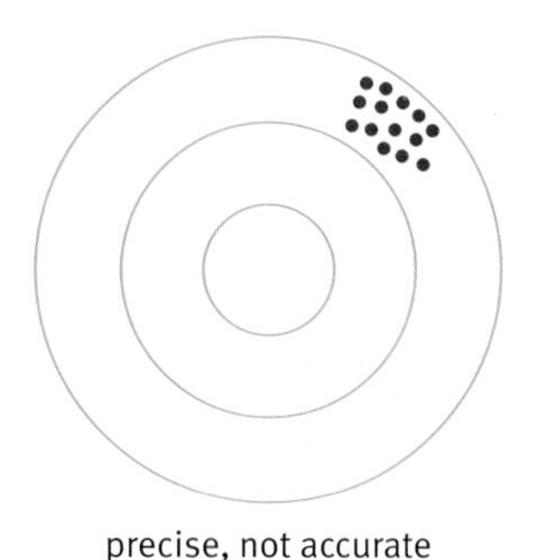
precise, not accurate

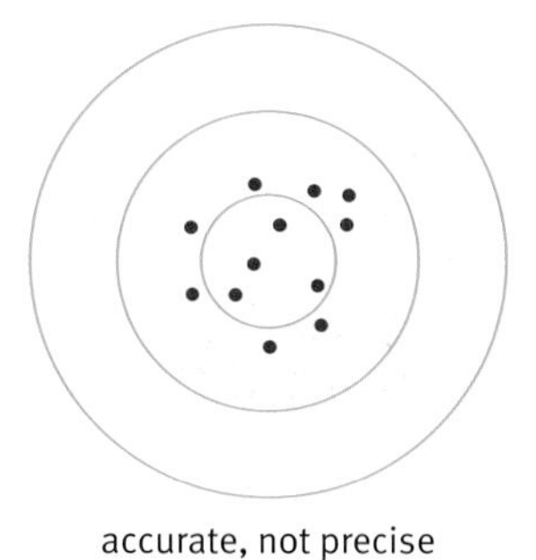
accurate, not precise

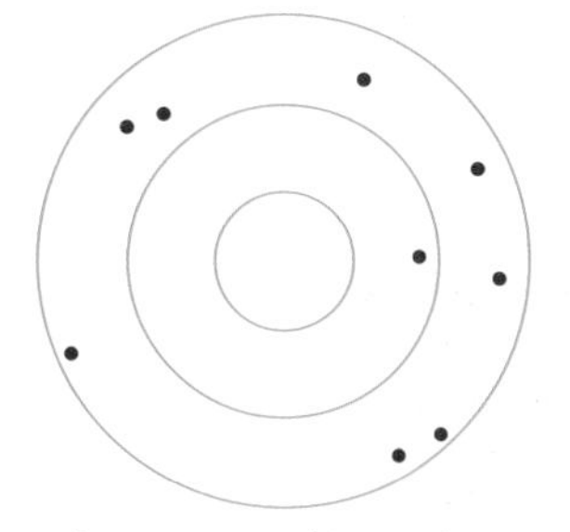
inaccurate and imprecise

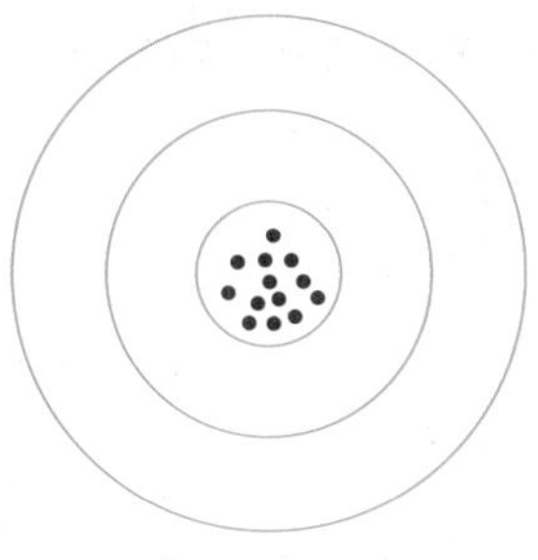
precise and accurate

Accuracy and precision are not the same thing.

Conclusions

The conclusions scientists draw from their work must be valid and justifiable.

- Valid means that the techniques and procedures used were suitable for what was being analysed or tested.
- Justifiable means that conclusions reached are backed by sound, reliable evidence.

Questions

1 An analyst determined the percentage of potassium in three brands of plant fertiliser for house plants by making five measurements for each brand.

These are the results of measuring the percentage by mass of potassium in three brands:

A 4.93, 4.89, 4.71, 4.81, 4.74
B 6.76, 7.91, 6.94, 6.71, 6.86
C 4.72, 4.76, 4.68, 4.70, 4.69

  a Determine the mean and range for each brand.
  b What conclusions can you draw about the three brands?

2 Why would the results be inaccurate if an analyst used hot solutions in graduated glassware?

Module Summary

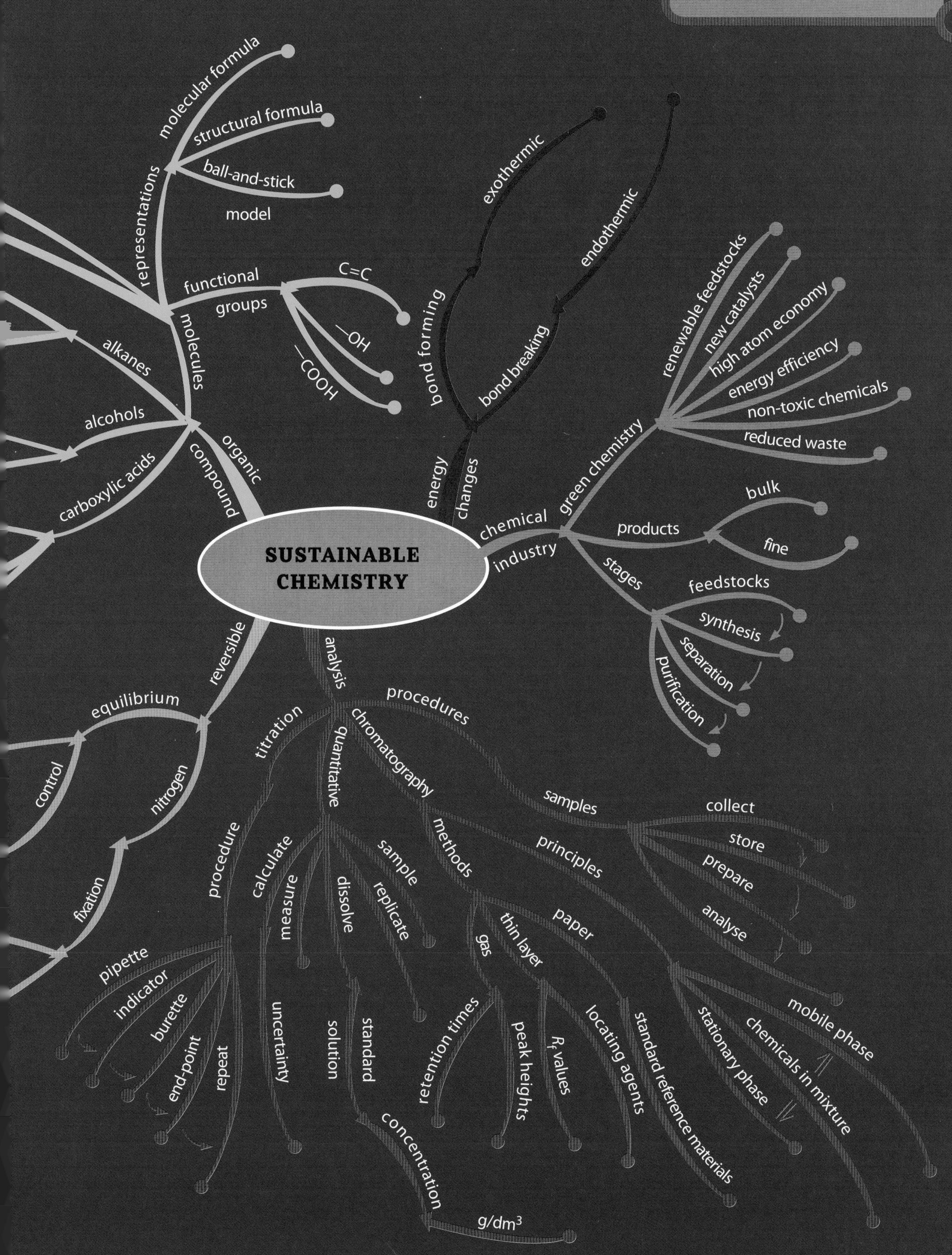

SUSTAINABLE CHEMISTRY
organic compound
molecules
representations
molecular formula
structural formula
ball-and-stick model
functional groups
C=C
—OH
—COOH
alkanes
alcohols
carboxylic acids
energy changes
bond forming
exothermic
bond breaking
endothermic
chemical industry
green chemistry
renewable feedstocks
new catalysts
high atom economy
energy efficiency
non-toxic chemicals
reduced waste
products
bulk
fine
stages
feedstocks
synthesis
separation
purification
reversible
equilibrium
control
nitrogen
fixation
analysis
procedures
samples
collect
store
prepare
analyse
titration
procedure
pipette
indicator
burette
end-point
repeat
quantitative
calculate
measure
uncertainty
dissolve
solution
standard
concentration
g/dm³
sample
replicate
chromatography
methods
gas
retention times
peak heights
thin layer
R_f values
paper
locating agents
standard reference materials
principles
stationary phase
chemicals in mixture
mobile phase

Science Explanations

Green chemistry

The chemical industry is reinventing many of the processes used to convert raw materials into useful products.

You should know:

- how to distinguish between bulk and fine chemicals produced by the chemical industry
- the importance of research to develop new products and processes
- the importance of regulations to control the chemical industry and the uses of its products
- the main stages in the industrial production of useful chemicals
- how the principles of green chemistry help to make the industry more sustainable
- why catalysts are important and how they affect the rates of reactions
- the contribution that enzymes can make to green chemistry
- how to use balanced symbol equations to calculate theoretical yields and atom economies.

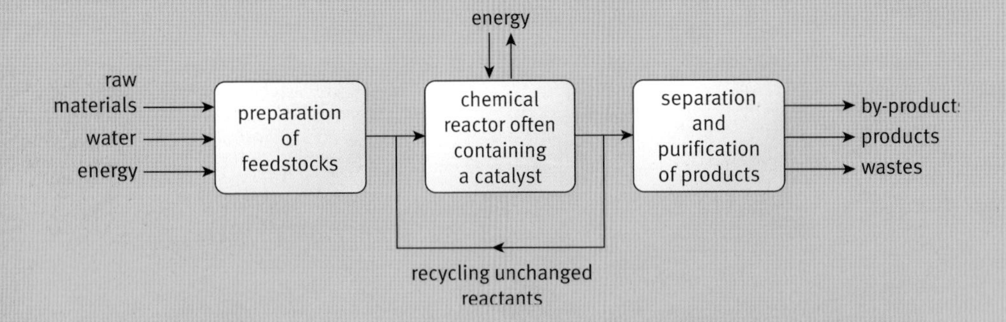

Alcohols, carboxylic acids, and esters

Chemists make sense of this great variety of compounds by classifying them according to their functional groups.

You should know:

- how to translate between molecular, structural, and ball-and-stick representations of simple molecules
- the names, formulae, and structures of the simpler examples of alkanes, alcohols, and carboxylic acids

- how to write and interpret symbol equations for organic reactions
- that the characteristic properties of organic molecules arise from their functional groups
- the functional groups of alcohols and carboxylic acids
- the difference between saturated and unsaturated compounds
- that some organic compounds, such as carboxylic acids and esters, have distinctive odours and tastes
- what happens when alkanes burn in air and why alkanes are unreactive towards aqueous reagents
- how the properties of ethanol compare with the properties of water and with those of alkanes
- the reactions of alcohols with air and with sodium
- that ethanol can be made by fermentation and is then concentrated by distillation
- which factors control the optimum conditions for making ethanol by fermentation
- the conditions that favour the use of genetically modified bacteria to make ethanol from biomass
- that there is a synthetic route from oil or natural gas to ethanol, and be able to evaluate the sustainability of this route compared with processes based on fermentation
- that carboxylic acids react like other acids with metals, alkalis, and carbonates
- why carboxylic acids are described as weak acids
- the conditions under which carboxylic acids and alcohols form esters
- the techniques used, and the reasons for each stage, in the procedure for making a simple ester from a carboxylic acid and an alcohol
- how fats are related to glycerol and fatty acids, and how fats differ from vegetable oils
- why living organisms make fats
- examples of the uses of some organic compounds such as methanol, ethanol, ethanoic acid, and simple esters.

Energy changes in chemistry

Chemists explain exothermic and endothermic reactions in terms of the energy changes when chemical bonds break and form.

You should know:

- how to draw and interpret energy-level diagrams for exothermic and endothermic reactions
- that energy is needed to break chemical bonds, and that energy is given out as bonds form
- how to use data on the energy needed to break bonds to determine the overall energy change for a reaction between simple molecules
- what is meant by the term activation energy.

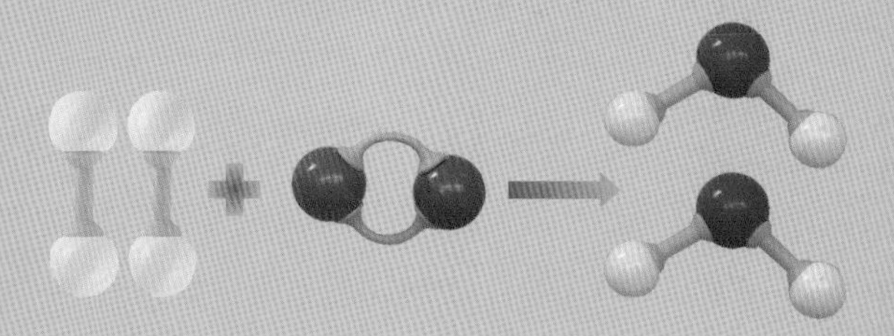

Two H—H bonds and one O=O bond break when hydrogen reacts with oxygen. The atoms recombine to make water as four new O—H bonds form.

Reversible reactions and equilibria

Many important processes are reversible and reach a state of equilibrium, which can be controlled by varying the conditions of concentration, temperature, and pressure.

You should know:

- that reversible reactions can reach a state of dynamic equilibrium
- why it is important that there are natural and artificial ways to fix nitrogen
- how and why the conditions for the Haber process are chosen to give the optimum yield
- why it is desirable to find new ways to manufacture ammonia.

Analysis

Analysis is important in checking the quality of food, in detecting pollution, in diagnosing diseases, and in gathering evidence that can help to convict criminals.

You should know:

- why standard procedures are needed for the collection, storage, and preparation of samples for analysis
- how different methods of chromatography separate mixtures as a mobile phase moves through a stationary phase
- why standard reference materials and locating agents are used in chromatography
- the similarities and differences between paper and thin-layer chromatography
- how to calculate and interpret R_f values
- the procedure for gas chromatography, and how to interpret simple gas chromatograms
- the key stages in a quantitative analysis
- how to make up a standard solution and calculate its concentration in g/dm^3
- how to carry out a titration, interpret the data, and assess the degree of uncertainty in the calculated result.

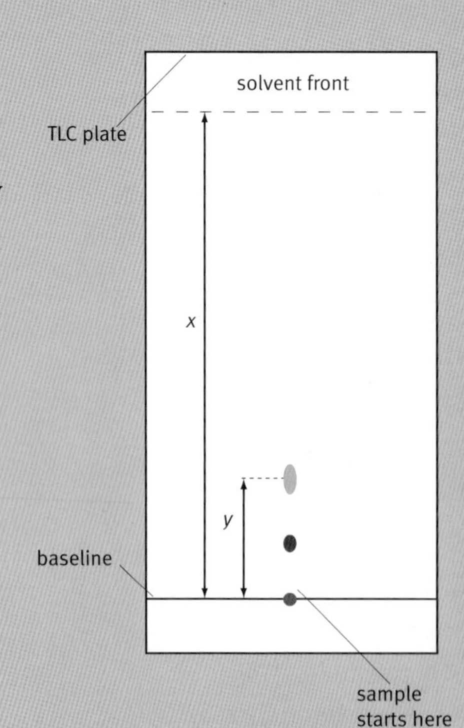

Retardation factors (R_f) can be calculated by measuring the distances travelled by chemicals in the sample and by the solvent.

Ideas about Science

Data: its importance and limitations

Scientists can never be sure that a measurement tells them the true value of the quantity being measured. Data is more reliable if it can be repeated. In the context of chemical analysis by chromatography or titration, you should be able to:

- explain the importance of repeating measurements
- estimate the true value from a set of repeated measurements and suggest the range in which the true value probably lies
- discuss and defend the decision to discard or retain an outlier.

Developing scientific explanations

Scientific explanations are based on data but they go beyond the data and are distinct from it. You should be able to:

- distinguish statements about energy changes, rates, and equilibria, which report data from statements of explanatory ideas
- recognise data or observations that are accounted for by (or conflict with) an explanation
- identify where creative thinking is involved in the development of an explanation illustrated, for example, by explanations of the effect of catalysts on rates or of dynamic equilibrium.

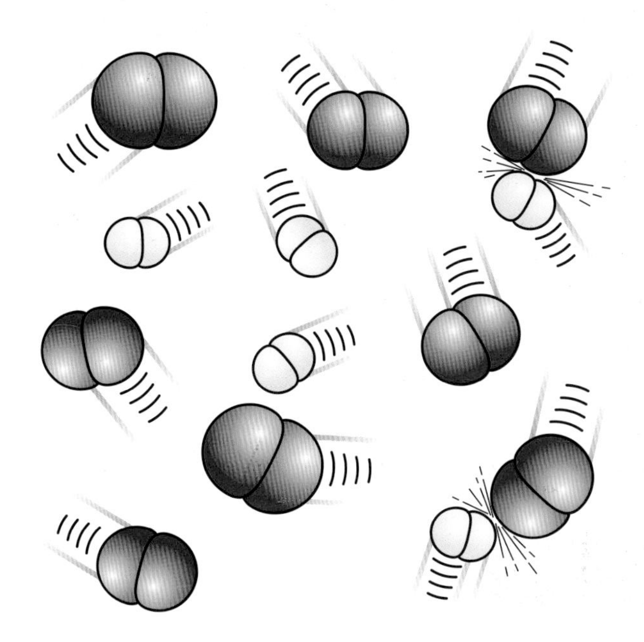

Making decisions about science and technology

Science helps us find ways of using natural resources in a more sustainable way. You should be able to:

- identify benefits and costs of making chemicals such as ammonia and ethanol
- suggest reasons why the choice of method for making a chemical depends on the social or economic context
- explain how the principles and practices of green chemistry contribute to sustainable development
- use data, such as atom economies and yields, to compare the sustainability of alternative processes.

Review Questions

1 Below are the structural formulae of three organic compounds.

A
```
    H   H   H
    |   |   |
H — C — C — C — H
    |   |   |
    H   H   H
```

B
```
    H   H
    |   |
H — C — C — OH
    |   |
    H   H
```

C
```
    H    O
    |   //
H — C — C
    |    \
    H     OH
```

a Write the molecular formula of each of the compounds.

b From the compounds above, identify:

i an alkane

ii an alcohol

iii a compound that reacts with calcium carbonate to produce carbon dioxide gas

iv a compound that is used as a solvent and a fuel

v a compound that is made by fermentation

vi two compounds that combine to make an ester.

2 This question is about two alkanes, butane and propane. Both of these alkanes are useful fuels.

a The molecular formula of butane is C_4H_{10}. Draw its structural formula.

b A chemist bubbles butane gas through acidic and alkaline solutions. Explain why butane does not react with the substances in the solutions.

c Butane burns in air to produce carbon dioxide and water. Write a balanced symbol equation for the burning reaction. Include state symbols.

d The equation for the reaction of propane with oxygen is given below.

$$C_3H_8 + 5O_2 \longrightarrow 3CO_2 + 4H_2O$$

Calculate the mass of carbon dioxide produced when 58 g of butane is burned in a good supply of air.

3 This question is about carboxylic acids.

a Give one use of ethanoic acid.

b Ethanoic acid reacts with magnesium to make magnesium ethanoate and one other product. Give the name of the other product.

c Copy and complete the word equation below for the reaction of butanoic acid with sodium hydroxide.

butanoic acid + sodium hydroxide $\longrightarrow$

d Explain why carboxylic acids are called weak acids.

e The table below gives the pH values of samples of hydrochloric acid and ethanoic acid. The concentration of each acid sample is the same. Identify which of the two acids – X or Y – is ethanoic acid, and give a reason for your choice.

Acid	pH
X	1.0
Y	2.9

4 Every year, UK chemical companies produce more than 1 million tonnes of ammonia by the Haber process.

a **i** Explain why ammonia is manufactured on such a large scale.

ii The feedstocks for the Haber process are hydrogen and nitrogen. State the source of each of these gases.

b The equation for the Haber process reaction is:

$$N_2(g) + 3H_2(g) \rightleftharpoons 2NH_3(g)$$

i Use ideas about molecules and equilibrium to explain why increasing the pressure increases the yield of ammonia.

ii The lower the temperature, the higher the yield of ammonia. Explain why a relatively high temperature of 450 °C is chosen for the Haber process.

5 Gas chromatography (GC) is a useful analytical technique. The diagram below outlines the instrument used.

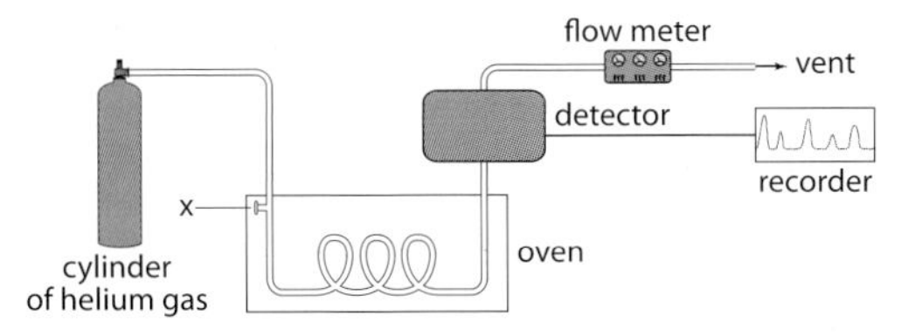

a Name the substance that is the mobile phase in the diagram above.

b A sample of a mixture is injected at X. Describe what happens to the sample as it moves through the instrument to the detector.

c A student injects a sample into a GC instrument. He obtains the chromatogram below.

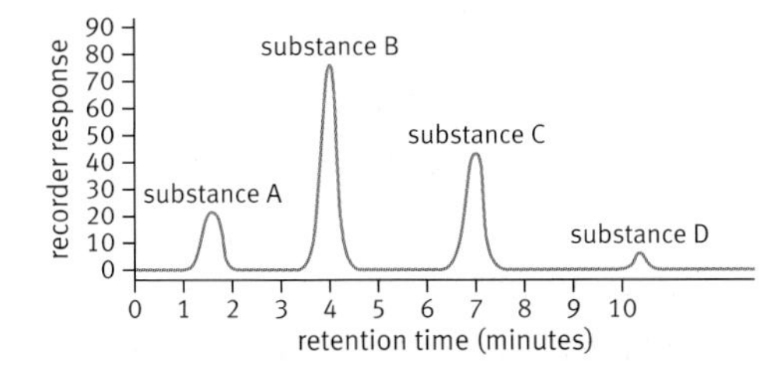

i How many substances were present in the sample?

ii Which substance passes through the column most quickly?

iii Which substance has the shortest retention time?

iv Which substance is present in the mixture in the smallest amount?

6 A technician makes up some solutions.

a First, she dissolves exactly 1.00 g of sodium hydroxide in water and makes up exactly 250 cm^3 of solution. Calculate the concentration of the solution, in g/dm^3.

b Next, the technician needs to make up 500 cm^3 of sodium carbonate solution of concentration 10.6 g/dm^3. What mass of sodium carbonate does she need?

7 Hydrogen reacts with fluorine. The equation for the reaction is:

$$H_2(g) + F_2(g) \longrightarrow 2HF(g)$$

Use the data in the table to calculate the energy change for the reacting masses in the equation above.

Process	Energy change for the formula masses (kJ)
Breaking one H–H bond	434 needed
Breaking one F–F bond	158 needed
Forming one H–F bond	562 given out

8 The sustainability of a chemical process depends on many factors. Describe and explain six of these factors.

Glossary

abundant Abundance measures how common an element is. Silicon is abundant in the lithosphere. Nitrogen is abundant in the atmosphere.

accumulate To collect together and increase in quantity.

accuracy How close a quantitative result is to the true or 'actual' value.

acid A compound that dissolves in water to give a solution with a pH lower than 7. Acid solutions change the colour of indicators, form salts when they neutralise alkalis, react with carbonates to form carbon dioxide, and give off hydrogen when they react with a metal. An acid is a compound that contains hydrogen in its formula and produces hydrogen ions when it dissolves in water.

activation energy The minimum energy needed in a collision between molecules if they are to react. The activation energy is the height of the energy barrier between reactants and products in a chemical change.

actual yield The mass of the required chemical obtained after separating and purifying the product of a chemical reaction.

alcohols Alcohols are organic compounds containing the reactive group —OH. Ethanol is an alcohol. It has the formula C_2H_5OH.

alkali A compound that dissolves in water to give a solution with a pH higher than 7. An alkali can be neutralised by an acid to form a salt. Solutions of alkalis contain hydroxide ions.

Alkali Acts Acts of Parliament passed in the UK in order to control levels of pollution. They led to the formation of an Alkali Inspectorate, which checked that at least 95% of acid fumes were removed from the chimneys of chemical factories.

alkali metal An element in Group 1 of the periodic table. Alkali metals react with water to form alkaline solutions of the metal hydroxide.

alkane Alkanes are hydrocarbons found in crude oil. All the C—C bonds in alkanes are single bonds. Ethane is an alkane. It has the formula C_2H_6.

alkene Alkenes are hydrocarbons that contain a C=C double bond. Ethene is an alkene. It has the formula C_2H_4.

alloy A mixture of metals. Alloys are often more useful than pure metals.

aqueous An aqueous solution is a solution in which water is the solvent.

atmosphere The layer of gases that surrounds the Earth.

atom The smallest particle of an element. The atoms of each element are the same as each other and are different from the atoms of other elements.

atom economy A measure of the efficiency of a chemical process. The atom economy for a process shows the mass of product atoms as a percentage of the mass of reactant atoms.

attractive forces (between molecules) Forces that try to pull molecules together. Attractions between molecules are weak. Molecular chemicals have low melting points and boiling points because the molecules are easy to separate.

balanced equation An equation showing the formulae of the reactants and products. The equation is balanced when there is the same number of each kind of atom on both sides of the equation.

best estimate When measuring a variable, the value in which you have most confidence.

biodegradable Materials that are broken down in the environment by microorganisms. Most synthetic polymers are not biodegradable.

biomass Plant material and animal waste that can be used as a fuel. A renewable energy source.

bleach A chemical that can destroy unwanted colours. Bleaches also kill bacteria. A common bleach is a solution of chlorine in sodium hydroxide.

bond strength A measure of how much energy is needed to break a covalent bond between two atoms. It is measured in joules.

branched chains Chains of carbon atoms with short side branches.

brine A solution of sodium chloride (salt) in water. Brine is produced by solution mining of underground salt deposits.

bulk chemicals Chemicals made by industry on a scale of thousands or millions of tonnes per year. Examples are sulfuric acid, nitric acid, sodium hydroxide, ethanol, and ethanoic acid.

burette A graduated tube with taps or valves used to measure the volume of liquids or solutions during quantitative investigations such as titrations.

by-products Unwanted products of chemical synthesis. By-products are formed by side-reactions that happen at the same time as the main reaction, thus reducing the yield of the product required.

carbonate A compound that contains carbonate ions, CO_3^{2-}. An example is calcium carbonate, $CaCO_3$.

carboxylic acid Carboxylic acids are organic compounds containing the reactive group —COOH. Ethanoic acid (acetic acid) is an example. It has the formula CH_3COOH.

carrier gas The mobile phase in gas chromatography.

catalyst A chemical that speeds up a chemical reaction but is not used up in the process.

catalytic converter A device fitted to a vehicle exhaust that changes the waste gases into less harmful ones.

cause When there is evidence that changes in a factor produce a particular outcome, then the factor is said to cause the outcome. For example, increases in the pollen count cause increases in the incidence of hay fever.

centrifuge A piece of equipment used to separate a mixture of liquids and solids by spinning the mixture very fast.

ceramic Solid materials such as pottery, glass, cement, and brick.

chemical change/reaction A change that forms a new chemical.

chemical equation A summary of a chemical reaction showing the reactants and products with their physical states (see balanced chemical equation).

chemical formula A way of describing a chemical that uses symbols for atoms. It gives information about the number of different types of atom in the chemical.

chemical industry The industry that converts raw materials such as crude oil, natural gas, and minerals into useful products such as pharmaceuticals, fertilisers, paints, and dyes.

chemical properties A chemical property describes how an element or compound interacts with other chemicals, for example, the reactivity of a metal with water.

chemical species The different chemical forms that an element can take. For example, chlorine has three chemical species: atom, molecule, and ion. Each of these forms has distinct properties.

chemical synthesis Making a new chemical by joining together simpler chemicals.

chlorination The process of adding chlorine to water to kill microorganisms, so that it is safe to drink.

chlorine A greenish toxic gas, used to bleach paper and textiles, and to treat water.

chromatogram The resulting record showing the separated chemicals at the end of a chromatography experiment.

chromatography An analytical technique in which the components of a mixture are separated by the movement of a mobile phase through a stationary phase.

collision theory The theory that reactions happen when molecules collide. The theory helps to explain the factors that affect the rates of chemical change. Not all collisions between molecules lead to reaction.

combustion When a chemical reacts rapidly with oxygen, releasing energy.

compression A material is in compression when forces are trying to push it together and make it smaller.

concentration The quantity of a chemical dissolved in a stated volume of solution. Concentrations can be measured in grams per litre.

condensed The change of state from a gas to a liquid, for example, water vapour in the air condenses to form rain.

conservation of atoms All the atoms present at the beginning of a chemical reaction are still there at the end. No new atoms are created and no atoms are destroyed during a chemical reaction.

conservation of mass The total mass of chemicals is the same at the end of a reaction as at the beginning. No atoms are created or destroyed and so no mass is gained or lost.

convection The movement that occurs when hot material rises and cooler material sinks.

correlation When an outcome happens if a specific factor is present, but does not happen when it is absent, or if a measured outcome increases (or decreases steadily) as the value of a factor increases, there is a correlation between the two. For example, a matching pattern in the variation of pollen count and the incidence of hay fever is evidence of a correlation.

corrosive A corrosive chemical may destroy living tissue on contact.

covalent bonding Strong attractive forces that hold atoms together in molecules. Covalent bonds form between atoms of non-metallic elements.

cross-links Links or bonds joining polymer chains together.

crude oil A dark, oily liquid found in the Earth, which is a mixture of hydrocarbons.

crust (of the Earth) The outer layer of the lithosphere.

crystalline A material with molecules, atoms, or ions lined up in a regular way as in a crystal.

crystalline polymer A polymer with molecules lined up in a regular way as in a crystal.

crystallise To form crystals, for example, by evaporating the water from a solution of a salt.

denatured When the shape of an enzyme has been changed, usually as a result of external temperatures being too high or pH changes. The enzyme no longer works.

density A dense material is heavy for its size. Density is mass divided by volume.

diamond A gemstone. A form of carbon. It has a giant covalent structure and is very hard.

diatomic A molecule with two atoms, for example, N_2, O_2, Cl_2.

displacement reaction A more reactive halogen will displace a less reactive halogen, for example, chlorine will displace bromide ions to form bromine and chloride ions:

$$Cl_2(aq) + Br^-(aq) \rightarrow 2Cl^-(aq) + Br_2(aq)$$

dissolve Some chemicals dissolve in liquids (solvents). Salt and sugar, for example, dissolve in water.

distillation A method of separating a mixture of two or more substances with different boiling points.

double bond A covalent bond between two atoms involving the sharing of two pairs of electrons. They are found in alkenes and unsaturated hydrocarbons.

drying agent A chemical used to remove water from moist liquids or gases. Anhydrous calcium chloride and anhydrous sodium sulfate are examples of drying agents.

durable A material is durable if it lasts a long time in use. It does not wear out.

dynamic equilibrium Chemical equilibria are dynamic. At equilibrium the forward and back reactions are still continuing but at equal rates so that there is no overall change.

efficiency The percentage of energy supplied to a machine that is usefully transferred by it.

electrode A conductor made of a metal or graphite through which a current enters or leaves a chemical during electrolysis. Electrons flow into the negative electrode (cathode) and out of the positive electrode (anode).

electrolysis Splitting up a chemical into its elements by passing an electric current through it.

electrolyte A chemical that can be split up by an electric current when molten or in solution is the electrolyte. Ionic compounds are electrolytes.

electron arrangement The number and arrangement of electrons in an atom of an element.

electrons Tiny particles in atoms. Electrons are found outside the nucleus. Electrons have negligible mass and are negatively charged, 1–.

electrostatic attraction The force of attraction between objects with opposite electric charges. A positive ion, for example, attracts a negative ion.

emission Something given out by something else, for example, the emission of carbon dioxide from combustion engines.

end point The point during a titration at which the reaction is just complete. For example, in an acid–alkali titration, the end point is reached when the indicator changes colour. This happens when exactly the right amount of acid has been added to react with all the alkali present at the start.

endothermic An endothermic process takes in energy from its surroundings.

energy level The electrons in an atom have different energies and are arranged at distinct energy levels.

energy-level diagram A diagram to show the difference in energy between the reactants and the products of a reaction.

enzyme A biological catalyst.

equilibrium A state of balance in a reversible reaction when neither the forward nor the backward reaction is complete. The reaction appears to have stopped. At equilibrium reactants and products are present and their concentrations are not changing.

erosion The movement of solids at the Earth's surface (for example, soil, mud, and rock) caused by wind, water, ice, gravity, and living organisms.

esters An organic compound made from a carboxylic acid and an alcohol. Ethyl ethanoate is an ester. It has the formula $CH_3COOC_2H_5$.

evaporate The change of state from a liquid to a gas.

exothermic An exothermic process gives out energy to its surroundings.

extraction (of metals) The process of obtaining a metal from a mineral by chemical reduction or electrolysis. It is often necessary to concentrate the ore before extracting the metal.

extruded A plastic is shaped by being forced through a mould.

factor A variable that changes and may affect something else.

fat Fats are esters of glycerol with long-chain carboxylic acids (fatty acids). The fatty acids in animal fats are mainly saturated compounds.

fatty acids Another name for carboxylic acids.

feedstocks A chemical, or mixture of chemicals, fed into a process in the chemical industry.

fermentation The conversion of carbohydrates to alcohols and carbon dioxide using yeast.

fibres Long thin threads that make up materials such as wool and polyester. Most fibres used for textiles consist of natural or synthetic polymers.

filter To separate a solid from a liquid by passing it through a filter paper.

fine chemicals Chemicals made by industry in smaller quantities than bulk chemicals. Fine chemicals are used in products such as food additives, medicines, and pesticides.

flame colour A colour produced when a chemical is held in a flame. Some elements and their compounds give characteristic colours. Sodium and sodium compounds, for example, give bright yellow flames.

flavouring Mixtures of chemicals that give food, sweets, toothpaste, and other products their flavours.

flexible A flexible material bends easily without breaking.

formula (chemical) A way of describing a chemical that uses symbols for atoms. A formula gives information about the numbers of different types of atom in the chemical. The formula of sulfuric acid, for example, is H_2SO_4.

fossils The stony remains of an animal or plant that lived millions of years ago, or an imprint it has made (for example, a footprint) in a surface.

fractional distillation The process of separating crude oil into groups of molecules with similar boiling points called fractions.

fractions A mixture of hydrocarbons with similar boiling points that have been separated from crude oil by fractional distillation.

functional group A reactive group of atoms in an organic molecule. The hydrocarbon chain making up the rest of the molecule is generally unreactive with common reagents such as acids and alkalis. Examples of functional groups are —OH in alcohols and —COOH in carboxylic acids.

giant covalent structure A giant, three-dimensional arrangement of atoms that are held together by covalent bonds. Silicon dioxide and diamond have giant covalent structures.

giant ionic structure The structure of solid ionic compounds. There are no individual molecules, but millions of oppositely charged ions packed closely together in a regular, three-dimensional arrangement.

glycerol Glycerol is an alcohol with three —OH groups. Its chemical name is propan-1,2,3-triol. Its formula is $CH_2OH—CHOH—CH_2OH$.

grain Relatively small particle of a substance, for example, grains of sand.

graphite A form of carbon. It has a giant covalent structure. It is unusual for a non-metal in that it conducts electricity.

group Each column in the periodic table is a group of similar elements.

Haber Process The reaction between nitrogen and hydrogen gas used to make ammonia on an industrial scale.

halogens The family name of the Group 7 elements.

hard A material that is difficult to dent or scratch.

harmful A harmful chemical is one that may cause damage to health if swallowed, breathed in, or absorbed through the skin.

heat under reflux Heating a reaction mixture in a flask fitted with a vertical condenser. Vapours escaping from the flask condense and flow back into the reaction mixture.

hydrocarbon A compound of hydrogen and carbon only. Ethane, C_2H_6, is a hydrocarbon.

hydrogen chloride gas An acid gas that is toxic and corrosive, and is produced by the Leblanc process.

hydrogen ion A hydrogen atom that has lost one electron. The symbol for a hydrogen ion is H^+. Acids produce aqueous hydrogen ions, $H^+(aq)$, when dissolved in water.

hydrogen sulfide gas A poisonous gas that smells of rotten eggs.

hydrosphere All the water on Earth. This includes oceans, lakes, rivers, underground reservoirs, and rainwater.

hydroxide ion A negative ion, OH^-. Alkalis give aqueous hydroxide ions when they dissolve in water.

incinerator A factory for burning rubbish and generating electricity.

indicator A chemical that shows whether a solution is acidic or alkaline. For example, litmus turns blue in alkalis and red in acids. Universal indicator has a range of colours that show the pH of a solution.

ionic bonding Very strong attractive forces that hold the ions together in an ionic compound. The forces come from the attraction between positively and negatively charged ions.

ionic compounds Compounds formed by the combination of a metal and a non-metal. They contain positively charged metal ions and negatively charged non-metal ions.

ionic equation An ionic equation describes a chemical change by showing only the reacting ions in solution, for example,

$$Ba^{2+}(aq) + SO_4^{2-}(aq) \rightarrow BaSO_4(s)$$

ions An electrically charged atom or group of atoms.

irreversible change A chemical change that can only go in one direction, for example, changes involving combustion.

landfill Disposing of rubbish in holes in the ground.

latitude The location of a place on Earth, north or south of the equator.

leach The movement of the plasticisers in a polymer into water, or another liquid, that is flowing past the polymer or is contained by it.

Leblanc process A process that used chalk (calcium carbonate), salt (sodium chloride) and coal to make the alkali, sodium carbonate. The Leblanc process was highly polluting.

Le Chatelier's principle The principle that the position of an equilibrium will respond to oppose a change in the reaction conditions.

life cycle assessment A way of analysing the production, use, and disposal of a material or product to add up the total energy and water used and the effects on the environment.

limiting factor The factor that prevents the rate of growth of living things.

line spectrum A spectrum made up of a series of lines. Each element has its own characteristic line spectrum.

lithosphere The rigid outer layer of the Earth, made up of the crust and the part of mantle just below it.

locating agent A chemical used to show up colourless spots on a chromatogram.

long-chain molecule Polymers are long-chain molecules. They consist of long chains of atoms.

macroscopic Large enough to be seen without the help of a microscope.

magnetic A material that is attracted to a magnet. For example, iron is magnetic.

mantle The layer of rock between the crust and the outer core of the Earth. It is approximately 2900 km thick.

material The polymers, metals, glasses, and ceramics that we use to make all sorts of objects and structures.

mean value A type of average, found by adding up a set of measurements and then dividing by the number of measurements. You can have more confidence in the mean of a set of measurements than in a single measurement.

measurement uncertainty Variations in analytical results owing to factors that the analyst cannot control. Measurement uncertainty arises from both systematic and random errors.

melting point The temperature at which something melts.

metal Elements on the left side of the periodic table. Metals have characteristic properties: they are shiny when polished and they conduct electricity. Some metals react with acids to give salts and hydrogen. Metals are present as positive ions in salts.

metal hydroxide A compound consisting of metal positive ions and hydroxide ions. Examples are sodium hydroxide, NaOH, and magnesium hydroxide, $Mg(OH)_2$.

metal oxide A compound of a metal with oxygen.

metallic bonding Very strong attractive forces that hold metal atoms together in a solid metal. The metal atoms lose their outer electrons and form positive ions. The electrons drift freely around the lattice of positive metal ions and hold the ions together.

microorganisms Living organisms that can only be seen by looking at them through a microscope. They include bacteria, viruses, and fungi.

mineral A naturally occurring element or compound in the Earth's lithosphere.

mixture Two or more different chemicals, mixed but not chemically joined together.

mobile phase The solvent that carries chemicals from a sample through a chromatographic column or sheet.

molecular model Models to show the arrangement of atoms in molecules, and the bonds between the atoms.

molecule A group of atoms joined together. Most non-metals consist of molecules. Most compounds of non-metals with other non-metals are also molecular.

molten A chemical in the liquid state. A chemical is molten when the temperature is above its melting point but below its boiling point.

monomer A small molecule that can be joined to others like it in long chains to make a polymer.

nanometre A unit of length 1000000000 times smaller than a metre.

nanoparticle A very tiny particle, whose size can be measured in nanometres.

nanotechnology The use and control of matter on a tiny (nanometre) scale.

natural A material that occurs naturally but may need processing to make it useful, such as silk, cotton, leather, and asbestos.

negative ion An ion that has a negative charge (an anion).

neutralisation A reaction in which an acid reacts with an alkali to form a salt. During neutralisation reactions, the hydrogen ions in the acid solution react with hydroxide ions in the alkaline solution to make water molecules.

neutrons An uncharged particle found in the nucleus of atoms. The relative mass of a neutron is 1.

nitrogen cycle The continual cycling of nitrogen, which is one of the elements that is essential for life. By being converted to different chemical forms, nitrogen is able to cycle between the atmosphere, lithosphere, hydrosphere, and biosphere.

nitrogen fixation The conversion of nitrogen gas into compounds either industrially or by natural means.

nitrogenase The enzyme system that catalyses the reduction of nitrogen gas to ammonia.

non-aqueous A solution in which a liquid other than water is the solvent.

nucleus The tiny central part of an atom (made up of protons and neutrons). Most of the mass of an atom is concentrated in its nucleus.

ore A natural mineral that contains enough valuable minerals to make it profitable to mine.

organic chemistry The study of carbon compounds. This includes all of the natural carbon compounds from living things and synthetic carbon compounds.

organic matter Material that has come from dead plants and animals.

outcome A variable that changes as a result of something else changing.

outlier A measured result that seems very different from other repeat measurements, or from the value you would expect, which you therefore strongly suspect is wrong.

oxidation A reaction that adds oxygen to a chemical.

oxide A compound of an element with oxygen.

particulate Tiny bit of a solid.

percentage yield A measure of the efficiency of a chemical synthesis.

period In the context of chemistry, a row in the periodic table.

periodic In chemistry, a repeating pattern in the properties of elements. In the periodic table one pattern is that each period starts with metals on the left and ends with non-metals on the right.

persistent organic pollutants (POPs) POPs are organic compounds that do not break down in the environment for a very long time. They can spread widely around the world and build up in the fatty tissue of humans and animals. They can be harmful to people and the environment.

petrochemical Chemicals made from crude oil (petroleum) or natural gas.

photosynthesis A chemical reaction that happens in green plants using the energy in sunlight. The plant takes in water and carbon dioxide, and uses sunlight to convert them to glucose (a nutrient) and oxygen.

pH scale A number scale that shows the acidity or alkalinity of a solution in water.

phthalate A chemical that is used as a plasticiser, added to polymers to make them more flexible.

physical properties Properties of elements and compounds such as melting point, density, and electrical conductivity. These are properties that do not involve one chemical turning into another.

pipette A pipette is used to measure small volumes of liquids or solutions accurately. A pipette can be used to deliver the same fixed volume of solution again and again during a series of titrations.

pilot plant A small-scale chemical processing facility. A pilot plant is used to test processes before scaling up to full-scale production.

plant A chemical plant is an industrial facility used to manufacture chemicals.

plasticiser A chemical (usually a small molecule) added to a polymer to make it more flexible.

pollutant Waste matter that contaminates the water, air, or soil.

polymer A material made of very long molecules formed by joining lots of small molecules, called monomers, together.

polymerise The joining together of lots of small molecules called monomers to form a long-chain molecule called a polymer.

positive ions Ions that have a positive charge (cations).

precipitate An insoluble solid formed on mixing two solutions. Silver bromide forms as a precipitate on mixing solutions of silver nitrate and potassium bromide.

precision A measure of the spread of quantitative results. If the measurements are precise all the results are very close in value.

preservative A chemical added to food to stop it going bad.

product The new chemicals formed during a chemical reaction.

properties Physical or chemical characteristics of a chemical. The properties of a chemical are what make it different from other chemicals.

proportional Two variables are proportional if there is a constant ratio between them.

proton number The number of protons in the nucleus of an atom (also called the atomic number). In an uncharged atom this also gives the number of electrons.

protons Tiny particles that are present in the nuclei of atoms. Protons are positively charged, 1+.

qualitative Qualitative analysis is any method for identifying the chemicals in a sample. Thin-layer chromatography is an example of a qualitative method of analysis.

quantitative Quantitative analysis is any method for determining the amount of a chemical in a sample. An acid–base titration is an example of quantitative analysis.

range The difference between the highest and the lowest of a set of measurements.

rate of reaction A measure of how quickly a reaction happens. Rates can be measured by following the disappearance of a reactant or the formation of a product.

reactant The chemicals on the left-hand side of an equation. These chemicals react to form the products.

reacting mass The masses of chemicals that react together, and the masses of products that are formed. Reacting masses are calculated from the balanced symbol equation using relative atomic masses and relative formula masses.

reactive metal A metal with a strong tendency to react with chemicals such as oxygen, water, and acids. The more reactive a metal, the more strongly it joins with other elements such as oxygen. So reactive metals are hard to extract from their ores.

real difference The difference between two mean values is real if their ranges do not overlap.

recycling A range of methods for making new materials from materials that have already been used.

reducing agent A chemical that removes oxygen from another chemical. For example, carbon acts as a reducing agent when it removes oxygen from a metal oxide. The carbon is oxidised to carbon monoxide during this process.

reduction A reaction that removes oxygen from a chemical.

reference materials Known chemicals used in analysis for comparison with unknown chemicals.

regulations Rules that can be enforced by an authority, for example, the government. The law that says that all vehicles that are three years old and older must have an annual exhaust emission test is a regulation that helps to reduce atmospheric pollution.

relative atomic mass The mass of an atom of an element compared to the mass of an atom of carbon. The relative atomic mass of carbon is defined as 12.

relative formula mass The combined relative atomic masses of all the atoms in a formula. To find the relative formula mass of a chemical, you just add up the relative atomic masses of the atoms in the formula.

renewable resource Resources that can be replaced as quickly as they are used. An example is wood from the growth of trees.

repeatable A quality of a measurement that gives the same result when repeated under the same conditions.

replicate sample Two or more samples taken from the same material. Replicate samples should be as similar as possible and analysed by the same procedure to help judge the precision of the analysis.

representative sample A sample of a material that is as nearly identical as possible in its chemical composition to that of the larger bulk of material sampled.

reproducible A quality of a measurement that gives the same result when carried out under different conditions, for example, by different people or using different equipment or methods.

retardation factor A retardation factor, R_f, is a ratio used in paper or thin-layer chromatography. If the conditions are kept the same, each chemical in a mixture will move a fixed fraction of the distance moved by the solvent front. The R_f value is a measure of this fraction.

retention time In chromatography, the time it takes for a component in a mixture to pass through the stationary phase.

risk The chance that a hazardous substance or process will harm someone.

risk assessment A check on the hazards involved in a scientific procedure. A full assessment includes the steps to be taken to avoid or reduce the risks from the hazards identified.

rock A naturally occurring solid, made up of one or more minerals.

rubber A material that is easily stretched or bent. Natural rubber is a natural polymer obtained from latex, the sap of a rubber tree.

salt An ionic compound formed when an acid neutralises an alkali or when a metal reacts with a non-metal.

sample A small portion collected from a larger bulk of material for laboratory analysis (such as a water sample or a soil sample).

saturated In the molecules of a saturated compound, all of the bonds are single bonds. The fatty acids in animal fats are all saturated compounds.

scale up To redesign a synthesis to produce a chemical in larger amounts. A process might be scaled up first from a laboratory method to a pilot plant, then from a pilot plant to a full-scale industrial process.

sedimentary rock Rock formed from layers of sediment.

shell A region in space (around the nucleus of an atom) where there can be electrons.

small molecules Particles of chemicals that consist of small numbers of atoms bonded together. Chemicals made up of one or more non-metallic elements and that have low boiling and melting points consist of small molecules.

soft A material that is easy to dent or scratch.

solution Formed when a solid, liquid, or gas dissolves in a solvent.

solvent front The furthest position reached by the solvent during paper or thin-layer chromatography.

spectroscopy The use of instruments to produce and analyse spectra. Chemists use spectroscopy to study the composition, structure, and bonding of elements and compounds.

standard solution A solution whose concentration is accurately known. They are used in titrations.

stationary phase The medium through which the mobile phase passes in chromatography.

stiff A material that is difficult to bend or stretch.

strong A material that is hard to pull apart or crush.

strong acid A strong acid is fully ionised to produce hydrogen ions when it dissolves in water.

subatomic particle The particles that make up atoms. Protons, neutrons, and electrons are subatomic particles.

subsidence The sinking of the ground's surface when it collapses into a hole beneath it.

surface area How much exposed surface a solid object has.

surface area (of a solid chemical) The area of a solid in contact with other reactants that are liquids or gases.

sustainable Using the Earth's resources in a way that can continue in future, rather than destroying them.

sustainable development A plan for meeting people's present needs without spoiling the environment for the future.

synthetic A material made by a chemical process, not naturally occurring.

tap funnel A funnel with a tap to allow the controlled release of a liquid.

tarnish When the surface of a metal becomes dull or discoloured because it has reacted with the oxygen in the air.

tectonic plates Giant slabs of rock (about 12, comprising crust and upper mantle) that make up the Earth's outer layer.

tension A material is in tension when forces are trying to stretch it or pull it apart.

theoretical yield The amount of product that would be obtained in a reaction if all the reactants were converted to products exactly as described by the balanced chemical equation.

theory A scientific explanation that is generally accepted by the scientific community.

titration An analytical technique used to find the exact volumes of solutions that react with each other.

toxic A chemical that may lead to serious health risks, or even death, if breathed in, swallowed, or taken in through the skin.

toxin A poisonous chemical produced by a microorganism, plant, or animal.

trend A description of the way a property increases or decreases along a series of elements or compounds, which is often applied to the elements (or their compounds) in a group or period.

triple bond A covalent bond between the two atoms involving the sharing of three pairs of electrons, for example, nitrogen gas. It makes the molecule very stable and unreactive.

unsaturated There are double bonds in the molecules of unsaturated compounds. There is no spare bonding. The fatty acids in vegetable oils include a high proportion of unsaturated compounds.

vegetable oil Vegetable oils are esters of glycerol with fatty acids (long-chain carboxylic acids). More of the fatty acids in vegetable oils are unsaturated when compared with the fatty acids in animal fats.

vinegar A sour-tasting liquid used as a flavouring and to preserve foods. It is a dilute acetic (ethanoic) acid made by fermenting beer, wine, or cider.

vulcanisation A process for hardening natural rubber by making cross-links between the polymer molecules.

weak acids Weak acids are only slightly ionised to produce hydrogen ions when they dissolve in water.

wet scrubbing A process used to remove pollutants from flue gases.

word equation A summary in words of a chemical reaction.

Index

Appendices

Useful relationships

You will need to be able to carry out calculations using these mathematical relationships.

C6 Chemical synthesis and C7 Further chemistry

percentage yield = (actual yield / theoretical yield) × 100%

C7 Further chemistry

concentration of a solution = mass of solute / volume of solution

chromatography:

retardation factor (R_f) = distance travelled by solute / distance travelled by solvent

Units that might be used in the Chemistry course

length: kilometres (km), metres (m), centimetres (cm), millimetres (mm), micrometres (μm), nanometres (nm)

mass: kilograms (kg), grams (g), milligrams (mg)

time: seconds (s), milliseconds (ms)

temperature: degrees Celsius (°C)

area: cm^2, m^2

volume: cm^3, dm^3, m^3, litres (l), millilitres (ml)

Prefixes for units

nano	micro	milli	kilo	mega	giga	tera
one thousand millionth	one millionth	one thousandth	× thousand	× million	× thousand million	× million million
0.000000001	0.000001	0.001	1000	1000 000	1000 000 000	1000 000 000 000
10^{-9}	10^{-6}	10^{-3}	$\times 10^3$	$\times 10^6$	$\times 10^9$	$\times 10^{12}$

Chemical Formulae

C1

carbon dioxide	CO_2
carbon monoxide	CO
sulfur dioxide	SO_2
nitrogen monoxide	NO
nitrogen dioxide	NO_2
water	H_2O

C4

water	H_2O
hydrogen	H_2
chlorine	Cl_2
bromine	Br_2
iodine	I_2
lithium chloride	$LiCl$
lithium bromide	$LiBr$
lithium iodide	LiI
sodium chloride	$NaCl$
sodium bromide	$NaBr$
sodium iodide	NaI
potassium chloride	KCl
potassium bromide	KBr

potassium iodide	KI
lithium hydroxide	$LiOH$
sodium hydroxide	$NaOH$
potassium hydroxide	KOH

C5

nitrogen	N_2
oxygen	O_2
argon	Ar
carbon dioxide	CO_2
sodium chloride	$NaCl$
magnesium chloride	$MgCl_2$
sodium sulfate	Na_2SO_4
magnesium sulfate	$MgSO_4$
potassium chloride	KCl
potassium bromide	KBr

C6

chlorine	Cl_2
hydrogen	H_2
nitrogen	N_2
oxygen	O_2
hydrochloric acid	HCl
nitric acid	HNO_3
sulfuric acid	H_2SO_4
sodium hydroxide	$NaOH$
sodium chloride	$NaCl$
sodium carbonate	Na_2CO_3
sodium nitrate	$NaNO_3$
sodium sulfate	Na_2SO_4
potassium chloride	KCl
magnesium oxide	MgO
magnesium hydroxide	$Mg(OH)_2$
magnesium carbonate	$MgCO_3$
magnesium chloride	$MgCl_2$
magnesium sulfate	$MgSO_4$
calcium carbonate	$CaCO_3$
calcium chloride	$CaCl_2$
calcium sulfate	$CaSO_4$

C7

methanol	CH_3OH
ethanol	C_2H_5OH
methanoic acid	$HCOOH$
ethanoic acid	CH_3COOH

Qualitative analysis data

Tests for negatively charged ions

ion	test	observation
carbonate CO_3^{2-}	add dilute acid	effervesces, and carbon dioxide gas produced (the gas turns lime water milky)
chloride (in solution) Cl^-	acidify with dilute nitric acid, then add silver nitrate solution	white precipitate
bromide (in solution) Br^-	acidify with dilute nitric acid, then add silver nitrate solution	cream precipitate
iodide (in solution) I^-	acidify with dilute nitric acid, then add silver nitrate solution	yellow precipitate
sulfate (in solution) SO_4^{2-}	acidify, then add barium chloride solution or barium nitrate solution	white precipitate

Tests for positively charged ions

ion	test	observation
calcium Ca^{2+}	add sodium hydroxide solution	white precipitate (insoluble in excess)
copper Cu^{2+}	add sodium hydroxide solution	light-blue precipitate (insoluble in excess)
iron(II) Fe^{2+}	add sodium hydroxide solution	green precipitate (insoluble in excess)
iron(III) Fe^{3+}	add sodium hydroxide solution	red–brown precipitate (insoluble in excess)
zinc Zn^{2+}	add sodium hydroxide solution	white precipitate (soluble in excess, giving a colourless solution)

OXFORD
UNIVERSITY PRESS

Great Clarendon Street, Oxford OX2 6DP

Oxford University Press is a department of the University of Oxford.
It furthers the Universityís objective of excellence in research,
scholarship, and education by publishing worldwide in

Oxford New York

Auckland Cape Town Dar es Salaam Hong Kong Karachi
Kuala Lumpur Madrid Melbourne Mexico City Nairobi
New Delhi Shanghai Taipei Toronto

With offices in
Argentina Austria Brazil Chile Czech Republic France Greece
Guatemala Hungary Italy Japan Poland Portugal Singapore
South Korea Switzerland Thailand Turkey Ukraine Vietnam

First published 2011.

British Library Cataloguing in Publication Data.

Data available.

ISBN 978-0-19-913837-1

10 9 8 7 6 5 4 3

Printed in China by Printplus

Paper used in the production of this book is a natural, recyclable product made
from wood grown in sustainable forests. The manufacturing process conforms to
the environmental regulations of the country of origin.

Acknowledgements
The publisher and authors would like to thank the following for their permission to reproduce photographs and other copyright material:
P13l: Sam Ogden/Science Photo Library; **P13r**: Martyn F. Chillmaid/Science Photo Library; **P14**: Kelly Redinger/Design Pics/Corbis; **P16t**: Charles D. Winters/Science Photo Library; **P16b**: NASA/Zooid Pictures; **P17**: M.T. Mangan/USGS; **P18l**: David Hardy/Science Photo Library; **P18r**: Christian Darkin/Science Photo Library; **P19t**: Russell Shively/Shutterstock; **P19m**: George Steinmetz/Science Photo Library; **P19b**: Dirk Wiersma/Science Photo Library; **P20t**: John Wilkinson/Ecoscene/Corbis; **P20b**: Harvey Pincis/Science Photo Library; **P21**: Victor De Schwanberg/Science Photo Library; **P28t**: Raoux John/Orlando Sentinel/Sygma/Corbis; **P28m**: Mate 3rd Class Daniel Scott/U.S. Navy photo; **P28b**: Cordelia Molloy/Science Photo Library; **P30**: Nick Hawkes/Ecoscene/Corbis; **P31**: Andrew Lambert Photography/Science Photo Library; **P32t**: Burkard Manufacturing Co. Limited; **P32m**: Andy Harmer /Science Photo Library; **P32b**: Philippe Plailly/Eurelios/Science Photo Library; **P33**: Garo/Phanie/Rex Features; **P34**: Ian Hooton/Science Photo Library; **P35**: Action Press/Rex Features; **P36**: RPL Carburettor and Injection Centre; **P37**: Spencer Grant/Science Photo Library; **P38**: Simon Fraser/Science Photo Library; **P42**: George Steinmetz/Science Photo Library; **P44**:FotografiaBasica/Istockphoto; **P46**: Chris Hellier/Science Photo Library; **P47**: Catherine Yeulet/Istockphoto; **P48tl**: Danish Khan/Istockphoto; **P48tm**: PeskyMonkey/Istockphoto; **P48tr**: Lee Torrens/Istockphoto; **P48bl**: Yuri Arcurs/Shutterstock; **P48bm**: Igor Terekhov/Bigstock; **P48br**: Kenneth Sponsler/Shutterstock; **P49tl**: PhotoCuisine/Corbis; **P49tm**: David Constantine/Science Photo Library; **P49tr**: Empics; **P49bl**: Taryn Cass/Zooid Pictures; **P49br**: Yves Forestier/Sygma/Corbis; **P50tl**: David Keith Jones/Images of Africa Photobank/Alamy; **P50tr**: Dennis Gilbert/VIEW Pictures Ltd/Alamy; **P50bl**: Tom Tracy Photography/Alamy; **P50br**: Rich Carey/Shutterstock; **P51t**: Masterfile; **P51b**: Duncan Moody/Istockphoto; **P52t**: Instron® Corporation; **P52b**: J & P Coats Ltd; **P53**: Studio 1One/Shutterstock; **P54t**: Tina Chang/Photolibrary; **P54ml-r**: Andrew Syred/Science Photo Library, Eye Of Science/Science Photo Library; **P54b**: Eye Of Science/Science Photo Library; **P57**: Science & Society Picture Library; **P58t**: Dan Sinclair/Zooid Pictures; **P58b**: Taryn Cass/Zooid Pictures; **P59**: Tim Pannell/Corbis; **P60l**: Zooid Pictures; **P60r**: Abaca/Empics: **P62**: W. L. Gore & Associates, Ltd.; **P63t**: Du Pont (UK) Ltd; **P63b**: Eye Of Science/Science Photo Library; **P64l**: Paul Rapson/Science Photo Library; **P64r**: Paul Rapson/Science Photo Library; **P66t**: Robert Wisdom/Dreamstime; **P66b**: Eric Isselée/Fotolia; **P67t**: David Buffington/Photolibrary; **P67b**: Dr P. Marazzi/Science Photo Library; **P68t**: Irabel8/Shutterstock; **P68bl**: Charles M. Ommanney/Rex Features; **P68br**: Back Page Images/Rex Features; **P69**: Fact Fact/Photolibrary; **P72**: PeskyMonkey/Istockphoto; **P74**: FotografiaBasica/Istockphoto; **P78t**: Kaido Kärner/Shutterstock; **P78b**: Vincent Lowe/Alamy; **P79tl**: Andrew J. Martinez/Science Photo Library; **P79tr**: Martin Bond/Photolibrary; **P79bl**: Dirk Wiersma/Science Photo Library; **P79br**: James King-Holmes/Science Photo Library; **P80t**: Unclesam/Fotolia; **P80bl**: Pascal Goetgheluck/Science Photo Library; **P80br**: Winsford Rock Salt Mine; **P81**: Geographical; **P82**: benicce/Shutterstock; **P84t**: Patrick Frilet/Rex Features; **P84b**: Timur Kulgarin/Shutterstock; **P85t**: Richard Watson/Photolibrary; **P85b**: Martyn F. Chillmaid/Science Photo Library; **P86**: © Catalyst; **P88t**: Samrat35/Dreamstime; **P88b**: Sean Sprague/Photolibrary; **P89**: American Chemistry Council, Inc.; **P90**: Robert Brook/Science Photo Library: **P92**: Tobias Schwarz/Reuters; **P94**: Kodda/Shutterstock; **P95**: Bob Edwards/Science Photo Library; **P96l**: Craig Holmes Premium/Alamy; **P96m**: Aj Photo/Science Photo Library; **P96r**: aikotel/Shutterstock; **P97**: Peter Ryan/Science Photo Library; **P98t**: Erika Craddock/Science Photo Library; **P98b**: ginosphotos/Istockphoto; **P99t**: Gordon Ball LRPS/Shutterstock; **P99b**: Ton Kinsbergen/Science Photo Library; **P102:** Kaido Kärner/Shutterstock; **P104:** Dirk Wiersma/Science Photo Library; **P106t:** The Print Collector/Photolibrary; **P106b:** Andrew Lambert Photography/Science Photo Library; **P108:** Andrew Lambert Photography/Science Photo Library: **P109:** Charles D. Winters/Science Photo Library; **P112:** Herve Berthoule/Jacana/Science Photo Library; **P113:** Andrew Lambert Photography/Science Photo Library; **P114t:** William B. Jensen/Oesper Collections: University of Cincinnati; **P114bl:** Martyn F. Chillmaid; **P114br:** Dept. Of Physics, Imperial College/Science Photo Library; **P115:** Roger Ressmeyer/Corbis; **P117:** David Parker/Science Photo Library; **P118:** Dept. Of Physics, Imperial College/Science Photo Library; **P122tl:** Arnold Fisher/Science Photo Library; **P122ml:** José Manuel Sanchis Calvete/Corbis; **P122bl:** Dirk Wiersma/Science Photo Library; **P122tm:** Andrew Lambert Photography/Science Photo Library, Charles D. Winters/Science Photo Library; **P122tr:** Andrew Lambert Photography/Science Photo Library; **P123t:** Andrew Lambert Photography/Science Photo Library; **P123b:** Charles D. Winters/Science Photo Library; **P124:** Science Photo Library; **P128t:** Charles D. Winters/Science Photo Library; **P128b:** NASA/Science Photo Library; **P129t:** Andrew Lambert Photography/Science Photo Library; **P129m:** Arnold Fisher/Science Photo Library; **P129b:** Martyn F. Chillmaid/Science Photo Library; **P132:** Roger Ressmeyer/Corbis; **P134:** Dane Steffes/Istockphoto; **P139:** Nina Towndrow/Nuffield Curriculum Centre; **P140:** Alen Rowell/Corbis; **P142l:** Joe Gough/Shutterstock; **P142m:** Martin Garnham/Dreamstime; **P142r:** Dogi/Shutterstock; **P143:** Dirk Wiersma/Science Photo Library; **P144:** Andrew Lambert Photography/Science Photo Library; **P145t:** Andrew Lambert Photography/Science Photo Library; **P145b:** Kesu/Shutterstock; **P146l:** George Bernard/Science Photo Library; **P146m:** Roberto de Gugliemo/Science Photo Library; **P146r:** Jos E Manuel Sanchis Calvete/Corbis; **P147l:** Pascal Goetgheluck/Science Photo Library; **P147r:** Peter Falkner/Science Photo Library; **P148:** Charles D. Winters/Science Photo Library; **P149t:** Mikeuk/Istockphoto; **P149bl:** Sinclair Stammers/Science Photo Library; **P149br:** Richard Megna/Fundamental/Science Photo Library; **P150t:** Layne Kennedy/Corbis; **P150b:** James L. Amos/Corbis; **P151:** Kevin Fleming/Corbis; **P156tl:** H. David Seawell/Corbis; **P156tm:** Alexis Rosenfeld/Science Photo Library; **P156tr:** John Van Hasselt/Sygma/Corbis; **P156b:** Charles E. Rotkin/Corbis; **P159:** Nik Wheeler/Corbis; **P162t:** Dirk Wiersma/Science Photo Library; **P162b:** H. David Seawell/Corbis; **P164:** Michael Rosenfeld/Science Faction/Corbis; **P166t:** Maximilian Stock Ltd/Science Photo Library; **P166b:** Alexander Raths/Shutterstock; **P167l:** Geoff Tompkinson/Science Photo Library; **P167r:** William Taufic/Corbis; **P168l:** Dave Bartruff/Corbis, Andrew Lambert Photography/Science Photo Library; **P168r:** Georgy Markov/Shutterstock, Martyn F. Chillmaid/Science Photo Library; **P169t:** Lurgi Metallurgie/Outokumpu; **P169bl:** Martyn F. Chillmaid/Science Photo Library; **P169br:** Fotex/Rex Features; **P170l:** Charles D. Winters/Science Photo Library; **P170m:** Marketa Mark/Shutterstock; **P170r:** Kojiro/Shutterstock; **P170b:** Zooid Pictures; **P171:** Andrew Lambert Photography/Science Photo Library; **P175:** Martyn F. Chillmaid/Science Photo Library; **P176t:** John Casey/Fotolia; **P176b:** AJ Photo/Science Photo Library; **P177:** Martyn F. Chillmaid/Science Photo Library; **P178:** Peter Bowater/Alamy; **P182:** Gary Banks/BP Saltend; **P184:** Bsip/Beranger/Science Photo Library; **P185:** Holt Studios International; **P187:** Sidney Moulds/Science Photo Library; **P192:** Peter Bowater/Alamy; **P194:** Y.Beaulieu, Publiphoto Diffusion/Science Photo Library; **P196:** Maximilian Stock Ltd./Science Photo Library; **P197:** R Estall/Robert Harding Picture Library Ltd/Alamy; **P198:** Steve Bicknell/The Steve Bicknell Style Library/Alamy; **P199t:** Mark Thomas/Science Photo Library; **P199b:** William Taufic/Corbis; **P200l:** Laurance B. Aiuppy/Stock Connection/Alamy; **P201:** Du Pont (UK) Ltd; **P202t:** Martyn F. Chillmaid/Science Photo Library; **P202b:** chas53/Fotolia; **P204:** Nigel Cattlin/Hot Studios International; **P205:** Alamy; **P207:** Leslie Garland/Leslie Garland Picture Library/Alamy; **P208:** Fred Hendriks/Shutterstock; **P210t:** Christine Osborne/Corbis; **P210b:** Corbis UK Ltd.; **P211:** Nina Towndrow/Nuffield Curriculum Centre; **P212:** Andrew Lambert Photography/Science Photo

Library; **P214:** Nina Towndrow/Nuffield Curriculum Centre; **P216t:** Alex Bartel/Science Photo Library; **P216b:** David R. Frazier/Science Photo Library; **P217:** Jamie Jones/Rex Features; **P218t:** North Dakota Department of Commerce Division of Community Services; **P218b:** Kathy Collins/Getty Images; **P219:** Patrick Wallet/Eurelios/Science Photo Library; **P220t:** Imagebroker/Alamy; **P220b:** Helene Rogers/Art Directors & Trip Photo Library; **P221:** Nina Towndrow/Nuffield Curriculum Centre; **P222:** Nina Towndrow/ Nuffield Curriculum Centre; **P224-225:** Fred Hendriks/Shutterstock; **P226:** Nina Towndrow/Nuffield Curriculum Centre; **P228:** Zooid Pictures;**P228-229:** Fred Hendriks/Shutterstock; **P230t:** The Metropolitan Council; **P230b:** Crown Copyright Health & Safety Laboratory/Science Photo Library; **P231:** Getty Images; **P237:** Fred Hendriks/Shutterstock; **P238t:** Fred Hendriks/ Shutterstock; **P238b:** Charles D. Winters/Science Photo Library; **P239:** Eulenblau/Istockphoto; **P241:** Science Photo Library; **P246:** Emilio Segre Visual Archives/American Institute Of Physics/Science Photo Library; **P248:** Dr Jeremy Burgess/Science Photo Library; **P251t:** Clive Freeman, The Royal Institution/Science Photo Library; **P251b:** Laguna Design/Science Photo Library; **P252t:** HR Bramaz, ISM/Science Photo Library; **P252b:** Photo courtesy of LGC; **P253t:** Pullman; **P253b:** Nick laham/Getty Images; **P254:** Bayer AG; **P255:** www.ars.usda.gov; **P256:** Zooid Pictures; **P257t:** Adrian Arbib/Corbis; **P257b:** Zooid Pictures; **P258:** Environmental Health Department, London Borough of Camden/Oxford University Press; **P258-259:** Fred Hendriks/ Shutterstock; **P259l:** BBC Photograph Library; **P259r:** David Stoecklein/Corbis; **P261:** Maximilian Stock Ltd/Science Photo Library; **P262:** Analtech Inc.; **P263:** Analtech Inc.; **P264:** Wellcome Trust; **P265:** James Holmes/Thomson Laboratories/Science Photo Library; **P266-267:** Fred Hendriks/Shutterstock; **P270:** Anna Grayson; **P270-271:** Fred Hendriks/Shutterstock; **P271t:** Anna Grayson; **P272t:** Anna Grayson, **P272:** Fred Hendriks/Shutterstock.

Illustrations by IFA Design, Plymouth, UK, Clive Goodyer, and Q2A Media.

Project Team acknowledgements
These resources have been developed to support teachers and students undertaking the OCR suite of specifications GCSE Science Twenty First Century Science. They have been developed from the 2006 edition of the resources.

We would like to thank David Curnow and Alistair Moore and the examining team at OCR, who produced the specifications for the Twenty First Century Science course.

Authors and editors of the first edition
We thank the authors and editors of the first edition, David Brodie, Anna Grayson, John Holman, Andrew Hunt, John Lazonby, Allan Mann, Peter Nicolson, Cliff Porter, and Charles Tracy.

Many people from schools, colleges, universities, industry, and the professions contributed to the production of the first edition of these resources. We also acknowledge the invaluable contribution of the teachers and students in the pilot centres.

The first edition of Twenty First Century Science was developed with support from the Nuffield Foundation, The Salters Institute, and the Wellcome Trust.

A full list of contributors can be found in the Teacher and Technician Resources.

The continued development of *Twenty First Century Science* is made possible by generous support from:
- The Nuffield Foundation
- The Salters' Institute